Lecture Notes in Computer Science 4250

Commenced Publication in 1973
Founding and Former Series Editors:
Gerhard Goos, Juris Hartmanis, and Jan van Leeuwen

H. Jaap van den Herik   Shun-Chin Hsu
Tsan-sheng Hsu   H.H.L.M. Donkers (Eds.)

Advances
in Computer Games

11th International Conference, ACG 2005
Taipei, Taiwan, September 6-9, 2005
Revised Papers

 Springer

Volume Editors

H. Jaap van den Herik
H.H.L.M. (Jeroen) Donkers
MICC, Universiteit Maastricht
Institute for Knowledge and Agent Technology (IKAT)
P.O. Box 616, 6200 MD, Maastricht, Netherlands
E-mail: herik@micc.unimaas.nl
donkers@micc.unimaas.nl

Shun-Chin Hsu
Chang Jung Christian University
Department of Information Management
G396 Chang Jung Road, Section 1, Kway Jen, Tainan, Taiwan
E-mail: schsu@mail.cju.edu.tw

Tsan-sheng Hsu
Academia Sinica
Institute of Information Science
128 Academia Road, Section 2, Nankang, Taipei, 115, Taiwan
E-mail: tshsu@iis.sinica.edu.tw

Library of Congress Control Number: 2006936974

CR Subject Classification (1998): G, I.2.1, I.2.6, I.2.8, F.2, E.1

LNCS Sublibrary: SL 1 – Theoretical Computer Science and General Issues

ISSN   0302-9743
ISBN-10   3-540-48887-1 Springer Berlin Heidelberg New York
ISBN-13   978-3-540-48887-3 Springer Berlin Heidelberg New York

Springer is a part of Springer Science+Business Media

springer.com

© Springer-Verlag Berlin Heidelberg 2006
Printed in Germany

Typesetting: Camera-ready by author, data conversion by Scientific Publishing Services, Chennai, India
Printed on acid-free paper   SPIN: 11922155   06/3142   5 4 3 2 1 0

Preface

This book contains the papers of the 11th conference on Advances in Computer Games (ACG11) held in Taipei, Taiwan. The conference took place during September 6-8, 2005, in conjunction with the 10th Computer Olympiad. It was the first time that this conference took place in Asia. The Advances in Computer Games conference series is a major international forum for researchers and developers interested in all aspects of artificial intelligence and computer-game playing. The Taipei conference was definitively characterized by new games and new ideas.

The Programme Committee (PC) received 32 submissions. Each paper was initially sent to two referees. If conflicting views on a paper were reported, it was sent to a third referee. Out of the 32 submissions, 2 were withdrawn before the final acceptance decision was made. With the help of many referees (see after the preface), the PC accepted 20 papers for presentation at the conference and publication thereafter provided that the authors submitted their contribution to a post-conference editing process. The second refereeing process was meant (a) to give authors the opportunity to include in the paper the results of the fruitful discussion after the lecture and (b) to maintain the high-quality threshold of the ACG series. The authors enjoyed this procedure.

Moreover, the PC was able to invite three key-note speakers who each opened one of the three conference days. The first invited speaker was Tony Marsland (University of Alberta), a former ICCA President. He opened the first day with the presentation "Trials and Tribulations of a Programmer." The second day was opened by Hiroyuki Iida (Japan Advanced Institute of Science and Technology (JAIST)). His talk was titled "Towards Dynamics of Intelligence in the Field of Games." The last day of the conference started with the invited speaker Feng-hsiung Hsu, who is well-known as the main programmer of the DEEP BLUE project. He is currently affiliated to Microsoft and so the title of his lecture was: "Hardware-Related Research at Microsoft Research Asia."

The above-mentioned set of 20 papers covers a wide range of computer games. There are 13 games that are popular among humans too, viz., 13 Western Chess, Chinese Chess, Japanese Chess, Checkers, Lose Checkers, Amazons, Go, Poker, LOA, Mastermind, Awari, Ataxx, and Pool. Moreover, there are two theoretical games, viz., Connect and Sumbers. The games also cover a wide range of research topics, including automatic generation, optimization, opponent modeling, search, knowledge representation, and graph history interaction. The ever-reiterating choices between publication per game domain (e.g., Chess and Go) or per research topic (e.g., learning) was solved this time in favor of the game domain with an open eye to the clustering of the research topics. We start the book with a paper on opening books in Western Chess, followed by a paper on endgame databases in Checkers. Then two papers follow that present methods to generate automatically search engines or to learn parameters for search engines in multiple domains. After a paper on a search algorithm in opponent modeling is a paper on Amazons. A group of three papers on Go is followed by two papers on Shogi. After papers on King Race, Chinese Chess, Connect, and Mastermind, a paper on multi-player chess is included. Subsequently two papers from the combinatorial game theory are presented. The sequence ends with two papers on robotic pool.

We hope that our readers will enjoy reading the efforts of the researchers. Below we provide a brief characterization of the 20 contributions, in the order in which they are printed in the book. It is a summary of their abstracts, yet it provides a fantastic three-page overview of the progress in the field.

"Innovative Opening-Book Handling" by Chrilly Donninger and Ulf Lorenz presents a heuristic in which the opening-book database is explored during the game. The main point is to avoid playing the "bad" grandmaster moves that are incidentally included in this book. The paper combines the chess expertise of a computer with some (partially dirty) statistical information. The technique is currently used in the chess program HYDRA.

"Partial Information Endgame Databases" is written by Yngvi Björnsson, Jonathan Schaeffer, and Nathan Sturtevant. The paper describes a method to build selective portions of end game databases, without fully computing portions of the database that will almost never be needed. It presents a new win-loss-draw value algorithm that can build endgame databases when unknown (partial information) values are present. The paper shows that significant portions of these databases can be resolved using these methods.

"Automatic Generation of Search Engines" by Markian Hlynka and Jonathan Schaeffer introduces PILOT, a system for automatically selecting enhancements for the $\alpha\beta$ search. PILOT generates its own test data and then uses a greedy search to explore the space of possible enhancements. Experiments in multiple domains show different enhancement selections. Tournament results further indicate that automatically generated $\alpha\beta$ search performs at least on a par with what is achievable by hand-crafted search engines. Moreover, the automatic generation involves many orders of magnitude less effort.

"RSPSA: Enhanced Parameter Optimization in Games" is written by Levente Kocsis, Csaba Szepesvári, and Mark Winands. The authors describe an algorithm for optimization of search parameters, which combines Simultaneous Perturbation Stochastic Approximation (SPSA) with Resilient Backpropagation (RPROP). The algorithm is tested in two domains: Poker and LOA. Experiments indicate that using RSPSA is a viable approach.

"Similarity Pruning in PrOM Search" by Jeroen Donkers, Jaap van den Herik, and Jos Uiterwijk introduces a new pruning mechanism for Probabilistic Opponent-Model search. The mechanism imposes a bound on the differences between the values that the opponent models may return for each position. The authors prove two properties of PrOM-search game trees: the *bound-conservation property* and the *bounded-gain property*. These properties lead to Similarity pruning in PrOM search. Experiments on random game trees show that Similarity Pruning increases the efficiency of PrOM search considerably.

"Enhancing Search Efficiency by Using Move Categorization Based on Game Progress in Amazons" is authored by Yoshinori Higashiuchi and Reijer Grimbergen. They propose a new method for improving the search in Amazons by using move categories to order moves. The categories are based on the likelihood of the move actually being selected as the best move, but also depend on the progress of the game. Self-play experiments show that using move categories significantly improves the strength of an Amazons program.

"Recognizing Seki in Computer Go," by Xiaozhen Niu, Akihiro Kishimoto, and Martin Müller, presents a new method for deciding whether an enclosed area is or can become a seki. The method combines local search with global-level static analysis. Experimental results show that a safety-of-territory solver enhanced by this method can successfully recognize a large variety of local-scale and global-scale test positions related to seki.

"Move-Pruning Techniques for Monte-Carlo Go," written by Bruno Bouzy, yields two new Monte-Carlo pruning techniques: Miai Pruning (MP) and Set Pruning (SP). In MP the second move of the random games is selected at random among a set of candidate moves. SP consists of gathering statistics about "good" and "bad" moves, pruning the latter when statistically inferior to the former. Both enhancements speed up the search at 9×9 boards. MP slightly improves the playing level. At 19×19 boards, MP results in a 30% speed-up enhancement and in a four-point improvement on average.

"A Phantom-Go Program" contains Tristan Cazenave's contribution on the relatively new computer game Phantom-Go. The new technique introduced is based on a Monte-Carlo approach. The program called ILLUSION plays Phantom Go at an intermediate level. The emphasis is on strategies, tactical search, and specialized knowledge. The paper provides a better understanding of the fundamentals of Monte-Carlo search in Go.

"Dual Lambda Search and Shogi Endgames" is a joint effort by Shunsuke Soeda, Tomoyuki Kaneko, and Tetsuro Tanaka. The authors propose a new threat-base search algorithm which takes into account threats by both players. They applied λ-search mutually recursively so that it searches the best move by taking into account threats by both players. The search algorithm, called *dual λ-search*, is implemented with DF-PN as the driver. Experiments with difficult Shogi-endgame problems show the effectiveness of the algorithm. It solves problems that even one of the strongest Shogi programs could not solve correctly.

"Chunking in Shogi: New Findings" is a contribution by Takeshi Ito, Hitoshi Matsubara, and Reijer Grimbergen. The paper focuses on cognitive experiments with expert Shogi players. The authors repeated the chess experiments by Chase and Simon with a set of next-move problem Shogi positions. The experiments show that expert Shogi players (1) search more moves, (2) search deeper and

(3) search faster than non-expert players. The experiments also show that expert Shogi players memorize the patterns of the positions and recognize move sequences before and after the position. The results suggest that Shogi players become stronger when they acquire "temporal chunks" of meaningful move sequences.

"King Race," by Alejandro González Romero, presents the results of semi-automated rule discovery in a small chess game, called King Race. From a manually devised set of attributes and a set of test positions, a decision tree is learned. The author then derives some rules from the decision tree and proves these rules to be correct. The author believes that these techniques could be used in more complex games as well.

"The Graph-History Interaction Problem in Chinese Chess" by Kuang-che Wu, Shun-Chin Hsu, and Tsan-sheng Hsu reports an improved implementation of Chines-chess rules in a computer program. The contribution focuses in particular on the rules concerning cycles. The authors present an algorithm that deals with most of the GHI problems encountered in Chinese chess. They allow an acceptable performance degradation only. On average, 3.5% more search time is needed, but the accuracy is improved substantially. Experiments show that the algorithm can solve many of the cases that could not be solved previously.

"A New Family of k-in-a-row Games" is written by I-Chen Wu and Dei-Yen Huang. The paper introduces the game family Connect(m, n, k, p, q) in which two players alternately place p stones on an $m \times n$ board, except for the first turn when the first player places q stones on the board. The player who first obtains k consecutive stones of their own color, wins. The authors analyze the family of games for fairness. Moreover, the paper proposes a threat-based strategy to play Connect(∞, ∞, k, p, q). Finally, the authors illustrate a new null-move search approach by solving Connect(∞, ∞, 6, 2, 3).

"Exact-Bound Analyses and Optimal Strategies for Mastermind with a Lie" is a contribution by Li-Te Huang, Shan-Tai Chen, and Shun-Shii Lin. This paper presents novel and systematic algorithms to solve a variant of the Mastermind game, which is called "Mastermind with a lie." First, a k-way-branching algorithm is used to get an upper bound of the number of guesses for the problem. Then a fast backtracking algorithm, based on the pigeonhole principle, is used to get a lower bound of the number of guesses. The authors show that the lowest upper bound and the highest lower bound are both 7, which means that the problem is solved completely.

"Player Modeling, Search Algorithms and Strategies in Multi-Player Games," by Ulf Lorenz and Tobias Tscheuschner, investigates a four-person chess variant in order to understand the peculiarities of multi-player games without chance components. In this contribution, player models and search algorithms are presented that have been tested in the four-player chess world. From the result follows that the more successful player models can benefit from more efficient algorithms and speed, because searching more deeply leads to better results. Moreover, a meta-strategy is presented that beats a paranoid $\alpha\beta$ player, the best known player so far in multi-player games.

"Solving Probabilistic Combinatorial Games" is a contribution by Ling Zhao and Martin Müller. It discusses Probabilistic Combinatorial Games (PCG) in which terminal positions in each subgame are evaluated by a probability distribution. The distribution expresses the uncertainty in the local evaluation. The paper focuses on the analysis and solution methods for a special case, 1-level binary PCG. Monte-Carlo analysis is used for move ordering in an exact solver that can compute the winning probability of a PCG efficiently. Monte-Carlo interior evaluation is used in a heuristic player. Experimental results show that both types of Monte-Carlo methods work well in this special case.

"On Colored Heap Games of Sumbers" by Kuo-Yuan Kao deals with sumbers. These heap games are a special type of combinatorial games. Sumbers can describe the positions of many partisan infinitesimal game. In this paper, the author elaborates further on previously obtained results on sumbers and presents three variations of colored heap games; each of them can be solved by sumbers.

"An Event-Based Pool Physics Simulator," written by Will Leckie and Michael Greenspan, presents a method to simulate the physics of the game of pool. The method is based on a parametrization of ball motion which allows the time of occurrence of events, such as collisions and transitions between motion states, to be solved analytically. It is shown that the occurrences of all possible events are determined as the roots of polynomials up to the fourth order, for which closed-form solutions exist. The method is both *accurate*, i.e., returning continuous space solutions for both time and space parameters, and *efficient*, i.e., requiring no iterative numerical methods. It is suitable for use within a game-tree search, which requires a great many potential shots to be modeled efficiently, and within a robotic pool system, which requires a high accuracy in predicting shot outcomes.

"Optimization of a Billiard Player – Position Play," is a paper by Jean-Pierre Dussault and Jean-François Landry. It describes optimization principles to produce a computer pool player that is good both technically and in planning. The authors provide optimization models to compute the shots to sink a given ball as well as to bring the cue ball at a specified target. Some hints on planning optimization strategies are given.

This book would not have been produced without the help of many persons. In particular we would like to mention the authors and the referees. Moreover, the organizers of the events in Taipei contributed quite substantially by bringing the researchers together. Then we would like to thank Ms. Tons van den Bosch for her assistance in making the manuscript fit for publication. Without much emphasis, a special word of thanks goes to the Program Committee of the ACG 11. At the same time, the editors gratefully acknowledge the expert assistance of all our referees. Finally, we happily recognize the generous sponsors Acer Inc, FunTown, Taiwan National Science Council, and ChessBase.

August 2006   Jaap van den Herik
                                               Shun-Chin Hsu
                                               Tsan-sheng Hsu
                                               Jeroen Donkers

Organization

Executive Committee

Editors   H. Jaap van den Herik
                       Shun-Chin Hsu
                       Tsan-sheng Hsu
                       H. (Jeroen) H. L. M. Donkers
Program Co-chairs   H. Jaap van den Herik
                       Shun-Chin Hsu
Honorary Co-chairs   D.T. Lee, Director, IIS, Academia Sinica
                       Chin-Sheng Chen, President, Chang Jung
                           Christian University

Organizing Committee

Johanna W. Hellemons (Chair)   Shun-Chin Hsu
H. (Jeroen) H. L. M. Donkers   H. Jaap van den Herik
Michael Greenspan   Martine Tiessen
Tsan-sheng Hsu

Sponsors

Acer Inc.
FunTown (TWP Corporation, a member of the Acer Group)
National Science Council, Taiwan
Chessbase, Hamburg, Germany

Program Committee

Ingo Althöfer	Michael Greenspan	Tony Marsland
Yngvi Björnsson	Reijer Grimbergen	Martin Müller
Bruno Bouzy	Jaap van den Herik	Jonathan Schaeffer
Michael Buro	Shun-Chin Hsu	Pieter Spronck
Tristan Cazenave	Tsan-sheng Hsu	Jos Uiterwijk
Keh-Hsun Chen	Hiroyuki Iida	I-Chen Wu
Paolo Ciancarini	Akihiro Kishimoto	Shi-Jim Yen
Jeroen Donkers	Yoshiyuki Kotani	
Aviezri Fraenkel	Shun-Shii Lin	

Referees

Victor Allis
Ingo Althöfer
Yngvi Björnsson
Bruno Bouzy
Mark G. Brockington
Michael Buro
Tristan Cazenave
Keh-Hsun Chen
Paolo Ciancarini
Rémi Coulom
Jeroen Donkers
Markus Enzenberger
David Fotland
Aviezri Fraenkel
Michael Greenspan
Reijer Grimbergen
Dap Hartmann
Guy Haworth

Bernard Helmstetter
Shun-Chin Hsu
Tsan-sheng Hsu
Robert Hyatt
Hiroyuki Iida
Graham Kendall
Akihiro Kishimoto
Jelle Kok
Yoshiyuki Kotani
Hans Kuijf
Shun-Shii Lin
Richard J. Lorentz
Ulf Lorenz
Shaul Markovitch
Frans Morsch
Martin Müller
Teigo Nakamura
Jan Paredis

Jacques Pitrat
Christian Posthoff
Jonathan Schaeffer
Nathan Sturtevant
Pascal Tang
Gerald Tesauro
Thomas Thomsen
Edward Trice
John Tromp
Jos Uiterwijk
Jan Willemson
Mark Winands
Christoph Wirth
I-Chen Wu
Shi-Jim Yen
Jan van Zanten

Table of Contents

Innovative Opening-Book Handling

Chrilly Donninger[1] and Ulf Lorenz[1,2]

[1] HydraChess.com
{chrilly, ulf}@hydrachess.com
[2] Department of Computer Science,
Universität Paderborn, Germany
flulo@uni-paderborn.de

Abstract. The best chess programs have reached the level of top players in the human chess world. The so-called opening books, which are databases containing thousands of Grandmaster games, have been seen as a big advantage of programs over humans, because the computers do never forget a variation. Interestingly, it is this opening phase which causes most problems for the computers. Not because they do not understand openings in general, but because the opening books contain too much rubbish. We introduce a heuristic which explores the database during a game. Without that, the computer repeats failures of weaker players. Our contribution presents best practice.

1 Introduction

Game playing is one of the core topics in Artificial Intelligence (AI). Unequivocal success stories come along with this topic. In some of the most popular parlor games, the computer players have already conquered the crown, such as in Checkers [19] or Othello [3]. Some games like Connect-4 have even been completely solved [21] which means that there exists an algorithm that for all positions of the game is able to play the perfect move. The game of Go is still out of the machine's reach, but in computer chess we nowadays see the leading programs crossing the borderline to the human top players. For instance, the current human vs. machine team-chess world-championship is in the hands of the machines [13]. Although our new heuristic is certainly powerful in any 2-person zero-sum parlor game, we restrict our description for the sake of clarity to the example of Chess, the most famous board game.

1.1 How a Chess Programs Works

The key feature, which enables current chess programs to play as strong as, or even stronger than the best human beings, is their search algorithm. The programs perform a forecast. Given a certain position, the line of reasoning is: what can I do, what can my opponent do next, what can I do thereafter? Modern programs use some variants of the so-called Alphabeta algorithm [12] in order to examine the resulting game tree. This algorithm is efficient in the sense that in most cases it will examine only $O(b^{d/2})$ leaves instead of b^l leaves, assuming a game-tree depth d and a uniform branching factor b. With the help of upper and lower bounds, the algorithm uses information that

H.J. van den Herik et al. (Eds.): ACG11, LNCS 4250, pp. 1–10, 2006.
© Springer-Verlag Berlin Heidelberg 2006

it collects during the search process in order to keep the remaining search tree small. This results in a sequential procedure which is difficult to parallelize, and naive approaches waste resources.

Although the Alphabeta algorithm is efficient, we are not able to compute true values for all positions in games like Chess, because the game tree is far too large. Therefore, computers apply tree search as an approximation procedure. This works as follows. First of all, a partial tree rooted near the top of the complete game tree is selected for examination. Usually, this selection is based on a maximum depth parameter. Then they assign heuristic values (e.g., one side having a Queen more means that this side will probably win) to the artificial leaves of the pre-selected partial tree and propagate these values up to the root of the tree as if they were true ones. When Shannon [20] proposed a first architecture for a computer chess program – which is in its core the same architecture as still used today [14] – it seemed quite natural that deeper searching leads to better results. This, however, is by far not self-evident, as theoretical analyses by Nau [16], Althöfer [1], or Beal [2] show. Nevertheless, the key observation over the last 40 years in most games like Chess is that a game tree acts as an *error filter*. The larger the tree which we can examine, and the more sophisticated its shape, the better is its error-filter property. New interesting ideas like in [11] have not reached practice yet.

1.2 The Race Between the Top-Four

Let us briefly summarize the breathtaking modern computer-chess history over the last 35 years. Before, between 1940 and 1970, programmers attempted to mimic human chess style, but the resulting programs were weak. Then in the 1970s, CHESS 4.5 was the first 'strong' program, emphasizing tree search. It was the first one that could win a tournament against humans, in the Minnesota Open 1977. In 1983, the chess-machine BELLE became National Master, with a rating of 2100 Elo[1]. In 1988, HITECH won for a first time against a Grandmaster, and in the same year DEEP THOUGHT already played on Grandmaster level itself. The world changed in 1992 when the CHESSMACHINE, a conventional PC program by Ed Schröder became World Champion. IBM's DEEP BLUE [10] beat Kasparov in a 6-game match, five years later [18].

Interestingly, since 1992 only PC programs have been World Champions. They have dominated the world, increasing their playing strength by about 30 Elo points per year. Nowadays, the computer-chess community is highly developed. Everything desirable exists, even a Virtual Reality with special machine rooms, closed and open tournament rooms etc. Anybody can play against Grandmasters or strong machines via Internet. At the moment, programs are at the point that they supersede their human counterparts. Four chess programs have a race for the chess crown. (1) SHREDDER, by Stefan Meyer-Kahlen, has been the dominating program over the last decade. (2) FRITZ, by Frans Morsch is the best-known program, (3) JUNIOR, by Amir Ban and Shay Bushinsky, is the current Computer-Chess World Champion, and last but not least (4) HYDRA [6], which is certainly the strongest program at the moment. These

---

[1] Elo: statistical measure. 100 Elo points difference correspond to 64% winning chance. Beginner ~1000 Elo, Int. Master ~2400, Grandmaster ~2500, World Champion ~2830.

four programs (indeed not with HYDRA, but some predecessor) scored more than 95% of the points against their opponents on the World Championship 2003.

1.3 Organization of This Paper

Besides search the existence of opening books is important in computer chess. We start emphasizing the importance. Then we discuss the techniques, which are traditionally used to generate opening databases. Thereafter, we present our own ideas and, last but not least, we show the effectiveness of the new heuristics. Our main idea is that the number of how often a move has been played in a certain position gives us a good hint of how risky it is to play this move; and that the success, with which a move has been played in the past, gives us a good hint on its potential. How these hints can be related to a specific position should be computed by the best available chess expert: the chess program.

2 Opening Books

The opening books of chess programs are large databases which contain thousands or millions of early positions together with one or several proposed moves for each of the positions. It is called 'opening book', because it contains early positions of the game, all taken from previously played games. The databases are partially generated with the help of known grandmaster games, and partially consist of analyses of chess experts. They are of great importance because (1) they can be computed off-line and (2) the top Grandmasters make so very few mistakes that they can lead a clear opening advantage to victories (thus a program should prohibit that a Grandmaster has a clean advantage after the opening phase).

2.1 The Situation

The opening books have always been an important feature in chess programs. For instance, in 1995, the program FRITZ beat IBM's DEEP BLUE, and later won the World Championship, because DEEP BLUE's opening book ended earlier than FRITZ's, and DEEP BLUE played a short castling directly after the book was finished. This mistake was so bad in a long term sense that Fritz had no difficulties to win the game.

Within the last decade, many people tried to generate top-level books for chess programs, but only three of them have been successful and have become famous in the computer chess community.

Over the years, these book writers increased their importance any further. Most of the games between top programs were decided in the opening. In the World Championship in Graz, 2003, the situation became nearly absurd. When FRITZ played against SHREDDER, SHREDDER left the book with a difficult position. Subsequently, FRITZ won. Indeed, the pairing seemed not to be FRITZ vs. SHREDDER, but Mr. Necci vs. Mr. Kure (the two most successful book writers) in a remote fight. Later, SHREDDER played against BRUTUS and SHREDDER won because of the opening book. A strange detail was that Kure made the opening book both for FRITZ and for BRUTUS. In the game FRITZ vs. BRUTUS, the programmers of both programs had in use the same opening book by Kure, but FRITZ' programmer made the mistake to add a change.

This was the decisive change which brought BRUTUS an advantage. Subsequently, BRUTUS won. In another game, JUNIOR played against SHREDDER, and SHREDDER stayed in the opening database until the final draw! As a direct consequence, some people predicted the death of computer chess. They assumed the influence of the book writers so strong that they believed that the books would become the dominating factor when two programs would play each other.

In order to find out what the book writers really do we interviewed most of the involved people. They are definitely not the strongest chess players in the world, but range between 2200 and 2550 Elo themselves.

2.2 State of the Art

From the interviews we learned that the top book writers used a mixture of different heuristics in which the following components seemed to play an important role.

(1) **Personal intuition.** This mainly seems to be used (a) to weight the points (2), (3), and (4) and (b) to decide on the 'main systems' which the program should play. (In Chess, the opening phase is divided into so-called 'systems' which in the course of time have names such as 'Sicilian Defence' and others.)

(2) **Their own chess expertise.** The writers seem to be convinced that they are able to improve the decisions of any chess program in some positions, of which they say that the machine does 'not understand' them.

(3) **Test-games.** All computers play in a similar manner. So, if you find a way to beat one program, you will have good chances that another program will lose in the same way.

(4) **Statistical information.** For each position played in the past, the following information is available, (a) which moves have been played in the past, (b) how strong the players who played the move are, and (c) how the games after the moves ended. For example: after 1. e4 e6 2. d4 d5 Nc3 Bb4 4. e5 c5 5. dxc…, one of the commercial FRITZ books presents us the following data:

Move	#	%	Elo	performance
5. … Nc6	3	0	2505	1692
5. … Qc7	2	0	2512	1707
5. … Ne7	2	25	2548	2301

From a statistical point of view, the move 5. … Ne7 seems to be the strongest move, because it has been partially successful with 25% of the points and has been played by reasonable strong players with an average of 2548 Elo. Nevertheless, the chances for Black seem low, because even the strong players could only reach a performance of 2301 Elo with 5. … Ne7 in this position.

The fact that even the top book writers look at the statistical data may be astonishing. It is generally believed that Chess is dominated by strategic thoughts and, moreover, the fundamentals of statistics are severely violated in strategic games. Let us, e.g., assume that a program A plays some hundreds test games against another program B. Let B score 75% of the points. Do we now know that B is stronger than A? No, we only know that B's score is not the result of a random process based on some normal distribution with a peak at zero. It is still possible that the programmer observes that A played only 1. e4 and 1. d4 and that A lost all d4-games and won all e4-games. Because A can in principle choose whether it plays 1. e4 or 1. d4, A is the stronger program.

Moreover, there are many examples in Chess in which a certain move in a certain position could be proven to be a bad move, whereas it was previously believed that the move was a good one. In these examples, the statistical information will look favorable for all cases, because the move will never be played again.

3 The New Way

We are convinced that strategic information is more valuable in games such as Chess than statistical knowledge plus medium-player expertise can be. The best experts for strategic information are the top chess programs themselves. Nevertheless, the statistical information seems to possess some value, because our experiments show that playing without any opening book is not a good option. Instead, we want to make statistical information available to the chess program as a second-order evaluation criterion for a move, during its search process. We distinguish between risk information that we want to use in order to spare time, and bonus information in order to make the program tend to play moves which have been successful in the past.

Let a big game database be available, and let us be interested in the value of a certain move in a specific position. From Section 2.2 we know that commercial tools may give information such as (a) how often a move has been played, and (b) how strong the players who played the move are on the average, (c) how successfully the move has been played. However, there is more information in the database. For instance, it makes a difference whether (a) a leading player (a so called Top-GM) played the move, and (b) whether the performance of the move is mainly based on draw games. Moreover, may we may consider (c) whether the move has been played in recent games after 1999, and (d) whether it has been played by especially aggressive players. Because good players tend to play weaker moves against weaker players (in order to hide their knowledge), we also distinguish between good and bad games. A game is estimated the better, the higher the two protagonists are ranked. So-called top games are games with both players having at least BookTopElo Elo. After all, the following eight parameters which influence a move's bonus and its risk are used for a weighted sum.

- BookTopFac = 8   Weight of a Top-GM game.
- BookDrawFac = 4   Weight for draw-game.
- BookWinFac = 8   Weight for win-game.
- BookRecentBias = 1   A move's weight increases, the more often it has been played after 1999.
- BookTopBias = 8   Similar to BookTopFac, but independent of its results.

- BookAggrBias = 6   A move is the better, the more often it has been played by an aggressive player (e.g., by Kasparov, Tal, Larsen, or Vaganian).

- BookTopElo = 2700   A game is a top-GM game if both players of a game have at least BookTopElo Elo.

- BookMinElo = 2500   Only games in that both players have at least BookMinElo Elo are of any interest.

The numbers assigned to the parameters are the values that we have used for experiments in Section 4.

Moreover, a 'spam'-filter suppresses games. If the words 'Blitz', 'Rapid', 'Blind', 'Interrnet', or '.com' occur in the Event or Site field of the pgn-game-header, we do not consider games with these keywords.

Let $Games_p[i]$ describe how often move i has been played in position p. $TopGames_p[i]$ is the number of games in that players with more than BookTopElo Elo have played move i in position p. $Wins_p[i]$ are defined as the number of games which have been won with move i in position p, $Draws_p[i]$ as the number of games which have ended in a draw with move i in position p. It is analogous with $TopWins_p[i]$, $RecentGames_p[i]$, and $GamesAggr_p[i]$. Let Gamesum be defined as follows.

$$Gamesum_p[i] := (TopGames_p[i]*BookTopFac+Games_p[i])$$

For a move i in position p, we define its goodness as:

```
Goodness_p[i] := (
    Wins_p[i] * BookWinFac + Draws_p[i] * BookDrawFac          (a)
    + BookTopFac * (TopWins_p[i]*BookWinFac +
                    TopDraws_p[i]*BookDrawFac)                  (b)
    + RecentGames_p[i] * BookRecentBias                        (c)
    + TopGames_p[i] * BookTopBias                              (d)
    + GamesAggr_p[i] * BookAggrBias                            (e)
) / (Gamesum_p[i]),                                           (f)
```

and its risk as:

```
risk_p[i] := 10 / (3*sqrt(Gamesum_p[i])).
```

The goodness is then adjusted by

```
Goodness_p[i] += (Gamesum_p[i] *3) / totalgames,
```

totalgames being the number of how often position p occurred in the database.

The intuition behind terms (a) to (f) is the following. Term (a) weights winning games and draw games against each other. Term (b) does the same, but here only the games of grandmasters with a high Elo rating count. Term (c) gives a bias to games which are played after 1999. Concerning term (d), top-games, i.e., games in that both players had at least 2700 Elo, should have higher impact than other games. In order to avoid openings which are preferred by calm players, games of Kasparov, Tal, Larsen and Vaganian are preferred with the help of term (e). Last but not least, the weighted sum is normalized in (f). The risk of a move i in a position p says that we can trust a move more, it has been played more often, always under the assumption that the program acknowledges the move as being 'good'.

Let us now assume that our program is in position p, and that the desired time for the next move is m minutes. Our program starts its computations as if there was no database in the background, but it adds the 'Goodness' to the root moves. This is achieved by appropriately changing the Alphabeta window at the root. If move i becomes the preferred move by the chess program itself, we will additionally adjust the computing time by the factor $risk_p[i]$. The new desired computing time is $m*risk_p[i]$.

4 Experiments

Finding convincing metrics and methods to evaluate a new idea, is quite a challenging task. Two important obstacles are: (1) if one runs experiments with a public-domain chess program, one will be in the danger that the program will not be sufficiently strong to fulfil the expert's requirements, and (2) practitioners may argue that results against weaker programs are irrelevant for top programs. Moreover, the whole idea has been developed as an attack against the top-book writers. Top books, however, are not publicly available. The first obstacle is not our problem, because we can make all experiments with our program HYDRA [6], which is strong enough to fulfill the requirements. The second obstacle is more severe. All that we can do is to present the data that we have: many test games against the top PC-chess program SHREDDER with a commercially available opening book, and some few games of 'real' competitions from public events.

4.1 The Program Hydra

The HYDRA project [6] is internationally driven and financed by PAL Computer Systems in Abu Dhabi, United Arabian Emirates. The core team consists of programmer Chrilly Donninger (Austria), researcher Ulf Lorenz (Germany), Chess Grandmaster Christopher Lutz (Germany), and the Pakistani project manager Muhammad Nasir Ali (Abu Dhabi). The FPGA chips from XilinX USA are provided on PCI cards from AlphaData in the UK. The compute cluster is built by Megware in Germany, supported by the Paderborn Center for Parallel Computing.

The hardware architecture. HYDRA uses the CHESSBASE/FRITZ graphical user interface (GUI), running on a Windows-XP PC. It connects via the Internet, using the secure shell (ssh) to a Linux cluster, which itself consists of 8 Dual PC server nodes, being able to handle two PCI busses simultaneously. Each PCI bus contains one FPGA accelerator card. One MPI process is mapped onto each of the processors and one of the FPGA cards is associated with it as well. A Myrinet network interconnects the server nodes.

The software architecture. The software is partitioned into two parts: (i) the distributed search algorithm, running on the Pentium nodes of the cluster and (ii) the 'soft-coprocessor' on the Xilinx FPGAs, related to [5].

The distributed search algorithm: Similar to [7,8,15], the basic idea of the parallelization is to decompose the search tree in order to search parts of the search tree in parallel and to balance the load dynamically with the help of the work-stealing concept. First, a special processor P_0 receives the search problem and starts performing the forecast algorithm as if it would act sequentially. At the same time, the other processors send requests for work to other randomly chosen processors. When a processor P_i that is already supplied with work, catches such a request, it checks whether there are unexplored parts of its search tree ready for evaluation. These unexplored parts are all rooted at the right siblings of the nodes of P_i's search stack. P_i sends back either that it cannot serve with work, or it sends a work packet (a chess position with bounds etc.) to the requesting processor P_j. Thus, P_i becomes a master itself, and P_j starts a sequential search on its own. The processors can be master and worker at the same time. The relationship dynamically changes during the computation. When P_j has finished its work (possibly with the help of other processors), it sends an answer message to P_i. The master/worker relationship between P_i and P_j is released, and P_j becomes idle. It again starts sending requests for work into the network. When a processor P_i finds out that it has sent a wrong Alphabeta window to one of its workers P_j, it makes a so-called window-message follow to P_j. P_j stops its search, corrects the window, and starts its old search from the beginning. If the message contained a so-called cutoff which indicates superfluous work, P_j just stops its work. HYDRA achieves speed-ups of 18 on the 32 cluster entities [17].

The soft-coprocessors: At a certain level of branching, the remaining subproblems are small enough that they can be solved with the help of a 'configware' coprocessor benefiting from the fine-grain parallelism inside the application. There is a complete chess program on-chip, consisting of modules for the search, the evaluation, the generating of moves and doing or taking back moves. The three main advantages of our configware solution are as follows.

1. Implementation of more knowledge requires additional space, but nearly no additional time.
2. FPGA code can be debugged and changed like software, without the long ASIC development cycles.
3. A large amount of fine-grain parallelism can be used.

At present, we use 67 BlockRAMs, 9879 Slices, 5308 TBUFs, 534 Flip-Flops, and 18403 LUTs. The longest path consists of 51 logic levels, and the design runs at 50MHz. An upper bound for the number of cycles per search node is 9 cycles.

4.2 Results

We played three different matches with HYDRA 1.09 against SHREDDER 8.0, with 72 games in each match. Before each match, we erased the `Hydra.plr` and the `Shredder.plr` learning files in order to keep these influences fair over all experiments. In the first match, SHREDDER played with its commercial opening book, without any learning, and HYDRA played without any book at all. In the second match, HYDRA used the opening book heuristic, as described above, called 'autobook'. In the third match, HYDRA used the autobook, and SHREDDER used the 'optimal' parameters, including book-learning. We do not exactly know how SHREDDER's book learning works. One possibility is described in [4].

Table 1. Results of the experiments

Match	Outcome	Strength
HYDRA 1.09, no book vs. SHREDDER 8.0, commercial book, no book-learning	+27,−25,=20	+9 Elo
HYDRA 1.09, autobook vs. SHREDDER 8.0, commercial book, no book-learning	+37,−20,=15	+83 Elo
HYDRA 1.09, autobook vs. SHREDDER 8.0, commercial book, optimal parameters, full book-learning	+33,−21,=18	+58 Elo

In Table 1 we see a clear positive influence of our autobook. For instance, the first line +27,−25,=20 means that HYDRA won 27 of the 72 games, lost 25, and drawed 20. We were impressed sufficiently to test this heuristic in tournament games as well.

In an 8-game match against SHREDDER in Abu Dhabi, HYDRA scored 5.5 out of 8 points and won at least the first 2 games because SHREDDER's opening book was outperformed. Moreover, we have never had to experience that we came with a bad position out of the opening. The technique was also successful against humans. In Abu Dhabi, HYDRA could beat GM Vladimirov (2650 Elo) with 3.5 out of 4, and in the Man vs. Machine World Championship in Bilbao in 2004, HYDRA scored 3.5 out of 4 points against Topalov (2757 Elo), Ponomariov (2710 Elo, 2x), and Karjakin (2576 Elo) [9].

5 Conclusions

We proposed a new heuristic how to combine the chess expertise of a computer with partially dirty statistical information. The technique is so successful that it is now used in the leading world chess program HYDRA [6].

References

1. I. Althöfer. Root Evaluation Errors: How they Arise and Propagate. *ICCA Journal*, 11(3):55–63, 1988. ISSN 0920-234X.
2. D.F. Beal. An Analysis of Minimax. *Advances in Computer Chess 2* (ed. M.R.B. Clarke), pages 103–109, Edinburgh University Press, 1980.
3. M. Buro. The Othello Match of the Year: Takeshi Murakami vs. Logistello. *ICCA Journal*, 20(3):189–193, 1997. ISSN 0920-234X.
4. M. Buro. Towards Opening Book Learning. *ICCA Journal*, 22(2):98–102, 1999. ISSN 0920-234X.
5. J.H. Condon and K. Thompson. Belle Chess Hardware. *Advances in Computer Chess 5* (ed. D.F. Beal), pages 44–54, Pergamon Press, 1982.
6. C. Donninger and U. Lorenz. The Chess Monster HYDRA. *In Proc. of 14th International Conference on Field-Programmable Logic and Applications*, Antwerp, Belgium, LNCS 3203, pages 927–932, 2004.
7. R. Feldmann, B. Monien, P. Mysliwietz, and O. Vornberger. *Distributed Game Tree Search. Parallel Algorithms for Machine Intelligence and Vision*, pages 66–101, Springer-Verlag, 1990.
8. C. Ferguson and R.E. Korf. Distributed Tree Search and Its Applications to Alpha-Beta Pruning. In: *Proceedings of the 7th National Conference Artificial Intelligence*, pages 128–132, 1988.
9. H.J. van den Herik. The World Chess Team Championship Man vs. Machine. *ICGA Journal*, 27(4):246–248, 2004. ISSN 1389-6911.
10. F. H. Hsu. Large Scale Parallelization of Alphabeta Search: An Algorithmic and Architectural Study with Computer Chess. *PhD thesis*, Carnegie-Mellon University, Pittsburgh, 1990.
11. H. Iida, J.W.H.M Uiterwijk, H.J. van den Herik, and I.S. Herschberg. Potential Applications of Opponent-Model Search. Part 2: Risks and Strategies, *ICCA Journal*, 17(1):10–15, 1994. ISSN 0920-234X.
12. D.E. Knuth and R.W. Moore. An Analysis of Alphabeta Pruning. *Artificial Intelligence*, 6(4):293–326, 1975.
13. D.N.L. Levy. The 2nd Bilbao Man vs. Machine Team Championship. *ICGA Journal*, 28(4):254–255, 2005. ISSN 1389-6911.
14. D. Levy and M. Newborn. *How Computers Play Chess*. W.H. Freeman and Company, New York, 1990.
15. T.A. Marsland, A. Reinefeld, and J.Schaeffer. Multiprocessor Tree-Search Experiments. *Artificial Intelligence*, 31(1):185–199, 1987.
16. D.S. Nau. An Investigation of the Causes of Pathology in Games. *Artificial Intelligence*, 19(3):257–278, 1982.
17. Personal communication with S. Meyer-Kahlen, M. Feist, F. Morsch, A. Kure, E. Günes, C. Lutz, C. de Gorter, 2000–2003.
18. A. Plaat and J. Schaeffer. Kasparov vs. Deep Blue: The Rematch. *ICCA Journal*, pp. 95–101, 1997.
19. J. Schaeffer, R. Lake, P. Lu, and M. Bryant. Chinook: The World Man-Machine Checkers Champion, *AI Magazine*, 17(1):21–29, 1996.
20. Claude E. Shannon. Programming a Computer for Playing Chess, *Philosophical Magazine*, 41(314):256–275,1950.
21. J.W.H.M. Uiterwijk, H.J. van den Herik, and L. V. Allis. A Knowledge-Based Approach to Connect-4. The Game is Over: White To Move Wins. *Heuristic Programming in Artificial Intelligence I*, pages 113–133, 1989.

Partial Information Endgame Databases

Yngvi Björnsson[1], Jonathan Schaeffer[2], and Nathan R. Sturtevant[2]

[1] Department of Computer Science,
Reykjavik University, Reykjavik, Iceland
yngvi@ru.is
[2] Department of Computing Science,
University of Alberta, Edmonton, Alberta, Canada
{jonathan, nathanst}@cs.ualberta.ca

Abstract. Endgame databases have previously been built based on complete analysis of endgame positions. In the domain of Checkers, where endgame databases consisting of 39 *trillion* positions have already been built, it would be beneficial to be able to build select portions of even larger databases, without fully computing portions of the database that will almost never be needed. We present a new win-loss-draw value algorithm that can build endgame databases when unknown (partial information) values are present, showing that significant portions of these databases can be resolved using these methods.

1 Introduction

Endgame databases were pioneered over 20 years ago [8,9]. The basic idea, computing the value of positions at the end of the game and backing up the values towards the start of the game, is both simple and powerful. These databases have provided numerous insights to human analysts (e.g., chess [3]), have been useful for solving games (e.g., awari [4], nine men's morris [1]), and have been instrumental in building super-human programs (e.g., CHINOOK [5]).

The biggest (and longest) endgame database computation is that from the CHINOOK checkers project. The databases currently contain 39 *trillion* (3.9×10^{13}) positions—all positions with 10 or fewer pieces on the board. These databases have been instrumental in an on-going effort to solve the game of checkers. In January 2005, the CHINOOK team achieved their first milestone, announcing a proof of the infamous White Doctor opening (it is a draw) [6]. Endgame databases introduce perfect knowledge into the proof, replacing the traditional heuristic evaluations. The attempt to solve checkers would be greatly accelerated if the 11-piece databases could be computed.

Endgame databases need to be computed in a certain order, to preserve the dependencies inherent in the calculation. For example, the database for all positions with two kings need to be computed before the 3-king database can be computed. Ideally, one would like to extend this computation to include *all* the pieces on the board (e.g., the 32-piece database for chess!). However, there are several limitations to extending the database calculations as far as possible. As

H.J. van den Herik et al. (Eds.): ACG11, LNCS 4250, pp. 11–22, 2006.

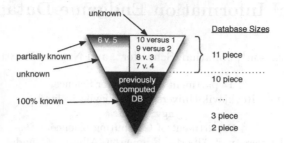

Fig. 1. Our goal: skip seldom used database and only compute more commonly used database positions

the number of pieces in the database increase, one encounters limits in data size (the databases become too large), computation time (there are more positions and longer winning sequences), memory (databases quickly become too big for RAM, resulting in costly I/O), and correctness (the computations need to be replicated to verify correctness).

A completed 11-piece checkers database would encompass 259 trillion positions, over six times larger than all existing checkers databases, which already took many years to build. The complete 6-piece vs. 5-piece subset (all combinations of checkers and kings) is a large portion of this, 118 trillion positions. Unfortunately, most of the computation is practically irrelevant for furthering the checkers proof. The first 11-piece database that must be computed by standard retrograde analysis, 6 kings vs. 5 kings, is never reached in tournament checkers games and never arises in our checkers proof trees. In effect, because of the computation dependencies, the least useful databases get computed before the most useful. The 11-piece database that would be most useful is the 6-checker vs. 5-checker database, but this is the very last calculation that gets done. This database has only 25 billion positions. Even the 6 vs. 5 endgames with a maximum of 1 king on the board come to 300 billion positions—easily doable using current technology. The problem is that we need to compute 118 trillion 6-piece vs. 5-piece positions before we get to the very small useful portion.

The goal of this research is illustrated in Fig. 1. Checkers databases that have been fully computed, such as the 2-piece through 10-pieces databases, are shaded in black. We would like to avoid computing the large and rarely used portions of the 11-piece database, and concentrate our efforts on a portion of the 6 vs. 5 database. Without calculating the full database this portion cannot be computed exactly, but even partial bounds on the values will be quite useful.

Instrumentation of the checkers prover used in [6] suggests that roughly 20% of non-database positions encountered in the search are 11-piece positions. Of these, over 90% are positions with 6 pieces vs. 5, with the total number of kings being 0 or 1. In the attempt to solve checkers, we do not need to prove the exact value of every position. In many cases a lower-bound or upper-bound on the value of a position will be adequate, for instance, to prove that a line of play is at least a draw. Thus, databases that only provide partial information about a set of positions are still useful. Similar methods to what we present here for backing up information in end-game databases are explored in [2].

This paper makes the following contributions.

1. A win-loss-draw value ordering when unknown (partial information) values are present.
2. The implementation of a retrograde analysis algorithm that correctly propagates partial information values.
3. Experimental results and validation for the 10-piece partial information databases.

2   Partial Information Endgame Databases

In this section we provide some background (2.1), introduce database values for partial information databases (2.2), explain the building of partial information databases (2.3), and sketch a proof of correctness (2.4).

2.1   Background: Perfection Information Endgame Databases

Endgame databases traditionally contain three possible outcomes, win (W), loss (L) or tie/draw (T). A new database is computed by repeatedly iterating through all positions in the database, filling in values for positions that are known, and repeating until all values are known. Note that we do not store the number of moves needed to resolve each position, which may be needed to successfully play out some positions.

Initially, all positions in the database are marked with an extra value, unknown (U), which does not remain in the database past the retrograde analysis. The only way a position can be assigned a value is if one of its children is a win, or if all children have been resolved to a loss or a tie.

There is an exception for games like Checkers which have rules for draw-by-repetition. It is possible that there are cycles of states for which the best move is a draw-by-repetition. When doing retrograde analysis these positions will never be resolved exactly because every node within the repeated cycle of moves will always have one unknown child. But, once no other changes can be propagated through the database, remaining positions with unknown values are given the value T, which effectively handles these positions.

This process of retrograde analysis, for which pseudo-code is shown in Table 1, relies on three operators, *completed*, *decide*, and *max*. In perfect-information endgame databases a position is *completed* as long as the value is not U, *decide* is the identity function except on input of U which becomes T, and *max* uses the following ordering on values: $L < T < U < W$.

The high-level pseudo-code and definitions above omit many practical considerations (e.g., for reducing I/O). For example, from a practical standpoint, the database values are always from the perspective of the player that has the move in the given position. Therefore, we must always negate the value of a child

Table 1. Pseudo-code for standard retrograde analysis

Retrograde_Analysis(database)
Initialize database to *unknown*, positionsUpdated = 1
while (positionsUpdated != 0)
  positionsUpdated = 0
  for each position **in** database
    if (!*completed*(position))
      if (*value*(position) != *max*(children of position))
        positionsUpdated += 1
        value(position) = *max*(children of position)
  for each position **in** database
    value(position) = *decide*(position)

---

before applying the max operator. In the above example, the negate operator maps L to W and vice verse, but T and U remain the same. Also, if no move is available for the player to move (player has no piece or all pieces are blocked) the position is labeled a loss for the player. However, this high-level view provides an elegant conceptual model of retrograde analysis, and allows us introduce our new ideas simply by redefining a few operators.

2.2 Database Values for Partial Information Databases

If we want to build partial information bounds into an endgame database, we will need more values than just $W/L/T$ in the database. Instead, we need an upper and lower-bound on the value of a position. So, we write possible values of each position as two letters, the lower-bound followed by the upper-bound.

First, the standard values of $W/L/T$ will be replaced with $WW/LL/TT$, meaning that the upper-bound and lower-bound for these positions is identical. Then, we introduce three new values, LT, LW and TW. LT means the position's value has a lower-bound of loss and an upper-bound of tie. Definitions for LW and TW follow similarly. We will refer to the new three values as partial-information values, because they do not provide perfect information about the value of a position, only bounds.

Given partial-information values, the first thing we need is an ordering on the values which can be used by the max operation. This ordering is defined in Fig. 2. A directed edge between any two states A and B means that B is strictly greater than A. This property is transitive, so the max of any two states A and B is the first common node on all paths between A and WW and B and WW. For most states this is straightforward. For instance, $max(LT, TW) = TW$. But, this is only a partial ordering, because TT is neither greater than or less than LW. So $max(TT, LW) = TW$. Another way to view this is that the lower-bound on the max of two values is the max of the lower-bounds on these values, and the upper-bound on the max of two values is the max of the upper-bounds.

These six bounded values are the only values that will be found in a completed partial information database. But, they are not adequate in themselves, because we need a richer definition for unknown value(s) when we actually build a partial information database.

2.3   Building Partial Information Databases

When we begin to build a partial information database there are several places where partial information values will enter the database. As before, we can initialize positions to a value of *unknown* before we begin. Positions on the fringe of the database being built can have children that are found in either a previously computed database, or a database that we have not computed. For instance, if we do not compute the 2-king 11-piece databases and we have a move that makes the second king on the board, we will not find that position in a database. Such positions are given the value *LW*, because we have no information about their values, and we will never attempt to calculate them.

Given the new definition of *max* and previous definitions of *decide* and *completed*, it may seem that we have enough information to build partial-information databases. But, there is one additional case that arises, which is the interaction of draw-by-repetition positions with other unknown values.

Consider the four positions which are part of a draw-by-repetition cycle in part (a) of Fig. 3. In a regular database, the player to move at the highlighted node would prefer a draw-by-repetition to the loss available as an alternate move. So, the final value of this state after retrograde analysis will be *T*.

Now consider part (b) of Fig. 3, where the alternate value to the draw-by-repetition is completely unknown, *LW*. In this situation, the first player is guaranteed at least a tie because of the draw-by-repetition, but there is a chance that the alternate move will lead to a win. Thus, the actual value of the state is *TW*. That is, there is one possible move that will lead to a tie by repetition, and there is another move that might lead to a win. At the parent of this position (using a nega-max formulation) the bound is then *LT*, and similarly in the rest of the cycle.

But this will not actually happen unless we modify the values used by retrograde analysis. If we have a single value for draw-by-repetition which is not

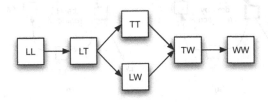

Fig. 2. The partial ordering of values used for database generation

resolved until after all passes through the database, none of the positions in this cycle will resolve. So, when all other positions have stopped updating, the *unknown* on the positions in the cycle will incorrectly be converted to TT. This process ignores the interactions between draw-by-repetition and positions that have partial information bounds.

Thus, in the same way that we introduced partial information bounds on the value of a state, we need to introduce new unknown values for use during retrograde analysis. These values are not bounds on the final value of a state, but instead represent the value a state would take if the retrograde analysis were to stop without further updates.

These new values are uLW, uTW, uTT and uLT. uTT is the same as the previous *unknown* value, meaning that a state is a draw-by-repetition. The additional values arise as a result of ways that uTT can combine with partial bounds in the game. We demonstrate this in Fig. 4.

In this figure we demonstrate how values are propagated through a draw-by-repetition loop. At the far left of this diagram we have the initial values in the database, with all values initialized to uTT. The only node that can update its value is the highlighted node, which is subsequently updated to uTW. In the next step, the parent of this state, which is highlighted, can now be updated to uLT. This process continues until no further updates are possible. If retrograde analysis ends at this point, all the unknown values will be converted to exact values.

Figure 4 shows how the values uTW and uLT can arise in the database. The other unknown value, uLW, is introduced when we have a state that has one child with a value of uLT and another child with the value LW.

We can now define the *completed*, *decide*, and *max* operators for partial information databases. A position is *completed* once it has a value of WW, TT, LL, LW, TW, or LT. We *decide* the final value of an unknown state by converting it to its corresponding known value. So, $decide(uLW) = LW$, and likewise for other unknown values. Finally, the maximization function is defined for all possible values by Table 2. Utilizing these new definitions, we can use

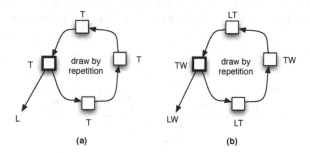

Fig. 3. Draw-by-repetition interactions

Table 2. The maximum of each possible combination of states that can be encountered in the process of building partial-information databases

max	WW	TT	LL	TW	LT	LW	uLW	uTW	uTT	uLT
WW	WW	WW	WW	WW	WW	WW	WW	WW	WW	WW
TT	WW	TT	TT	TW	TT	TW	uTW	uTW	uTT	uTT
LL	WW	TT	LL	TW	LT	LW	uLW	uTW	uTT	uLT
TW	WW	TW	TW	TW	TW	TW	uTW	uTW	uTW	uTW
LT	WW	TT	LT	TW	LT	LW	uLW	uTW	uTT	uLT
LW	WW	TW	LW	TW	LW	LW	uLW	uTW	uTW	uLW
uLW	WW	uTW	uLW	uTW	uLW	uLW	uLW	uTW	uTW	uLW
uTW	WW	uTW	uTW	uTW	uTW	uTW	uTW	uTW	uTW	uTW
uTT	WW	uTT	uTT	uTW	uTT	uTW	uTW	uTW	uTT	uTT
uLT	WW	uTT	uLT	uTW	uLT	uLW	uLW	uTW	uTT	uLT

the same methodology as shown in pseudo-code in Table 1 to compute partial information databases, except that the initialized *unknown* value is now uTT.

2.4   Proof of Correctness

In this section we sketch a proof of correctness; but many details are omitted for clarity. We first show that any positions which become *decided* during retrograde analysis must have correct bounds. Then, we show that draw-by-repetition values are computed correctly. Finally, we show that draw-by-repetition subsets of the database cannot improperly effect other parts of the database.

First, consider positions that will become *decided* during the process of retrograde analysis. This means that either all of the children of these positions are decided, or they have one child that leads to a win. If a position has one child that leads to a proven win, this will be a proven win no matter whether we are using partial information databases or perfect information databases. Similarly, if a position is resolved to exactly TT or LL during retrograde analysis, this value must have been calculated in the exact same manner as in a standard endgame database.

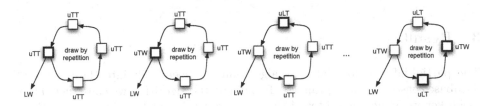

Fig. 4. The propagation of values combined with uTT

Now, consider positions that resolved to LW, LT or TW. There are two places that partial information bounds are introduced into the database. The first is when a child is found in a database that has not been computed, in which case it is given the value LW. This is the most general bound possible, so the actual value of this state must fall within these partial information bounds (loss to win). The second place we can get partial information bounds is from another partial information database, which we assume has already been calculated correctly. From Table 2 we can verify that the combination of LW, LT, and TW with any other *known* value correctly preserves upper- and lower-bounds on a particular state. Thus, any value that becomes decided during retrograde analysis is calculated correctly.

Next, consider positions involved in draw-by-repetition, which are not resolved until retrograde analysis completes. If they are part of a closed loop (all positions outside the loop lead to a loss or a tie) then these positions are handled no differently by retrograde analysis with partial information values then they would be handled by standard retrograde analysis.

In the case where a position near a draw-by-repetition loop has a partial information value that could possibly be a win, it will be incorporated into the bounds of positions in the draw-by-repetition loop, as in Fig. 4. Again, from Table 2 one can verify that there is no way to combine values in such a loop to eliminate the draw-by-repetition.

Finally, is it possible for unknown values from a draw-by-repetition to adversely effect other positions in the database, now that we have expanded the range of possible values that can occur in a draw-by-repetition loop? It is not possible because of the following observation: Any values which are not *known* besides uTT must trace to at least one position where there is a uTT value for one child and a partial information value (TW or LW) at the other child. The value of such states then will either be a draw-by-repetition or the result of the partial information value from the other child—but there is not enough information in the analysis to resolve which. In either case lower- and upper-bounds for such states will be correct. Presenting a formal proof of this would be quite detailed, but simply relies again on the values in Table 2.

While we have not shown in a completely formal manner that our partial information databases are correct, these description should give the reader a feel of the general correctness retrograde analysis using both partial information values and extended values for unknown positions.

3   Results

The partial information database algorithm has been built into the CHINOOK database construction program [7]. To validate the algorithm we constructed the 5-checker vs. 5-checker subset of the 10-piece databases. For this computation, we removed all the previously computed 10-piece databases. In other words, the database program only had access to the 2-piece through 9-piece databases. After constructing the partial information 10-piece databases we verified that

Table 3. Max table for partial databases, 5 vs. 5 checkers. Each entry is the percentage of resolved positions.

	7	6	5	4	3	2
7	16	25	33	37	44	70
6	26	38	49	56	68	84
5	35	48	62	71	81	88
4	35	53	66	80	86	88
3	40	63	79	86	91	95
2	62	81	84	91	97	100

all bounds were consistent with the actual values in the previously computed 10-piece databases.

Table 3 shows some results for the 5-checker vs. 5-checker database. The database is broken into smaller pieces based on the leading rank of the checker for each side. For example, the 7/7 table entry has both Black and White having a checker on the seventh rank. It is not possible to have a checker on the eighth rank (it already has become a king). The 6/4 entry has all of White's checkers on or before the sixth rank with at least one on the sixth rank, while all of Black's checkers are on the fourth rank or before (with at least one on the fourth rank). Note that there is no result for rank 1—since each side has 5 checkers, it is not possible to have them all on the first rank.

The table shows the fraction of (non-capture) positions that have been resolved. A position proven to be a win, loss or tie counts as one point, while a lower- or upper-bound of a tie counts as a half point. The total database score divided by the number of positions in the computation gives the score. For example, the 7/7 entry shows that when both sides have a leading checker on the seventh rank, then the database program was able to resolve 16% of the values. The 7/7 entry is low for two reasons.

1. Most positions in the 5 vs. 5 databases are drawn. This makes it difficult for the database construction algorithm, since a draw cannot be proven until the values (or bounds on values) of all children of a position are known.
2. This database computation has both sides "on the boundary". When the leading checker advances, it becomes a king—and there are no database results for that position. Hence many of the 7/7 computation are unresolved because they lead to unresolved (partial information) positions.

As one would expect, as one moves away from the boundary, a higher percent of positions are resolved. This is easily explained since when the leading checker in these databases advances, it moves into a database that has already been computed and has (partial) results. It is particularly noteworthy that 100% of the positions with both players leading checkers at the second rank are resolved correctly and completely. So, from one perspective, this portion of the databases is no longer a partial information database, as all values are exact.

Also of interest is that the database construction algorithm took as many as 36 passes over the data to compute the values in Table 3. This means that some of the values represent proven wins in 36 ply. If ever one of these positions comes up in the checkers proof, the perfect value from the partial information databases will replace 36-ply of search!

The above results point to an obvious way to improve the results. Since the boundary introduces the unknown values, an effort should be made to resolve as many boundary positions as possible. For example, one could do a small (say, 5-ply) search for each boundary position. This would increase the number of boundary positions that have useful data, either proven or partial values. These values, of course, get propagated into the database calculation, which would increase the percentage of overall positions resolved. We have not yet incorporated this into our program, but it is a point of future work.

So, how will this impact the 11-piece databases? The 11-piece databases are less likely to have draws—one side is up a checker, so we would expect the side with more pieces to win most positions. Thus, to provide additional insight into how the 11-piece database results will look, we also computed partial 4 vs. 5-checker 9-piece databases.

Table 4. Max table for 9 piece databases, 4 vs. 5 checkers

	7	6	5	4	3	2
7	7	10	13	15	25	74
6	46	55	57	61	73	98
5	64	71	78	80	91	97
4	74	81	87	93	96	99
3	82	88	94	97	98	100
2	86	93	97	99	99	100
1	87	94	95	99	100	100

Table 4 shows the 9-piece results when the player to move has four checkers against an opponent with five checkers. Table 5 shows the similar table when the player to move has five checkers against an opponent with four.

These numbers are best explained using specific examples from the table. In the first row, second column (7/6) of Table 4 we find the entry 10. This means that only 10% of the positions have been resolved when the player with 4 checkers is to move and has his most advanced checker on the seventh rank. But, in Table 5 in the first row, second column we find the entry 66, meaning 66% of positions have been resolved when the player with 5 checkers is to move with his most advanced checker on the seventh rank. So, each entry is the percent of resolved positions relative to the most advanced checker on the board, which player is to move next, and the number of checkers that each player has.

The first item of interest is that the ratio of resolved positions is noticeably higher than in the 10-piece database, which we expected, because of the uneven material value on the board. We see also that the stronger side can generally

Table 5. Max table for 9 piece databases, 5 vs. 4 checkers

	7	6	5	4	3	2	1
7	44	66	78	85	92	91	90
6	51	73	84	91	95	96	95
5	51	79	90	95	98	99	99
4	52	82	94	98	99	100	100
3	57	90	98	99	100	100	100
2	86	99	100	100	100	100	100

resolve a larger percentage of positions. The only exceptions to this is where the weaker side is just about to promote a checker to a king, because the databases containing the result of such a move has not been computed.

Also of interest is the asymmetric nature of the table. The entry in the sixth row and seventh (6/7) column for the weak player (46%) entry is much stronger for the weak side than the 7/6 entry (10%). This occurs because from the 6/7 entry the weak player can move his most advanced checker forward into the 7/7 database. Because it will then be the other player's turn, many of these positions (44%) will already be resolved.

4   Conclusions

The goal of this research has been to compute the important parts of the 11-piece checkers endgame databases. At the time of this writing the 6-piece vs. 5-piece endgames with one or fewer kings are currently being computed, followed by the 6-checker vs. 6-checker database. When complete, these will be added to the CHINOOK databases and used in the checkers prover. Although this computation may only cause roughly 10% of the positions examined by the prover to be resolved by the new databases, our experience is that in many cases this will result in many ply being eliminated from some of the sub-proofs. This will result in a substantial reduction in the computational effort needed to solve checkers.

We also plan to compute some 7-piece vs. 4-piece and 8-piece vs. 3-piece partial information endgames. Although these lopsided endgames are almost always won for the stronger side, having the proven database value will be very useful. Sadly, any checkers solver must always explore the path of maximum resistance, since it must prefer an unknown value over a loss or draw. Consequently, the prover tends to make moves to postpone resolving the value of a position—an "unknown" position down 5 pieces (it could be a win!) is preferred over a known draw.

In conclusion, partial information databases are not a complete set of resolved values, but they still provide useful information. They allow us to probe further forward in the game search space, allowing us to uncover new secrets of the game—without having to pay the full price of computational resources and time.

References

1. R. Gasser. Solving Nine Men's Morris. *Computational Intelligence*, 12:24–41, 1996.
2. T.R. Lincke. *Exploring the Computational Limits of Large Exhaustive Search Problems*. Ph.D. thesis, Swiss Federal Institute of Technology, 2002.
3. John Nunn. *Secrets of Rook Endings*. Gambit Books, B.T. Batsford Ltd., London, 1999.
4. J. Romein and H. Bal. Solving the Game of Awari Using Parallel Retrograde Analysis. *IEEE Computer*, 36(10):26–33, 2003.
5. J. Schaeffer. *One Jump Ahead*. Springer-Verlag, New York, 1997.
6. J. Schaeffer, Y. Bjornsson, N. Burch, A. Kishimoto, M. Müller, R. Lake, P. Lu, and S. Sutphen. Solving Checkers. In *International Joint Conference on Artificial Intelligence (IJCAI)*, pages 292–297, 2005.
7. J. Schaeffer, Y. Bjornsson, Neil Burch, Robert Lake, Paul Lu, and Steve Sutphen. Building the 10-piece Checkers Endgame Databases. In *10th Advances in Computer Games (ACG10), Many Games, Many Challenges* (eds. H. J. van den Herik, H. Iida, and E. A. Heinz), Kluwer Academic Publishers, Boston, pages 193–210, 2004.
8. T. Ströhlein. *Untersuchungen uber Kombinatorische Spiele*. Dissertation, Fakultät für Allgemeine Wissenschaften der Technischen Hochschule München, 1970.
9. K. Thompson. Retrograde Analysis of Certain Endgames. *ICCA Journal*, 9(3):131–139, 1986.

Automatic Generation of Search Engines

Markian Hlynka and Jonathan Schaeffer

Department of Computing Science
University of Alberta, Edmonton, Alberta, Canada
{markian, jonathan}@cs.ualberta.ca

Abstract. A plethora of enhancements are available to be used together with the $\alpha\beta$ search algorithm. There are so many, that their selection and implementation is a non-trivial task, even for the expert. Every domain has its specifics which affect the search tree Even seemingly minute changes to an evaluation function can have an impact on the characteristics of a search tree. In turn, different tree characteristics must be addressed by selecting different enhancements. This paper introduces PiLOT, a system for automatically selecting enhancements for $\alpha\beta$ search. PILOT generates its own test data and then uses a greedy search to explore the space of possible enhancements. Experiments with multiple domains show differing enhancement selections. Tournament results are presented for two games to demonstrate that automatically generated $\alpha\beta$ search performs at least on a par with what is achievable by hand-crafted search engines, but with orders of magnitude less effort in its creation.

1   Introduction

Programs which play two-player games of perfect information typically rely on two things to achieve high performance: an evaluation function and a search algorithm. The evaluation function estimates the desirability of a static board position. The search algorithm, using the evaluation function as a guide, acquires knowledge dynamically: it looks ahead through the possible lines of play and selects the one leading to the highest achievable evaluation given the search constraints. $\alpha\beta$ pruning [10] is the mainstay of this sort of game program. However, designers of high-performance search engines have long observed that while the pseudo code for $\alpha\beta$ is approximately 20 lines, an actual high-performance implementation of $\alpha\beta$ can run upwards of 20 pages of code. The reason for this disparity is straightforward. Since the development of $\alpha\beta$ pruning and the decision to pursue search as the cornerstone of two-player game research, most work in two-player games has been in the area of search enhancements [15]. While the choice of $\alpha\beta$ for a two-player perfect information game is trivial, the implementation for maximum performance on a particular problem is not. In practice, there is a plethora of enhancements which allow the basic search algorithm to focus its attention further on the areas of the search space most likely to produce useful information, while eliminating areas of the search which seem irrelevant.

H.J. van den Herik et al. (Eds.): ACG11, LNCS 4250, pp. 23–38, 2006.
© Springer-Verlag Berlin Heidelberg 2006

Search enhancements fall roughly into four categories: those that change the order in which successors are considered at each node in the tree; those that focus the search by dynamically extending or reducing the search depth; those that adjust the $\alpha\beta$ bounds to reduce the amount of search required; and those that employ some combination of memory or caching to avoid repeating work or to access knowledge acquired outside of the current search.

For each of these categories there are numerous enhancements in the literature. Some seem to work well in most domains. Others seem to be more specific, having their greatest effect on a class of domains, and less or detrimental impact on other domains. It is also pertinent to note that the word "domain" in this sense does not necessarily have a one-to-one correspondence with "game". That is, games typically pass through several phases in the course of normal play. These phases may differ significantly in their distinguishing features, and thus could be said to correspond to different domains. This is why high performance game programs often have multiple classes of settings corresponding to this changing nature of a game. Thus, different enhancements may be applicable in different phases.

When building a high-performance search engine for a particular game domain, it is not the choice of the algorithm that the programmer finds costly and time consuming. Rather, it is in the selection, implementation, testing, debugging, and interaction of the available search enhancements that the real work lies.

Consider the scenario of a programmer developing a high-performance search engine to play a new game. Currently, such a programmer is faced with a daunting task. The programmer must painstakingly sort through all the known search enhancements for the ones which will most benefit the new domain. The enhancements are sensitive to many potential differences between domains. They usually have parameters which require tuning for good performance. If the domain happens to be relatively new and unknown, this further exacerbates the difficulty of the task. Thus, the work is driven by intuition and experimentation, resulting in hundreds of hours of programmer and computer processing time [9,14].

The next problem our imaginary programmer faces is that different enhancements interact in various ways. Sometimes these interactions are well defined. However, more than enough enhancements exist to make a thorough list of interactions infeasible if not actually impossible. Of course, like the enhancements themselves, these interactions are also tied up with the domain to which they are applied. To make the problem interesting, the areas covered by each enhancement are not discrete and self-contained; different enhancements overlap in their goals as well as their methods.

The final problem is that most enhancements contain parameters that must be further tuned for maximum performance in the domain. Poor parameter selection could result in a highly beneficial enhancement being perceived as useless. Selecting the correct parameters is frequently at least as difficult as choosing the enhancement in the first place. And, of course, this must be done not just individually, but also in the context of all the other enhancements present:

their parameters may interact in increasingly complicated ways. For example, enhancements A and B may individually produce their best results with parameter sets A' and B' respectively. However, using A and B simultaneously with these same parameters may produce unsatisfactory results. Not only might the combined effect fail to increase performance, it may even decrease it.

Current methods (i.e.,"by hand"), however cumbersome, do work. Experts painstakingly "hill-climb", carefully implementing one enhancement at a time, evaluating it, and only accepting it if it yields better performance. That this approach works in practice is evidenced by such stunning successes as CHINOOK, DEEP BLUE, LOGISTELLO, and others.

Fürnkranz [7] identifies the learning of techniques to control search as an area which, "surprisingly... is more or less still open research in game-playing." Little work has been done to automate the task of devising a search algorithm along with its various enhancements for a particular purpose. Rather, most work has been focused in the area of search control, which refers to methods that allow a finer granularity of control over the search algorithm. The main forms of learning search control are focusing the search with extensions and reductions, modifying parameters which control the search, and using acquired knowledge to direct the search. These are exemplified by the work of [2] and [5].

Cook and Varnell [6] present an algorithm in which changing certain parameters modifies the basic behaviour of single agent parallel A* search. They call these variations search strategies, and their work is probably most similar to that presented herein. Indeed, Schaeffer et al. [18] posit that single-agent and two-player search are different sides of the same coin. Other related attempts have been made, such as that of ZILLIONS OF GAMES [19], which builds an evaluation function based on a game description. ZILLIONS's approach might complement the presented work, PILOT, nicely: we are working on high performance search, while ZILLIONS seems to focus on the evaluation. Unfortunately, further details on how ZILLIONS functions is unavailable.

The problem is simply stated: how does one find the right combination of enhancements, along with their associated parameters, to maximize the search performance of a game-playing program? Why not devise a system to automate this difficult task? Such a system could characterize the features of new search domains. It could then use this acquired knowledge to learn correlations between various features of the domain and search enhancements. The system would then test its hypotheses to confirm their accuracy. Upon completion of processing, the end result of the system would be a search algorithm, complete with enhancements, thoroughly tested, and specifically tailored to the required domain. Most importantly, all this would be accomplished without human intervention![1]

To this lofty goal this paper contributes the following. An implementation framework for two-player perfect-information games, wherein the search algorithm is separated from the domain-specific knowledge. This gives a game programmer two immediate advantages: implementing new games takes only one

---

[1] Save for the obvious human work required to define the new domain, determine the bounds, requirements, and evaluation function.

or two days, and each game immediately has access to the full set of available $\alpha\beta$ search enhancements.

A prototype program called PILOT is presented. PILOT is able to choose a set of search enhancements, compile a game program with these enhancements, automatically generate test sets for a specific domain (game), and perform a greedy search through the space of search enhancements to determine the best set for the particular domain in question.

Results are presented for two non-trivial games, Ataxx and Lose Checkers, as well as reports on some experiments with the (also non-trivial) games of Awari and Critical Mass.

2   PILOT

PILOT is the name given to a system which oversees the choice of $\alpha\beta$ search enhancements appropriate to a specific domain. PILOT is the control program responsible for *piloting* the various components which comprise the system. PILOT's job is (1) to generate a search engine by selecting search enhancements, (2) to create and execute tests to determine their effectiveness, and (3) to generate further search engines based on the data thus acquired. This process iterates until a user-imposed constraint is met, or until PILOT reaches a steady-state beyond which it cannot improve search performance.

2.1   Interfacing with PILOT

As stated in the introduction, PILOT includes an implementation framework which separates the search algorithm from game specific knowledge. This greatly speeds the process of developing a new game. To this end, a programming interface is provided to the developer of a new game. The programmer must provide the system with a number of standard functions and data types which are specific to the application. These fall into three main categories: rules, knowledge, and performance hooks.

Functions in the rules category implement the rules of the game. They include functions for making moves, unmaking moves, and move generation. The game evaluation is the main member of the knowledge category. While building a good evaluation function is difficult, this is not in the scope of our work. Search enhancements can be implemented in a generic fashion, but this can severely impact performance. Performance hooks rectify this by allowing the programmer to use domain-specific knowledge. Providing an incremental hash function for a transposition table is an example of this.

2.2   Components

PILOT's main modules are illustrated in Fig. 1. The modules are code generation, evaluation, analysis and learning, and enhancement selection.

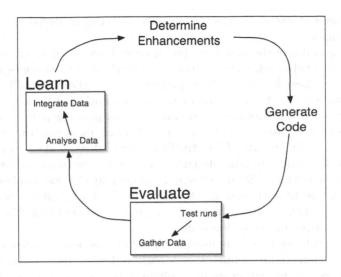

Fig. 1. The PILOT cycle

Code Generation. PILOT depends upon a code generator. Thus, the first task is to design a system in which the search algorithm and enhancements are written in a generic fashion, and can be easily enabled and disabled. It must also be easy to add new enhancements. Instrumentation must be embedded into this design to allow for the gathering of various measures of performance. Given such a framework, the process of designing a high-performance search engine for a specific application would be akin to fitting together pre-constructed blocks of code. Obviously there is a benefit in the fact that the code is already tested and the pieces are known to interact correctly. Nevertheless, there remains the problem of how to fit the provided pieces together in order to best accomplish a specific task.

PILOT currently generates code using the standard C/C++ preprocessor. Different enhancements are enabled and disabled via manipulation of a header file. The advantage of this approach is the speed of the generated code. While this approach seems the most straightforward, it is nevertheless challenging to code various enhancements independently such that any combination will produce correct code. This is not a simple matter of additive coding, as enhancements can have issues of dependence, precedence, mutual exclusion, and so forth. The prototype contains up to eight enhancements (with plans to increase this number), discussed in Subsection 2.3. Additionally, it has been instrumented to acquire a number of useful tree statistics.

The strength of this method is that it need not be merely a tool for automation with PILOT. The code generation can be leveraged for other tasks, such as more efficient evaluation of new enhancements, a starting point for expert programmers, or a tool for beginners with which to explore search algorithms.

Evaluation. Once the code generator produces a search engine for the task at hand, PILOT enters the evaluation phase. This phase consists of assessing the search engine and gathering data from which intelligent decisions can subsequently be made.

PILOT can evaluate the search engine in several ways. Currently, assessment is based on an evaluation suite. Suites are comprised of a number of move sequences. PILOT has the capability to generate these suites for itself as needed, use test cases supplied by a human, or use a combination of both.

The purpose of the experiments detailed in this paper was to verify the effectiveness of the entire cycle in the context of tournament performance against programs written by humans. Thus, the following settings might be seen as modest. Evaluation suites were generated using a plain $\alpha\beta$ search engine for the given game. From the initial position, the search engine played some number of moves into the game using some consistent search depth. The position thus reached becomes the starting position for a sequence of related positions.[2] Several such sequences comprise an evaluation suite.

When the evaluation suite is used, each initial position in a sequence is set up, and the search engine evaluates each position in the sequence. It is possible that the search engine will generate a different move from the one that leads to the next position in the sequence. PILOT notes this and continues with the next *predetermined* position in the test set.[3] The detection of a different move selection may be used in the evaluation process at a later date. This process repeats for every sequence in the evaluation suite.

In carrying out the evaluation, two sorts of numbers can result: performance metrics and search tree statistics. Performance metrics evaluate the effectiveness of a particular search engine; that is, how well it performs. Search tree statistics quantify the characteristics of a search tree. The combination of these two types of data is used to suggest what enhancements to try on the next iteration of PILOT's learning cycle. The line between these two aspects is not necessarily sharp: a search tree statistic might be used as a performance metric. Nevertheless, performance metrics compare different search engines with an eye to which will win more frequently. Search tree statistics are values which *may* correlate to the performance.[4] Thus, performance metrics indicate which search engine will win, while the search tree statistics may tell us why.

For the experiments reported in this paper, search tree size was chosen as the performance metric. This choice was made for its initial simplicity: a smaller tree indicates a more efficient search.[5] Thus, as a baseline, a smaller tree is better. Clearly there are trade-offs. For example, if generating a smaller tree takes twice as much time as generating a larger tree, then the larger tree may

---

[2] This is important because some enhancements require a context in which to operate, and testing them on randomly generated, unrelated positions would be ineffectual.

[3] That is, each search engine is tested on the same sequence of related positions. This keeps testing consistent.

[4] Though how they correlate may not be readily apparent.

[5] It is significant to note that this depends on the enhancements used. The tree may become smaller due to enhancements which induce more $\alpha\beta$ cutoffs; or search-reduction enhancements may directly reduce the size of the tree. These are not necessarily equivalent, and this is under investigation.

arrive at the desired result first! Initial experiments with a generic hash function for the transposition table showed precisely this disparity, and indicated that a domain-specific incremental hash function was an essential addition to the code if transposition tables were to be evaluated fairly. Similarly, adding a search extension increases the tree size for a search of some nominal depth. In this case, the size of the search tree does not convey any gain: the tree appears larger!

More interesting performance metrics include move quality and time elapsed. However, this makes comparisons a non-trivial matter, especially as the quality of a move is not easy to define. Tournament results or self-play may also prove a good performance indicator. Thus, while these are beyond the scope of the current experiments, there is clearly potential for future work.

Analysis and Learning. In Fig. 1, PILOT's next phase consists of analysis and learning from the experimental data. This involves making intelligent decisions from the statistics generated in the evaluation phase and integrating it with the acquired knowledge of the system. The current prototype of PILOT, however, is much more simplistic. It executes a straightforward hill-climbing algorithm based on search tree size, which is exactly what a human would do. While PILOT *is* currently able to gather detailed search tree statistics, this ability is not used in these initial experiments. Thus, this section is provided to indicate some of the intended directions of future work. PILOT's current hill-climbing approach is discussed in more detail in the next section.

Enhancement Selection. In the enhancement selection phase, PILOT determines what set of search enhancements to try in the next iteration. This phase is closely related to both the code generation and learning phases. In the initial stage of the search, PILOT can start either from a plain, unenhanced $\alpha\beta$ search or from a user specified starting point – a 'suggestion'.

Future incarnations of PILOT, upon arriving at the enhancement selection phase, will use acquired knowledge to direct the choice of what to try next. However, for the sake of an initial deterministic and easily understood baseline, PILOT currently performs a greedy search: from any state S, each unused enhancement is tested in turn (currently without the benefit of parameter tuning). The enhancement, e, which results in the greatest reduction in the size of the search tree compared to S, is enabled. That is, a new state $S' = S + e$ is created. This process is iterated with S' as the new starting state until no further improvement is possible. Thus, PILOT hill-climbs in much the way a human expert would,[6] but without human intervention, and with the potential to run many more tests than would be feasible for a human.

---

[6] With one difference. A human will sometimes *disable* an enhancement, even after it has been shown to be beneficial. Such testing might find a combination of lesser enhancements which might increase performance when interaction with another enhancement is removed. The current version of PILOT will miss these opportunities: PILOT sees the starting state only as something to which enhancements can be added, not from which they can be removed. It was felt that this was a reasonable decision in the context of the current set of enhancements. However, PILOT is rapidly outgrowing this decision, and this capability will be added.

Thus, while we hope to see good enhancements selected, we also hope that not *all* enhancements are selected, or the benefits will be unclear compared to simply selecting all enhancements. However, this is partly a function of the evaluation method used. Since we use search tree sizes, and most enhancements reduce the size of the tree, a certain amount of correlation must be expected. Nevertheless, early experiments with the games of Awari and Critical Mass confirmed the effectiveness of this method by demonstrating different enhancements being selected depending on both the domain and the evaluation technique.[7] For example, the transposition tables in Awari were implemented generically, and were therefore computationally expensive. Therefore, when PILOT evaluated this domain using timing data, transposition tables were seen to degrade search performance, and PILOT decided against using them.

Having determined which (if any) enhancements to use, PILOT generates a header file to enable the appropriate enhancements. Then, the code generator is commanded to build a new search engine with the indicated set of enhancements. It is PILOT's job to ensure that enhancement dependencies and prerequisites are met for any set of enhancements it passes to the code generator.

2.3   Enhancements

PILOT currently supports the following set of basic $\alpha\beta$ enhancements: transposition tables (TT), transposition table move ordering (TTO), history heuristic (HH), killer heuristic (KH), iterative deepening by 1 or 2 ply (ID1 or ID2), extending the search by 1 ply in positions with only a single move (EXT), and Principal Variation Search (PVS).

Transposition tables are the most commonly used method of information caching [13]. They are an important enhancement because a transposition table entry can serve three main purposes: narrowing of $\alpha\beta$ bounds, a source of move ordering, and a source of immediate cutoffs in the search. Parameters that can be learnt include the size of the table,[8] the replacement scheme of the hash table, and when to reset the table. These are parameters which future versions of PILOT will be able to address in more detail.

As just mentioned, transposition table move ordering is one use of a transposition table. At its simplest, it involves searching the move retrieved from the transposition table first. However, enhancements like ETC [17] and Presearch [8] provide information which might be used to further order moves at a node. Additionally, there exists the possibility of determining precedence between different move-ordering enhancements. Presently, PILOT uses a static precedence among move-ordering techniques, though we intend to alter all these parameters in future experiments.

---

[7] For completeness, experiments were also performed on a 'random' game where all positions and moves were uncorrelated. A random position generated a random number of random moves which led to random successor states. PILOT was unable to determine any enhancements to be useful in this 'game'.

[8] Larger is not necessarily better [11,17].

The history heuristic [13] is a mechanism whereby the moves in the search have an associated global score which is incremented when a move is determined to be the best in some position. When a new position is encountered, the history score of each move available at that position is used as an indicator of which ones are likely to be best. Whereas a transposition table stores the exact context of a move, the history heuristic is a table of moves with associated scores indicating the frequency of their 'goodness'. Parameters which might be generated automatically include the increment factor, the scaling factor, and the frequency of scaling.

By contrast, the killer heuristic [1] stores some number of refuting moves for each level of the search tree. The killer heuristic can be viewed as a special case of the history heuristic. However, the killer heuristic uses less memory and might be preferred in certain situations. Parameters to tune for the killer heuristic are the number of killer moves stored per ply, and the replacement scheme.

Iterative deepening's main purposes are for move ordering at the root of the search, and for operating under time constraints. While iterative deepening might be used to assist in move ordering at interior nodes, the main parameter of interest is the amount of the increment between iterations: should the search iterate by one ply at a time, two, or possibly more? Changing this value can help to balance odd-even effects of the evaluation function. Being able to determine it automatically when providing a new evaluation function would be useful. Currently, PILOT chooses between iterating either one or two ply at a time.

Search extensions are useful in situations when it is clear that a position is unresolved, or when the choices available are too minimal to be likely to provide useful information. In these situations, search is extended an additional ply in an effort to overcome the horizon effect. A straightforward search extension is to extend the search in any position where there is only a single available move. Since it takes no effort to choose this move, further effort can be spent looking deeper instead. However, two parameters are of utmost importance: the amount of the extension (i.e., extend the search by *how many* ply), and the choice of a maximum bound. That is, if an extended position leads to a position which is also extendable, at some point this process must be curtailed, lest the current line be searched to unreasonable depth at the cost of the rest of the search.

Principal Variation Search, also known as minimal-window, null-window, or zero-window search, is a basic windowing enhancement which improves on alpha-beta by using the smallest possible bounds to search moves which are not in the main line. Re-search may be required if the bounds do not hold, but the minimal bounds allow for small reductions in the tree. The more stable the main line is, the better PVS fairs.

3   Experiments

To evaluate the feasibility and effectiveness of PILOT's approach, several games were written within the PILOT search framework. These include Tictactoe, Hexpawn, Awari, Critical Mass, Halma, Lose Checkers, and Ataxx. Additionally, a random game was also created, as discussed briefly in Section 2.2. Of these, Tic-tactoe and Hexpawn are trivial games. The others, however, are non-trivial. It is also worth mentioning the increasing ease with which games can be added to this system. Each of the games here was implemented in day or two, and this is typically all that is now required to have a fully functioning game! The difficult part, of course, is coming up with a user-supplied evaluation function. For all games here, a simple evaluation function based on material difference was used. This allows PILOT to be evaluated solely in terms of the search it generates and not on the cleverness of the evaluation.

Awari is fairly well known in the games research community. It is played on a
2 × 6 board. Two players sit opposite the long sides of the board. Initially, each 'pit' contains 4 stones. Players take turns sowing their stones around the board in an effort to capture the most stones. Awari was solved by Romein and Bal [12].

Critical Mass is played on a 5 × 6 board, though other sizes are possible. Each square is assigned a 'critical mass' (CM) number which is equal to the number of adjacent horizontal and vertical squares. Starting with an empty board, players alternate in placing 'protons' on any square that is empty or already occupied by that player's own protons. When the number of protons on a square exceeds its CM, the square 'explodes': it becomes empty, and exactly one proton is added to each adjacent horizontal and vertical square. If any of the adjacent squares belonged to the opponent, they now belong to the current player. An explosion can lead to other squares exceeding their CM. Explosions are continued until no square is above critical mass. The game ends when one player owns all the squares on the board.

Lose Checkers is checkers where you play to lose all of your pieces.

Ataxx is played on an $n \times n$ board. Players take turns making either jump moves or cloning moves. The latter creates a new piece on an adjacent square, while the former moves a piece to a new location with a distance[9] of 2 from the starting square. The goal is to have the most pieces at the end of the game, and the game ends when there are no legal moves for either player, a position has repeated three times, or 50 jump moves have been made in a row [3].

3.1 Experiments Summary

The experiments reported here cover several phases in PILOT's development. The initial thrust was to test PILOT's greedy search strategy on two nontrivial games, Awari and Critical Mass. As PILOT was developed, more enhancements were added. After the initial stages of development, it was determined that it was necessary to evaluate PILOT in comparison to hand-crafted search engines. Assignments from a graduate course in heuristic search were made available for the games of Lose Checkers and Ataxx. These programs were hand-crafted for a course tournament, and thus represent an intensive month's work on the part of each student. Previous experience in this environment shows that the

---

[9] In at least one of the x or y components.

student programs typically range from extremely good (far outstripping any human players) through moderate to poor. Thus, it was determined that this was an excellent testbed in which to evaluate PILOT's automatic search generation.

We have two types of results for PILOT: the results of the enhancement selection process, and the results of the two tournaments. In the former, we would like to see different results for different games. In the latter, we would like confirmation that PILOT's automatic generation of a search algorithm does a comparable (or better) job than graduate students.

PILOT's results for the enhancement selection process are illustrated in Table 1. An interesting side-effect of the enhancement selection process is the order in which the enhancements are selected by the greedy search. This ranking, therefore, is an indication of the relative amount of benefit each enhancement adds to the particular domain. This has been indicated in the table. These experiments were produced by running PILOT through its cycle as discussed in Section 2. Specifically, PILOT generates and runs evaluation suites, as in Subsection 2.2. Evaluation suites were generated with fixed depth searches at one depth, and then used for search-engine assessment with an equal or deeper (but still fixed) depth. Since different domains can be searched to widely varying depths, for these experiments the depths were chosen manually for each domain in order to control the running time. As an example, Awari generated evaluation suites with a depth-5 search, and used the resulting suites with depth-10 searches.

Table 1. Enhancement selection by PILOT in a number of games and situations

Game	Enhancements							
	TT	ID1	ID2	TTO	HH	KH	EXT	PVS
Awari	1	3	—	2			—	—
Critical Mass	1	4	—	3	2	—	—	—
Critical Mass	2	5	—	4	3	1	—	—
LoseCheckers	1	+		3	2		—	—
LoseCheckers (PVS)	1		4	2	3			5
Ataxx (shallow)	3		2	4	1		5′	—
Ataxx (deep)	2	3		4	1	5	6′	—

Each row of Table 1 indicates a game. The columns indicate which enhancements are enabled. The enhancements are identified by the abbreviations noted in the list in Subsection 2.3. Each cell in the table has one of the following indicators:

$1 \cdots n$ A number indicates the order in which the enhancement was added to the search.

— Indicates an enhancement that was not yet implemented at the time of this experiment.

A blank space indicates an enhancement that was *not* selected for this experiment.

$+$ Indicates an enhancement not selected by PILOT, but turned on for the time-controlled tourments.

n' Indicates an enhancement that was chosen not because it caused an improvement, but because it was not detrimental in the tests. This is essentially an indication that the tests in these cases were not sufficiently broad.

It should be noted that there are multiple experiments reported for Critical Mass, Ataxx, and Lose Checkers. The difference in the Critical Mass experiments is that in the first one, the killer heuristic had not been implemented. Adding this enhancement to Critical Mass caused it to be selected first, which is noteworthy. Similarly, Lose Checkers (PVS) is the most recent addition, and in addition to the PVS enhancement it includes a re-written search engine, the combination of which account for the differences.

Due to Ataxx's high branching factor, a shallow run was done with PILOT where test positions were searched to a depth of 3. Later, a deeper test was done with positions searched to a depth of 6. The table shows two differences: a different iterative deepening increment, and the killer heuristic becoming useful in the deeper searches where it had been left out in the shallower tests. This is significant because it hints at the changing characteristics of the search tree as the depth increases, and also because the shallower search approximates the deeper one. (Only one enhancement is different, and the relative orders are the same.) This phenomenon bears further study, as it may be related to observations by [16] that for best performance the search depth at the time of learning should match the intended search depth (e.g., for tournaments).

It is also interesting to note that the transposition table is not always the first enhancement selected. When the games involved are considered, this may be explained. In the case of Ataxx, certain moves, or types of moves (jumps versus non-jumps) are extremely important. If the history heuristic is learning to favour one over the other, this total move ordering might be more effective than elimination of cycles, which are less prominent in Ataxx. Similarly in Critical Mass, transpositions are not as vital, since a single move can change the configuration of the entire board. Certain key moves will initiate a massive chain reaction across the entire board. Searching these moves first can save significant search effort.

By contrast, Lose Checkers is a more classical board game involving a diminishing number of moving pieces. We would expect transpositions to be prevalent, and indeed they are the first enhancement chosen. Since PILOT is selecting enhancements solely on their ability to reduce search-tree size (independent of time constraints), iterative deepening is at a disadvantage and therefore not selected.[10] Similarly, EXT could not be chosen by PILOT in "Lose Checkers (PVS)" because EXT typically increases the tree size in an effort to enhance the *quality* of the search; PILOT cannot yet understand this distinction.

---

[10] A related issue is that early versions of the search engine may not have ordered root moves effectively when interacting with ID. This would affect ID's perceived usefulness. However, this favorably reflects on PILOT's sensitivity to the minutiae of implementation details. Later experiments bear out this sensitivity.

Thus, we see how PILOT not only suggests different combinations of enhancements for different games, but might give insights into why these choices are appropriate.

3.2 Lose Checkers Tournament Results

Lose Checkers has proved to be a most interesting game. PILOT's search-engine configurations are summarized in the previous section. For the reported tournament the configuration determined in "Lose Checkers (PVS)" in Table 1 was used.

The Lose Checkers tournament was played against sixteen programs. These programs had already played in a full double-round-robin tournament using 20 seconds per move. The most recent experiment had the following results.

Against 16 programs, PILOT scored 9 wins, 10 losses, and 13 draws, searching to an average depth of 21.6 ($\sigma = 4.3$).[11] Scores were assigned using the formula of one point per win and half a point per loss. Thus, PILOT scores 15.5 points. These data were then combined with the full-tournament result to determine PILOT's relative ranking.

PILOT ranked in seventh place in this combined result. Looking at the spread of points across the tournament shows that there are distinct groups of programs. The most exceptional programs scored between 27 and 32 points. The next group scored between 20 and 23, and a third group from 13.5 to 15.5. The next two groups scored between 10 and 11, and below 4.

Observations. Subsequent discussions with a few of the students who wrote the best programs were illuminating. The top programs, for example, used a random evaluation along with a parity checker which could solve one-on-one piece endgames. Their experiences indicate that a material evaluation such as PILOT's was frequently detrimental. It also appears that in Lose Checkers, deeper search is not necessarily correlated with success, particularly when using material evaluation.

PILOT ranks at the top of the third group of programs. Note that the second and third groups consisted of programs that are moderately strong, as opposed to the highly tuned programs of group one. Given PILOT's simple material-only evaluation, and that the other programs had significant time to come up with a search and evaluation combination that would be better than simple material, PILOT's performance is commendable. An examination of the games shows that PILOT generates deeper search than the majority of its opponents, and its limiting factor appears to be the evaluation. However, for these very reasons, Lose Checkers is an ideal testbed for PILOT: a subsequent challenge is to develop improved analysis and evaluation suites which better reflect tournament conditions.

3.3 Ataxx Tournament Results

For Ataxx, a simple depth extension was available to PILOT. Table 1 shows the learned result. The second ('deep') result indicated in the table was used for

---

[11] Each program plays black once and white once.

the tournament. It is rare in Ataxx that only a single move is available, and if it happens, it is usually in the endgame. It is clear from the fact that PILOT detected no change when using the depth extension that our automatic test generation is not comprehensive enough.

The Ataxx tournament originally consisted of 21 programs. Unfortunately, only 8 were available for this experiment. However, these 8 programs were a representative sample across the entire gamut. Thus, it is with reasonable confidence that this result is presented as representative.

The procedure was similar to Lose Checkers. The original Ataxx tournament was run with 15 seconds per move. The tournament with PILOT also used the same 15 second move limit. No draws occurred in either the PILOT tournament or the original Ataxx tournament.

PILOT tied for fourth place among nine programs with 10 wins and 6 losses, searching to an average depth of 7.9 ply ($\sigma = 0.94$). The top program had 16 wins, while the next group of four programs had from 10 to 12 wins. The final four programs had between 0 and 5 wins. Thus, PILOT again achieved a ranking in the second grouping: not outstanding, but in with the best. Once again, it is important to remember that PILOT is competing against programs which had their search and evaluation functions tuned specifically to Ataxx. PILOT is purposely handicapped by a simplistic material-difference evaluation in order to gauge the effectiveness of the search enhancements.

The rankings of the 8 opponents in the original tournament from best to worst, were: 1, 3, 5, 6, 9, 10, 16, and 20. Thus, the stronger programs were clearly better represented. It was somewhat surprising, therefore, when PILOT won and lost one game each against the third and fifth ranked programs; we were quite pleased.

4   Conclusions and Future Work

It is encouraging that even with some of the restrictive limitations placed on the prototype of PILOT, it was able to generate a search powerful enough to rank with the second class of hand-crafted search programs. When this is combined with the fact that PILOT's code, of necessity, must sacrifice a certain amount of speed for flexibility of implementation, and that domain-specific tuning of the evaluation function was nonexistent, the results seem even more promising.

Additionally, it has been shown that even using a greedy search across the space of available enhancements, PILOT offers insights into individual domains both by what it selects, and the manner in which it does it. Some of the results presented here suggest that PILOT might eventually make it possible to generate not just a single search engine for a game, but multiple search engines tailored to different phases of the same game. This is a practise that has remained difficult for humans.

Thus, there are plenty of areas for future exploration. With the demonstration of the basic soundness and effectiveness of PILOT's approach, several avenues present themselves. The most pressing requirements follow. While eight enhancements are a good start, they are not yet representative of what a human programmer can use. More enhancements are needed. Also required is a new system of evaluation more powerful than search tree size. This new method might consider time, move quality, and tournament or self-play results. The current greedy search is effective, but simplistic. It is tempting to see how far it can go, but a better method of exploration must be found as the number and types of enhancements grows.

A second important area is parameter tuning. In the course of coding and experimenting, it has been observed that minor changes in one enhancement (intentional or otherwise) clearly impact upon the entire process of enhancement selection, and in the selection or discarding of *other* enhancements. Thus, even the current version of PILOT is able to effectively perceive these subtle differences. What currently lacks is a mechanism for explicitly testing such configurations.

Finally, PILOT has the potential to generate a new search engine on a per-evaluation-function basis. Current results have used minimalist evaluation functions, and it will be extremely interesting to see what is possible with an evaluation function that has been tuned for a specific game.

The ultimate goal is to free humans from the drudgery of creating a new search engine for every new game that comes along. Freed of the requirements of endless implementation and testing, we will be able to focus on an exciting new set of problems.

Acknowledgements

We would like to acknowledge the assistance of the following people and organizations, and express our sincere thanks: Akihiro Kishimoto, for his development of and assistance with the Generic Game Server clients for Lose Checkers and Ataxx; Michael Buro, for making his GGS code available to the world at large [4]. This research was supported by the Natural Sciences and Engineering Research council of Canada (NSERC) and Alberta's Informatics Circle of Research Excellence (iCORE).

References

1. A. G. Akl and M. M. Newborn. The Principle Continuation and the Killer Heuristic. In *ACM Annual Conference*, pages 466–473, 1977.
2. Y. Björnsson. *Selective Depth-First Game-Tree Search*. PhD thesis, University of Alberta, 2002.
3. M. Buro. Rules of Ataxx, 2004.
   http://www.cs.ualberta.ca/ ~mburo/ggsa/ax.rules.
4. M. Buro. Generic Game Server, 2005. http://www.cs.ualberta.ca/~mburo/.
5. M. Buro. ProbCut: An Effective Selective Extension of the $\alpha\beta$ Algorithm. *ICCA Journal*, 18(2):71–76, 1995.
6. D.J. Cook and R.C. Varnell. Adaptive Parallel Iterative Deepening Search. *Journal of Artificial Intelligence Research*, 9:167–194, 1999.

7. J. Fürnkranz. Machine Learning in Games: A Survey. In Johannes Fürnkranz and Miroslav Kubat, editors, *Machines that Learn to Play Games*, chapter 2, pages 11–59. Nova Science Publishers, Huntington, NY, 2001.
8. M. Hlynka and J. Schaeffer. Pre-Searching. *ICGA Journal*, 27(4):203–208, 2004.
9. P. Hoffman. *Archimedes' Revenge: The Joys and Perils of Mathematics*. W. W. Norton & Company, May 1988.
10. D.E. Knuth and R. Moore. An analysis of Alpha-Beta Pruning. *Artificial Intelligence*, 6(4):293–326, 1975.
11. A. Plaat. *Research Re: Search & Re-search*. PhD thesis, Erasmus University, Rotterdam, June 1996.
12. J.W. Romein and H.E. Bal. Solving Awari with Parallel Retrograde Analysis. *IEEE Computer*, 36(10):26–33, October 2003.
13. J. Schaeffer. The History Heuristic and Alpha-Beta Search Enhancements in Practice. *IEEE Transactions on Pattern Analysis and Machine Intelligence*, PAMI-11(11):1203–1212, 1989.
14. J. Schaeffer. *One Jump Ahead*. Springer-Verlag, New York, 1997. ISBN 0387949305.
15. J. Schaeffer. A Gamut of Games. *AI Magazine*, 22(3):29–46, Fall 2001.
16. J. Schaeffer, M. Hlynka, and V. Jussila. Temporal Difference Learning Applied to a High Performance Game. In *International Joint Conference on Artificial Intelligence (IJCAI-01)*, pages 529–534, 2001.
17. J. Schaeffer and A. Plaat. New Advances in Alpha-Beta Searching. In *ACM Computer Science Conference*, pages 124–130, 1996.
18. J. Schaeffer, A. Plaat, and A. Junghanns. Unifying Single-Agent and Two-Player Search. *Information Sciences*, 135(3-4):151–175, 2001.
19. Zillions Development Corporation. Zillions of Games – Unlimited Board Games & Puzzles, 1998–2005. http://www.zillionsofgames.com/.

RSPSA: Enhanced Parameter Optimization in Games

Levente Kocsis[1], Csaba Szepesvári[1], and Mark H.M. Winands[2]

[1] MTA Sztaki, Budapest, Hungary
{kocsis, szcsaba}@sztaki.hu
[2] Institute for Knowledge and Agent Technology,
MICC, Universiteit Maastricht, Maastricht, The Netherlands
m.winands@micc.unimaas.nl

Abstract. Most game programs have a large number of parameters that are crucial for their performance. Tuning these parameters by hand is rather difficult. Therefore automatic optimization algorithms in game programs are interesting research domains. However, successful applications are only known for parameters that belong to certain components (e.g., evaluation-function parameters). The SPSA (Simultaneous Perturbation Stochastic Approximation) algorithm is an attractive choice for optimizing any kind of parameters of a game program, both for its generality and its simplicity. Its disadvantage is that it can be very slow.

In this article we propose several methods to speed up SPSA, in particular, the combination with RPROP, using common random numbers, antithetic variables, and averaging. We test the resulting algorithm for tuning various types of parameters in two domains, Poker and LOA. From the experimental study, we may conclude that using SPSA is a viable approach for optimization in game programs, in particular if no good alternative exists for the types of parameters considered.

1 Introduction

Any reasonable game program has several hundreds if not thousands of parameters. These parameters belong to various components of the program, such as the evaluation function or the search algorithm. While it is possible to make educated guesses about "good" values of certain parameters, hand-tuning the parameters is a difficult and time-consuming task. An alternative approach is to find the "right" values by means of an automated procedure.

The use of parameter optimization methods for the performance tuning of game programs is difficult by the fact that the objective function is rarely available analytically. Therefore, methods that rely on the availability of an analytic expression for the gradient cannot be used. However, there exist several ways to tune parameters despite the lack of an analytic gradient. An important class of such algorithms is represented by temporal-difference (TD) methods that have been used successfully in tuning evaluation-function parameters [14]. Obviously, any general-purpose (gradient-free) global search method can be used for parameter optimization in games. Just to mention a few examples, in [3] genetic

H.J. van den Herik et al. (Eds.): ACG11, LNCS 4250, pp. 39–56, 2006.

algorithms were used to evolve a neural network to play checkers, whilst in [2] an algorithm similar to the Finite-Difference Stochastic Approximations (FDSA) algorithm was used successfully for tuning the search-extension parameters of CRAFTY. Nevertheless, we believe that automatic tuning of parameters remains a largely unexplored area of game programming.

In this article we investigate the use of SPSA (Simultaneous Perturbation Stochastic Approximation), a stochastic hill-climbing search algorithm for tuning the parameters of game programs. Since optimization algorithms typically exhibit difficulties when the objective function (performance measure) is observed in heavy noise, for one test domain we choose a non-deterministic game, namely Omaha Hi-Lo Poker, one of the most complex poker variants. For Texas Hold'em Poker several years of research has led to a series of strong programs: POKI, PSOPTI, and VEXBOT [1]. Our program, McRAISE, borrows several ideas from the above mentioned programs. The name of the program originates from the use of Monte-Carlo simulations and the program's aggressive style. In the second test domain, LOA, we use MIA, winner of the 8th and 9th Computer Olympiad.

The article is organized as follows. Section 2 describes the RSPSA algorithm that combines SPSA and RPROP. In Sect. 3, three ways to enhance the performance of RSPSA are proposed together with a discussion of the various trade-offs involved, supported by analytic arguments. Next, in Sect. 4 the test domains and the respective programs are described. Experiments with RSPSA in these domains are given in Sect. 5. Finally, we draw our conclusions in Sect. 6.

2   The RSPSA Algorithm

Below we start describing the basic setup (2.1) of the RSPSA algorithm. Then we provide details on the supporting algorithms SPSA (2.2) and RPROP (2.3). Finally in 2.4 we outline the RSPSA algorithm.

2.1   Basic Setup

Consider the task of finding a maximizer $\theta^* \in \mathbb{R}^d$ of some real-valued function f, i.e., find $\theta^* = \mathrm{argmax}_\theta f(\theta)$. In our case f may measure the performance of a player in some environment (e.g., against a fixed set of opponents), or it may represent an auxiliary performance index of interest that is used internally in the algorithm in such a way that a higher value of it might ultimately yield better play. In any case, θ represents some parameters of the game-playing program.

We assume that the algorithm of which the task is to tune the parameters θ can query the value of f at any point θ, but the value received by the algorithm will be corrupted by noise. The noise in the evaluation of f can originate from randomized decisions of the players or from the randomness of the environment. In a card game for instance the cards represent a substantial source of randomness in the outcomes of rounds. We shall assume that the value observed in the t-th step of the algorithm, when the simulation is run with parameter θ_t, is given by $f(\theta_t; Y_t)$ where Y_t is some random variable such that the expected value of $f(\theta_t; Y_t)$ conditioned on θ_t and given all past information equals to $f(\theta_t)$:

$$f(\theta_t) = \mathbb{E}\left[f(\theta_t; Y_t)\,|\,\theta_t, \mathcal{F}_t\right],\tag{1}$$

where $\mathcal{F}_t$ is the sigma-field generated by $Y_0, Y_1, \ldots, Y_{t-1}$ and $\theta_0, \theta_1, \ldots, \theta_{t-1}$. Stochastic gradient ascent algorithms work by changing the parameter θ in a gradual manner so as to increase the value of f on average:

$$\theta_{t+1} = \theta_t + \alpha_t \hat{g}_t(\theta_t).\tag{2}$$

Here θ_t is the estimate of θ^* in the t-th iteration (time step), $\alpha_t \geq 0$ is a learning rate parameter that governs the size of the changes to the parameters and $\hat{g}_t(\theta_t)$ is some approximation to the gradient of f such that the expected value of $\hat{g}_t(\theta_t)$ given past data is equal to the gradient $g(\theta) = \partial f(\theta)/\partial\theta$ of f and $(\hat{g}_t(\theta_t) - g(\theta))$ has finite second moments.

2.2 SPSA

When f is not available analytically then one must resort to some approximation of the gradient in order to use gradient ascent. One such approximation was introduced with the SPSA algorithm in [12]:

$$\hat{g}_{ti}(\theta_t) = \frac{f(\theta_t + c_t\Delta_t; Y_t^+) - f(\theta_t - c_t\Delta_t; Y_t^-)}{2c_t\Delta_{ti}}.\tag{3}$$

Here $\hat{g}_{ti}(\theta_t)$ is the estimate of the i-th component of the gradient, Δ_{ti}, Y_t^+, and Y_t^- are random variables: Y_t^+ and Y_t^- are meant to represent the sources of randomness of the evaluation of f, whilst Δ_t is a perturbation vector to be chosen by the user. Note that the numerator of this expression does not depend on the index i and therefore evaluating Eq. 3 requires only two (randomized) measurements of the function f. Still, SPSA provides a good approximation to the gradient: Under the conditions that (i) the random perturbations Δ_t are independent of the past of the process, (ii) for any fixed t, $\{\Delta_{ti}\}_i$ is an i.i.d. sequence[1], (iii) the distribution of Δ_{ti} is symmetric around zero, (iv) $|\Delta_{ti}|$ is bounded with probability one, and (v) $\mathbb{E}\left[\Delta_{ti}^{-1}\right]$ is finite, and assuming that f is sufficiently smooth, it can be shown that the bias of estimating that gradient $g(\theta_t)$ by $\hat{g}_t(\theta_t)$ is of the order $O(c_t^2)$. Further, the associated gradient ascent procedure can be shown to converge to a local optima of f with probability one [12].

A simple way to satisfy the conditions on Δ_t is to choose its components to be independent ± 1-valued Bernoulli distributed random variables with each outcome occurring with probability $1/2$. One particularly appealing property of SPSA is that it might need d times less measurements than the classical FDSA procedure and still achieve the same asymptotic statistical accuracy (see, e.g., [12]). FDSA works by evaluating f at $\theta_t \pm c_t e_i$ and forming the appropriate

---

[1] "i.i.d." is the shorthand of "independent, identically distributed".

differences – thus it requires $2d$ evaluations. For a more thorough discussion of SPSA, its variants, and its relation to other methods we refer to [12,13].

SPSA, like other stochastic approximation algorithms has quite a few tunable parameters. These are the gain sequences α_t, c_t and the distribution of the perturbations Δ_t. When function evaluation is expensive, as is often the case in games, small sample behavior of the algorithm becomes important. In that case the proper tuning of the parameters becomes critical.

In practice, the learning rate α_t and the gain sequence c_t are often kept at a fixed value. Further, in all previous works on SPSA known to us it was assumed that the perturbations Δ_{ti}, $i = 1, \ldots, d$, have the same distribution. When different dimensions have different scales (which we believe is a very common phenomenon in practice) then it does not make too much sense to use the same scales for all the dimensions. The issue is intimately related to the issue of scaling the gradient addressed also by second and higher-order methods. These methods work by utilising information about higher order derivatives of the objective function (see, e.g., [4,13]). In general, these methods achieve a higher asymptotic rate of convergence, but, as discussed, e.g., in [15], their practical value might be limited in the small sample size case.

2.3   RPROP

The RPROP ("resilient backpropagation") algorithm [11] and its variants are amongst the best performing first-order batch neural network gradient training methods and as such represent a viable alternative to higher-order methods.[2] In practice RPROP methods were found to be very fast and accurate, robust to the choices of their parameters, scale well with the number of weights. Further, RPROP is easy to implement, it is not sensitive to numerical errors and since the algorithm is dependent only on the sign of the partial derivatives of the objective function,[3] it is thought to be suitable for applications where the gradient is numerically estimated and/or is noisy.

A particularly successful variant is the iRPROP⁻algorithm [5]. The update equations of iRPROP⁻ for maximising a function $f = f(\theta)$ are as follows:

$$\theta_{t+1,i} = \theta_{t,i} + \text{sign}(g_{ti})\delta_{ti}, \quad t = 1, 2, \ldots; i = 1, 2, \ldots, d. \tag{4}$$

Here $\delta_{ti} \geq 0$ is the step size for the i-th component and g_t. is a gradient-like quantity:

$$g_{ti} = \mathbb{I}(g_{t-1,i}f_i'(\theta_t) \geq 0)f_i'(\theta_t), \tag{5}$$

i.e., g_{ti} equals the i-th partial derivative of f at θ except when a sign reversal is observed between the current and the previous partial derivative, in which case g_{ti} is set to zero.

---

[2] For a recent empirical comparison of RPROP and its variants with alternative, gradient optimization methods such as BFGS, CG and others see, e.g., [5].

[3] RPROP, though it was worked out for the training of neural networks, is applicable in any optimization problem where the gradient can be computed or approximated.

The individual step-sizes δ_{ti} are updated in an iterative manner based on the sign of the product $p_{t,i} = g_{t-1,i} f_i'(\theta_t)$:

$$\eta_{ti} = \mathbb{I}(p_{t,i} > 0)\eta^+ + \mathbb{I}(p_{t,i} < 0)\eta^- + \mathbb{I}(p_{t,i} = 0), \qquad (6)$$
$$\delta_{ti} = P_{[\delta^-, \delta^+]}(\eta_{ti}\delta_{t-1,i}), \qquad (7)$$

where $0 < \eta^- < 1 < \eta^+$, $0 < \delta^- < \delta^+$, $P_{[a,b]}$ clamps its argument to the interval $[a,b]$, and $\mathbb{I}(\cdot)$ is a $\{0,1\}$-valued function working on Boolean values and $\mathbb{I}(\mathcal{L}) = 1$ if and only if $\mathcal{L}$ is true, and $\mathbb{I}(\mathcal{L}) = 0$, otherwise.

2.4 RSPSA

Given the success of RPROP, we propose a combination of SPSA and RPROP (in particular, a combination with iRPROP$^-$). We call the resulting combined algorithm RSPSA ("resilient SPSA"). The algorithm works by replacing $f_i'(\theta_t)$ in Eq. 5 with its noisy estimates $\hat{g}_{t,i}(\theta_t)$. Further, the scales of the perturbation vector Δ_{ti} are coupled to the scale of the step sizes of δ_{ti}.

Before motivating the coupling let us make a few observations on the expected behavior of RSPSA. Since iRPROP$^-$ depends on the gradient only through the sign of it, it is expected that if the sign of $\hat{g}_{t,i}(\theta_t)$ coincides with that of $f_i'(\theta_t)$ then the performance of RSPSA will be close to that of iRPROP$^-$. This can be backed up by the following simple argument. Assuming that $|f_i'(\theta)| > \varepsilon$, applying Markov's inequality, we obtain that

$$\mathbb{P}(\text{sign}(\hat{g}_{t,i}(\theta)) \neq \text{sign}(f_i'(\theta))) \leq \mathbb{P}(|\hat{g}_{t,i}(\theta)) - f_i'(\theta)| \geq \varepsilon) \leq \frac{M_{t,i}}{\varepsilon^2}, \qquad (8)$$

where $M_{t,i} = \mathbb{E}\left[(\hat{g}_{t,i}(\theta) - f_i'(\theta))^2 | \mathcal{F}_t\right]$ denotes the mean square error of the approximation of $f_i'(\theta)$ by $\hat{g}_{t,i}(\theta)$, conditioned on past observations. In fact, this error can be shown to be composed of a bias term dependent only on f, θ and c, and a variance term dependent on the random quantities in $\hat{g}_{t,i}(\theta)$. Hence, it is important to make the variance of the estimates small.

Now, let us turn to the idea of coupling the scales of the perturbation vectors to the step sizes of iRPROP$^-$. This idea can be motivated as follows. On "flat areas" of the objective function, where the sign of the partial derivatives of the objective function is constant and where the absolute value of these partial derivatives is small, a perturbation's magnitude along the corresponding axis should be large or the observation noise will dominate the computed finite differences. In contrast, in "bumpy areas" where the sign of a partial derivative changes at smaller scales, smaller perturbations that fit the "scale" of desired changes can be expected to perform better. Since the step-size parameters of RPROP are larger in flat areas and are smaller in bumpy areas, it is natural to couple the perturbation parameters of SPSA to the step-size parameters of RPROP. A simple way to accomplish this is to let $\Delta_{ti} = \rho\,\delta_{ti}$, where ρ is some positive constant, to be chosen by the user.

3   Increasing Efficiency

In this section we describe three methods that can be used to increase the efficiency of RSPSA. The first method, known as the "Method of Common Random Numbers", was proposed earlier to speed up SPSA [9,6]. The second method, averaging, was proposed as early as in [12]. However, when averaging is used together with the method of common random numbers a new trade-off arises. By means of a formal analysis this trade-off is identified and resolved here for the first time. To the best of our knowledge, the third method, the use of antithetic variables has not been suggested earlier to be used with SPSA. All these methods aim at reducing the variance of the estimates of the gradient, which as noted previously should yield better performance. In this section we will drop the time index t in order to simplify the notation.

3.1   Common Random Numbers

In SPSA (and therefore also in RSPSA) the estimate of the gradient relies on differences of the form $f(\theta + c\Delta; Y^+) - f(\theta - c\Delta; Y^-)$. Denoting by F_i^+ the first term and by F_i^- the second term, elementary probability calculus gives $\mathrm{Var}\left(F_i^+ - F_i^-\right) = \mathrm{Var}\left(F_i^+\right) + \mathrm{Var}\left(F_i^-\right) - 2\mathrm{Cov}\left(F_i^+, F_i^-\right)$. Thus the variance of the estimate of the gradient can be decreased by introducing some correlation between F_i^+ and F_i^-, provided that this does not increase the variance of F_i^+ and F_i^-. The reason is that by our assumptions Y^+ and Y^- are independent and thus $\mathrm{Cov}\left(F_i^+, F_i^-\right) = 0$. Now, if $F_i^{\pm}$ is redefined to depend on the *same* random value Y (i.e., $F_i^{\pm} = f(\theta \pm c\Delta; Y)$) then the variance of $F_i^+ - F_i^-$ will decrease when $\mathrm{Cov}\left(f(\theta + c\Delta; Y), f(\theta - c\Delta; Y)\right) > 0$. The larger this covariance is the larger the decrease of the variance of the estimate of the gradient will be.

When f is the performance of a game program obtained by means of a simulation that uses pseudo-random numbers then using the same random series Y can be accomplished by using identical initial seeds when computing the values of f at $\theta + c\Delta$ and $\theta - c\Delta$.

3.2   Using Averaging to Improve Efficiency

A second method to reduce the variance of the estimate of the gradient is to average many independent estimates of it. However, the resulting variance reduction is not for free, since evaluating $f(\theta; Y)$ can be extremely CPU-intensive, as mentioned earlier. To study this trade-off let us define

$$\hat{g}_{q,i}(\theta) = \frac{1}{q} \sum_{j=1}^{q} \frac{f(\theta + c\Delta; Y_j) - f(\theta - c\Delta; Y_j)}{2c\,\Delta_i}, \tag{9}$$

where according to the suggestion of the previous section we use the same set of random number to evaluate f both at $\theta + c\Delta$ and $\theta - c\Delta$. Further, let $\hat{g}_{r,q,i}(\theta)$ be the average of r independent samples $\{\hat{g}_{q,i}^{(j)}(\theta)\}_{j=1,\dots,r}$ of $\hat{g}_{q,i}(\theta)$. By the Strong Law of Large Numbers $\hat{g}_{r,q,i}(\theta)$ converges to $f_i'(\theta) + O(c^2)$ as $q, r \to +\infty$ (i.e., its ultimate bias is of the order $O(c^2)$). It follows then that increasing $min(r,q)$ above the value where the bias term becomes dominating does not improve the finite sample performance. This is because increasing p decreases the frequency of updates to the parameters.[4]

In order to gain further insight into how to choose q and r, let us consider the mean squared error (MSE) of approximating the i-th component of the gradient by $\hat{g}_{r,q,i}$: $M_{r,q,i} = \mathbb{E}\left[(\hat{g}_{r,q,i}(\theta) - f_i'(\theta))^2\right]$. By some lengthy calculations, the following expression can be derived for $M_{r,q,1}$:[5]

$$M_{r,q,1} = \frac{1}{r}\mathbb{E}\left[\Delta_1^2\right]\mathbb{E}\left[1/\Delta_1^2\right]\sum_{j=2}^{d}\left\{\left(1 - \frac{1}{q}\right)\mathbb{E}\left[f_j'(\theta;Y_1)^2\right] + \frac{1}{q}\mathbb{E}\left[f_j'(\theta;Y_1)\right]^2\right\}$$

$$+ \frac{1}{rq}\mathbb{E}\left[(f_1'(\theta;Y_1) - f_1'(\theta))^2\right] + O(c^2). \qquad (10)$$

Here $f_i'(\theta;Y)$ is the partial derivative of $f(\theta;Y)$ w.r.t. θ_i: $f_i'(\theta;Y) = \frac{\partial f(\theta;Y)}{\partial \theta_i}$.

It follows from Eq. 10 that for a fixed budget of $p = qr$ function evaluations the smallest MSE is achieved by taking $q = 1$ and $r = p$ (disregarding the $O(c^2)$ bias term which we assume to be "small" as compared to the other terms).

Now the issue of choosing p can be answered as follows. Under mild conditions on f and Y (ensuring that the expectation and the partial derivative operators can be exchanged), $\sum_{j=2}^{d}\mathbb{E}\left[f_j'(\theta;Y_1)\right]^2 = \sum_{j=2}^{d}f_j'(\theta)^2$. Hence, in this case with the choices $q = 1$, $r = p$, $M_{p,1,1}$ becomes equal to

$$\frac{1}{p}\left\{\mathbb{E}\left[\Delta_1^2\right]\mathbb{E}\left[1/\Delta_1^2\right]\sum_{j=2}^{d}f_j'(\theta)^2 + \mathbb{E}\left[(f_1'(\theta;Y_1) - f_1'(\theta))^2\right]\right\} + O(c^2), \quad (11)$$

which is composed of two terms in addition to the bias term $O(c^2)$: the term $\sum_{j=2}^{d}f_j'(\theta)^2$ represents the contribution of the "cross-talk" of the derivatives of f to the estimation of the gradient, whilst the second term, $\mathbb{E}\left[(f_1'(\theta;Y_1) - f_1'(\theta))^2\right]$ gives the MSE of approximating $f_1'(\theta)$ with $f_1'(\theta;Y_1)$ (which is equal to the variance of $f_1'(\theta;Y_1)$ in this case). The first term can be large when θ is far from a stationary point of f, whilst the size of the second term depends on the amount of noise in the evaluations of f. When the magnitude of these two terms is larger than that of the bias term $O(c^2)$ then increasing p will increase the efficiency of the procedure, at least initially.

---

[4] In [12] it is shown that using decreasing gains $\alpha_t = a/t^\alpha$ and $c_t = c/t^\gamma$ with $\beta = \alpha - 2\gamma > 0$, $0 < \alpha \leq 1$, $0 < \gamma$, the optimal choice of p is governed by an equation of the form $p^{\beta-1}A + p^\beta B$, where $A, B > 0$ are some (unknown) system parameters. This equation has a unique minimum at $p = (1-\beta)A/(\beta B)$, however, since A, B are unknown parameters this result has limited practical value besides giving a hint about the nature of the trade-off in the selection of p.

[5] Without the loss of generality we consider the case $i = 1$.

3.3 Antithetic Variables

In Sect. 3.1 we have advocated the introduction of correlation between the two terms of a difference to reduce its variance. The same idea can be used to reduce the variance of averages: Let $U_1, U_2, \ldots, U_n$ be i.i.d. random variables with common expected value I. Then the variance of $I_n = 1/n \sum_{i=1}^{n} U_i$ is $1/n \mathrm{Var}\,(U_1)$. Now, assume that n is even, say $n = 2k$ and consider estimating I by

$$I_n^A = \frac{1}{k} \sum_{i=1}^{k} \frac{U_i^+ + U_i^-}{2}, \tag{12}$$

where now it is assumed that $\{U_1^+, \ldots, U_k^+\}$ are i.i.d., just like $\{U_1^-, \ldots, U_k^-\}$, $\mathbb{E}\left[U_i^+\right] = \mathbb{E}\left[U_i^-\right] = I$. Then $\mathbb{E}\left[I_n^A\right] = I$ and

$$\mathrm{Var}\left(I_n^A\right) = (1/k)\,\mathrm{Var}\left((U_1^+ + U_1^-)/2\right). \tag{13}$$

Using the elementary identity

$$\mathrm{Var}\left((U_1^+ + U_1^-)/2\right) = 1/4\left(\mathrm{Var}\left(U_1^+\right) + \mathrm{Var}\left(U_1^-\right) + 2\mathrm{Cov}\left(U_1^+, U_1^-\right)\right), \tag{14}$$

we get that if $\mathrm{Var}\left(U_1^+\right) + \mathrm{Var}\left(U_1^-\right) \leq 2\mathrm{Var}\,(U_i)$ and $\mathrm{Cov}\left(U_1^+, U_1^-\right) \leq 0$ then $\mathrm{Var}\left(I_n^A\right) \leq \mathrm{Var}\,(I_n)$. One way to achieve this is to let U_i^+, U_i^- be *antithetic* random variables: U_i^+ and U_i^- are called antithetic if their distributions are the same but $\mathrm{Cov}\left(U_i^+, U_i^-\right) < 0$.

How can we introduce antithetic variables in parameter optimization of game programs? Consider the problem of optimizing the performance of a player in a non-deterministic game. Let us collect all random choices external to the players into a random variable Y and let $f(Y; W)$ be the performance of the player in the game. Here W represents the random choices made by the players (f is a deterministic function of its arguments). For instance, in poker Y can be chosen to be the cards in the deck at the beginning of the play after shuffling. The idea is to manipulate the random variables Y in the simulations by introducing a "mirrored" version, Y', of it such that $f(Y; W)$ and $f(Y'; W')$ become antithetic. Here W' represents the player's choices in response to Y' (it is assumed that different random numbers are used when computing W and W').

The influence of the random choices Y on the outcome of the game is often strong. By this we mean that the value of $f(Y; W)$ is largely determined by the value of Y. For instance, it may happen in poker that one player gets a strong hand, whilst the other gets a weak one. Assuming two players, a natural way to mitigate the influence of Y is to reverse the hands of the players: the hand of the first player becomes that of the second and vice versa. Denoting the cards in this new scenario by Y', it is expected that $\mathrm{Cov}\left(f(Y; W), f(Y'; W')\right) < 0$. Since the distribution of Y and Y' are identical (the mapping between Y and Y' is a bijection), the same holds for the distributions of $f(Y; W)$ and $f(Y'; W')$. When the random choices Y influence the outcome of the game strongly then we often find that $f(Y; W) \approx -f(Y'; W')$. When this is the case then $\mathrm{Cov}\left(f(Y; W), f(Y'; W')\right) \approx -\mathrm{Var}\left(f(Y; W)\right)$ and thus $f(Y; W)$ and

$f(Y'; W')$ are "perfectly" antithetic and thus $\mathrm{Var}\left(I_n^A\right) \approx 0$. Of course, $f(Y; W)$ $= -f(Y'; W')$ will never hold and thus the variance of I_n^A will not be eliminated entirely – but the above argument shows that it can be reduced to a large extent.

This method can be used in the estimation of the gradient (and also when the performance of the players is evaluated). Combined with the previous methods we obtain

$$
\hat{g}_{p',i}(\theta) = \frac{1}{4cp'} \sum_{j=1}^{p'} \frac{1}{\Delta_i^{(j)}} \Big((f(\theta + c\Delta^{(j)}; Y) + f(\theta + c\Delta^{(j)}; Y'))
$$
$$
- \ (f(\theta - c\Delta^{(j)}; Y) + f(\theta - c\Delta^{(j)}; Y')) \Big), \tag{15}
$$

where $\Delta^{(1)}, \ldots, \Delta^{(p')}$ are i.i.d. random variables. In our experiments (see Sect. 5.1) we observed that to achieve the same accuracy in evaluating the performance of a player we could use up to 4 times less samples when antithetic variables were used. We expect similar speed-ups in other games where external randomness influences the outcome of the game strongly.

4 Test Domains

In this section we describe the two test domains, Omaha Hi-Lo Poker and Lines of Action. Together with the game-playing programs they are used in the experiments.

4.1 Omaha Hi-Lo Poker

The rules. Omaha Hi-Lo Poker is a card game played by two to ten players. At the start each player is dealt four private cards, and at later stages five community cards are dealt face up (three after the first betting round, and one after the second betting round and after the third betting round). In a betting round, the player on turn has three options: *fold*, *check/call*, or *bet/raise*. After the last betting round, the pot is split amongst the players depending on the strength of their cards. The pot is halved into a high side and a low side. For each side, players must form a hand consisting of two private cards and three community cards. The high side is won according to the usual poker hand ranking rules. For the low side, a hand with five cards with different numerical values from Ace to eight has to be constructed. The winning low hand is the one with the lowest high card.

Estimating the Expected Proportional Payoff. It is essential for a poker player is to estimate his winning chances, or more precisely to predict how much share one will get from the pot. Our program, McRAISE, uses the following calculations to derive an estimate of the expected proportional payoff.

N random card configurations, cc, are generated, each consisting of the opponent hands $h_{opp}(cc)$ and the community cards that still have to be dealt. Then, given the betting history *history* of the current game, the expected payoff as expressed as a proportion of the actual pot size (we call this quantity the expected proportional payoff or EPP) is approximated by

$$p_{win} = \frac{1}{N} \sum_{cc} win(cc) \prod_{opp} \frac{p(h_{opp}(cc)|history)}{p(h_{opp}(c))} . \qquad (16)$$

where $win(cc)$ is the percentage of the pot won for a given card configuration cc and $p(h_{opp}(cc)|history)$ is the probability of cc given the observed history of betting actions. Now, using Bayes' rule, we obtain

$$\frac{p(h_{opp}(cc)|history)}{p(h_{opp}(cc))} \propto p(history|h_{opp}(cc)), \qquad (17)$$

where the omission of $p(history)$ is compensated by changing the calculation of p_{win} by normalising the individual weights of $w(cc) = p(history|h_{opp}(cc))$ so as they sum to 1, i.e., p_{win} is estimated by means of weighted importance sampling.

The probability of a betting sequence given the hole cards, $p(history|h_{opp}(cc))$, is computed using the probability of a certain betting action given the game state, $p(a|h_{opp}(cc))$, which is the core of the opponent model. If we would assume independence among the actions of an opponent, then $p(history|h_{opp}(cc))$ would come down to a product over the individual actions. This is obviously not the case. A simple way to include the correlation among the betting actions inside a round is given by the following equation:

$$p(history|h_{opp}(cc)) = \prod_{rnd} \frac{\sum_{a \in history_{opp,rnd}} \frac{p(a|h_{opp}(cc))}{p(a)}}{na_{opp,rnd}} , \qquad (18)$$

where $na_{opp,rnd}$ is the number of actions of an opponent *opp* in a round *rnd*.

Estimating $p(a|h_{opp}(cc))$ can be done in various ways. Currently, we use a generic opponent model, fitted to a game database that includes human games played on IRC, and games generated by self-play.

Action Selection. MCRAISE's action selection is based on a straightforward estimate of the expected value of each actions followed by selecting the action with the highest value. Given the current situation s and the estimate of EPP, $p_{win} = p_{win}(s)$, the expected value of an action a is estimated by

$$Q(s, a) = p_{win} \Pi(a, s) - B(a, s), \qquad (19)$$

where $\Pi(a, s)$ is the estimated pot size provided that action a is executed and $B(a, s)$ is the contribution to the pot. For estimating $\Pi(a, s)$ and $B(a, s)$, we assume that every player checks from this point on.

4.2   Lines of Action

In this subsection we explain first the game of Lines of Action (LOA). Then, the tournament program MIA and its enhancement RPS are described briefly.

The rules. LOA is a chess-like game with a connection-based goal. The game is played on an 8×8 board by two sides, Black and White. Each side has twelve pieces at its disposal. The players alternately move a piece, starting with Black. A move takes place in a straight line, exactly as many squares as there are pieces of either colour anywhere along the line of movement. A player may jump over its own pieces. A player may not jump over the opponent's pieces, but can capture them by landing on them. The goal of a player is to be the first to create a configuration on the board in which all own pieces are connected in one unit. The connections within the unit may be either orthogonal or diagonal.

MIA. MIA 4++ is a world-class LOA program, which has won the LOA tournament at the eighth (2003) and ninth (2004) Computer Olympiad. It is considered as the best LOA-playing entity of the world [17]. Here we will focus on the program component optimized in the experiments, the realization-probability search (RPS) [16]. The RPS algorithm is a new approach to fractional plies. It performs a selective search similar to forward pruning and selective extensions. In RPS the search depth of the move under consideration is determined by the realization probability of its move category. These realization probabilities are based on the relative frequencies which are noticed in master games. In MIA, the move categories depend on center-of-mass, board position, and capturing. In total there are 277 weights eligible to be tuned. Levy [10] argues that it may be necessary for a computerized search process to have numbers for the categories that are different from the ones extracted from master games. Therefore, we also believe that there is still room to improve the algorithm's performance by tuning its weights.

5 Experiments

In poker, we tested the RSPSA algorithm by optimizing two components of McRaise, the opponent model and the action selection. For both components, we compare the performance resulting by using RSPSA with the performance given by an alternative (state-of-the-art) optimization algorithm. The experiments for the opponent-model optimization are described in Sect. 5.1 and for the move-selection optimization in Sect. 5.2. In LOA, the RSPSA algorithm is employed to tune the realization-probability weights in MIA. According to [7] these weights belong to a class of parameters (termed class-S search decisions) that can be evaluated using search trees. In Sect. 5.3 we show how this property is exploited for improving the efficiency of the learning.

5.1 Tuning the Opponent Model

The opponent model of McRaise is embodied in the estimation of $p(a|h_{opp}(cc))$ (see Sect. 4.1). The model uses in total six parameters.

For problems where the number of parameters is small, FDSA can be a natural competitor to SPSA. We combined both SPSA and FDSA with RPROP. The combined FDSA algorithm will be denoted in the following by RFDSA. Some preliminary experiments were performed with the standard SPSA, but they did not produce reasonable results (perhaps due to the anisotropy of the underlying optimization problem).

A natural performance measure of a player's strength is the average amount of money won per hand divided by the value of the small bet (sb/h). Typical differences between players are in the range of 0.05 to 0.2sb/h. For showing that a 0.05sb/h difference is statistically significant in a two-player game one has to play up to 20,000 games. It is possible to speed up the evaluation if *antithetic dealing* is used as proposed in Sect. 3.3. In this case, in every second game each player is given the cards which the opponent had the game before, while the community cards are kept the same. According to our experience, antithetic dealing reduces the necessary number of games by at least four. This technique is used throughout the poker experiments.

In the process of estimating the derivatives we employed the "Common Random Numbers" method: the same decks were used for the two opposite perturbations. Since many of the decks produced zero SPSA differences, thus producing zero contribution to the estimation of the sign of the derivatives, those decks that resulted in no-zero differences were saved for reuse. In subsequent steps, half of the decks used for a new perturbation were taken from those previously stored, whilst the other half was generated randomly. The idea of storing and reusing decks that 'make difference' can be motivated using ideas from importance sampling, a method known to decrease the variance of Monte-Carlo estimates.

The parameters of the algorithms are given in Table 1. Note that the performance corresponding to a single perturbation was evaluated by playing games in parallel on a cluster of sixteen computers. The number of evaluations (games) for a given perturbation was kept in all cases above 100 to reduce the communication overhead. The parameters of the opponent model were initialized to the original parameter settings of MCRAISE.

The evolution of performance for the two algorithms is plotted in Fig. 1(top) as a function of the number of iterations. The best performance obtained for RSPSA

Table 1. Learning parameters of RSPSA and RFDSA for opponent model (OM), RSPSA and TD for evaluation function (EF) and RSPSA for policy (POL)learning. $\eta+$, $\eta-$, δ_0 (the initial value of δ_{ti}), δ^- and δ^+ are the RPROP parameters; Δ is the SPSA (or FDSA) perturbation size, λ is the parameter of TD; *batchsize* is the number of performance evaluations (games) in an iteration which, for RSPSA and RFDSA, is equal to the product of the number of perturbations (q), the number of directions (2) and the number of evaluations per perturbation (r).

	$\eta+$	$\eta-$	δ_0	δ^-	δ^+	$\Delta(\lambda)$	*batchsize*
RSPSA (OM)	1.1	0.85	0.01	1e-3	1.0	δ	$40 \times 2 \times 250$
RFDSA (OM)	1.1	0.85	0.01	1e-3	1.0	δ	$6 \times 2 \times 1500$
RSPSA (EF)	1.2	0.8	0.05	1e-3	1.0	$\delta/0.7$	$100 \times 2 \times 100$
RSPSA (POL)	1.1	0.9	0.01	1e-3	1.0	$\delta/0.3$	$100 \times 2 \times 100$
TD (EF)	1.2	0.5	0.1	1e-6	1.0	0.9	10000

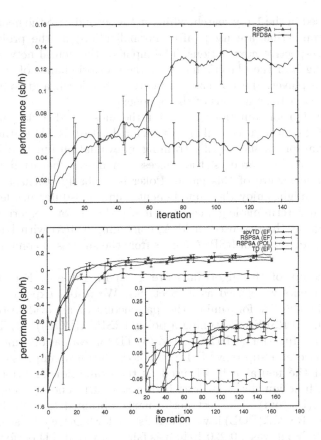

Fig. 1. Learning curves in poker: RSPSA and RFDSA for opponent-model learning (top) and RSPSA and TD for policy learning and evaluation-function learning (bottom). The graphs are obtained by smoothing the observed performance in windows of size 15. The error bars were obtained by dropping the smallest and largest values within the same windows centered around their respective coordinates.

was +0.170sb, whilst that of for RFDSA was +0.095sb. Since the performance resulting from the use of RSPSA is almost twice as good as that of resulting from the use of RFDSA, we may conclude that despite the small number of parameters, RSPSA is the better choice here.

5.2 Learning Policies and Evaluation Functions

The action-selection mechanism of MCRAISE is based on a simple estimation of the expected payoffs of actions and selecting the best action (see Sect. 4.1). This can be cast as a 1-ply search w.r.t. the evaluation function V if s', being the situation after action a is executed from situation s, and if we define $V(s') = Q(s, a)$. In the experiments we represent either V or Q with a neural network. In the first case the output of the neural network for a given situation s represents

$V(s)$ that is used in the 1-ply search, whilst in the second case the neural network has three outputs that are used (after normalization) as the probabilities of selecting the respective next actions. The input to the neural networks include EPP, the strength of the player's hand (as the a-priori chance of winning), the position of the player, the pot size, the current bet level, and some statistics about the recent betting actions of the opponent.

Learning evaluation functions is by far the most studied learning task in games. One of the most successful algorithm for this task is TD-learning and the best known example of successfully training an evaluation function is TDGammon [14]. By some researchers, the success can mostly be attributed to the highly stochastic nature of this game. Poker is similarly stochastic, therefore TD-algorithms might enjoy the same benefit. Temporal-difference learning had some success in deterministic games as well, e.g. [18]. In our experiment we use a similar design as the one used in MIA, combining TD(λ) with RRPOP (one source for the motivation of RSPSA comes from the success of combining TD(λ) and RPROP).

The parameters of the algorithms are given in Table 1. For RSPSA the same enhancements were used as in Sect. 5.1. We tested experimentally four algorithms: (1) RSPSA for tuning the parameters of an evaluation function (RSPSA(EF)), (2) RSPSA for tuning a policy (RSPSA(POL)), (3) TD for tuning an evaluation function (TD(EF)), and (4) TD for evaluation-function tuning with a supervised start-up (spvTD(EF)). For the latter a simple supervised algorithm tuned the neural network used as the evaluation function to match the evaluation function that was described in Sect. 4.1. The learning curves are given in Fig. 1(bottom). The best performance obtained for RSPSA(EF) was +0.194sb/h, for RSPSA(POL) it was +0.152sb/h, for TD(EF) it was +0.015sb/h and for spvTD(EF) it was +0.220sb/h. It is fair to say that TD performed better than RSPSA, which is a result one would expect given that TD uses more information about the gradient. However, we observe that for TD it was essential to start from a good policy and this option might not be always available. We note that although the two RSPSA algorithms did not reach the performance obtained by the combination of supervised and TD-learning, they did reach a considerable performance gain even though they were started from scratch.

5.3   Tuning the Realization-Probability Weights

Generally the parameters of a game program are evaluated by playing a number of games against a (fixed) set of opponents. In [7], it was noted that for parameters such as search extensions alternative performance measures exists as well. One such alternative is to measure the 'quality' of the move selected by the search algorithm (constrained by time, search depth or number of nodes). The quality of a move was defined in [8] as the negative negamax score returned by a sufficiently deep search for the position followed by the move. In the following, we describe two experiments. In the first, the performance is evaluated by game result. The result is averaged over 500 games played against five different opponents starting from 50 fixed positions with both colors. Each opponent is using

Table 2. Learning parameters of RSPSA for realization-probability weights using game result (GR) and move score (MS) for evaluation

	$\eta+$	$\eta-$	δ_0	δ^-	δ^+	$\Delta(\lambda)$	$batchsize$
RSPSA (GR)	1.2	0.8	0.005	1e-3	1.0	$\delta/0.7$	$500 \times 2 \times 1$
RSPSA (MS)	1.2	0.8	0.005	1e-3	1.0	$\delta/0.7$	$5000 \times 2 \times 1$

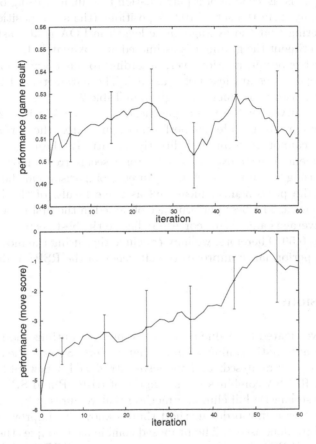

Fig. 2. Learning curves for RSPSA on game result (top) and moves score (bottom) as a function of the number of iteration. The graphs are obtained by smoothing the observed performance in windows of size 10. The error bars were obtained by dropping the smallest and largest values within the same windows centered around their respective coordinates.

a different evaluation function. Each player is searching a maximum of 250,000 nodes per move. In RSPSA the gradient is estimated with 500 perturbations, using one game per perturbation. The common random number technique is implemented in this case by using the same starting position, same opponent and same color for both the positive and the negative sides. In the second experiment the performance is evaluated by move score. The move score is averaged over a fixed set of 10,000 positions. For selecting the move the search is limited to 250,000 nodes. For evaluating the move a deeper search is used with a maximum of 10,000,000 nodes. Since the score of a move does not depend on the realization-probability weights, they are cached and reused when the same move is selected again (for the same position). So, the deeper search is performed far less frequently than the shallower search. The RSPSA gradient is estimated with 5,000 perturbations. Each side of a perturbation is evaluated using one position selected randomly from the set of 10,000 positions (the same position for both sides). Considering that the average game length in LOA is at least 40 ply, in the second experiment the gradient is estimated approximately four times faster than in the first experiment. Moreover, according to our observation, the estimates with move scores are less noisy as well. The parameters of the RSPSA algorithm for the two experiments are given in Table 2.

The learning curves for the two experiments are plotted in Fig. 2. Since the two experiments are using different performance measures, the performance for the two curves cannot be compared directly. Intuitively, the performance gain for the experiment using move scores (bottom) seems to be more significant than the one using game result. A more direct comparison can be performed by comparing the performance, measured as game result, of the best weight vector of each curve. The best performance obtained in the first experiment was 0.55, and the average game result corresponding to the best vector of the second experiment was 0.59. Therefore, we may conclude that using the move scores for estimating the performance improves the efficiency of the RSPSA algorithm.

6   Conclusions

This article investigated the value of a general purpose optimization algorithm, SPSA, for the automatic tuning of game parameters. Several theoretical and practical issues were analysed, and we have introduced a new SPSA variant, called RSPSA. RSPSA combines the strengths of RPROP and SPSA: SPSA is a gradient-free stochastic hill-climbing method that requires only function evaluations. RPROP is a first-order method that is known to improve the rate of convergence of gradient ascent. The proposed combination couples the perturbation parameter of SPSA and the step-size parameters of RPROP. It was argued that this coupling is natural.

Several other methods were considered to improve the performance of SPSA (and thus that of RSPSA). The effect of performing a larger number of perturbations was analyzed. An expression for the mean-square error of the estimate of the gradient was derived as the function of the number of (noisy) evaluations of the objective function per perturbation (q) and the number of perturbations (r). It was found that to optimize the mean-square error with a fixed budget $p = qr$, the number of perturbations should be kept at maximum. We suggested that besides using the method of "common random numbers", the method of antithetic variables should be used for the further reduction of the variance of the estimates of the gradient. These methods together are estimated to achieve a speed-up of factor larger than ten (since a smaller number of function evaluations is sufficient to achieve the same level of accuracy in estimating the gradient). The overall effect of these enhancements facilitated the application of SPSA for tuning parameters in our game programs McRAISE and MIA, whilst without the proposed modifications SPSA was not able to yield noticeable improvements.

The performance of RSPSA was tested experimentally in the games of Omaha Hi-Lo Poker and LOA. In poker, the optimization of two components of McRAISE were attempted: that of the opponent model and the action-selection policy. The latter optimization task was tried both directly when the policy was represented explicitly and indirectly via the tuning of the parameters of an evaluation function. In addition to testing RSPSA, for both components an alternative optimizer was tested (RFDSA and TD(λ), respectively). On the task of tuning the parameters of the opponent model, RSPSA led to a significantly better performance as compared with the performance obtained when using RFDSA. In the case of policy optimization, RSPSA was competitive with TD-learning, although the combination of supervised learning followed by TD-learning outperformed RSPSA. Nevertheless, the performance of RSPSA was encouraging on this second task as well. In LOA, the realization-probability weights of MIA were tuned by RSPSA. In the experiments, we have shown that using move scores for performance evaluation instead of game results can speed-up and improve the performance of RSPSA at the same time. In summary, from the experimental study we may conclude that the RSPSA algorithm using the suggested enhancements is a viable approach for optimizing parameters in game programs.

Acknowledgements

We would like to acknowledge support for this project from the Hungarian Academy of Sciences (Cs. Szepesvári, Bolyai Fellowship).

References

1. D. Billings, A. Davidson, T. Shauenberg, N. Burch, M. Bowling, R. Holte, J. Schaeffer, and D. Szafron. Game Tree Search with Adaptation in Stochastic Imperfect Information Games. In *Proceedings of Computers and Games (CG'04)*, 2004.
2. Y. Björnsson and T. A. Marsland. Learning Extension Parameters in Game-Tree Search. *Journal of Information Sciences*, 154:95–118, 2003.
3. K. Chellapilla and D. B. Fogel. Evolving Neural Networks to Play Checkers Without Expert Knowledge. *IEEE Transactions on Neural Networks*, 10(6):1382–1391, 1999.
4. J. Dippon. Accelerated Randomized Stochastic Optimization. *Annals of Statistics*, 31(4):1260–1281, 2003.
5. C. Igel and M. Hüsken. Empirical Evaluation of the Improved Rprop Learning Algorithm. *Neurocomputing*, 50(C):105–123, 2003.
6. N. L. Kleinman, J. C. Spall, and D. Q. Neiman. Simulation-based Optimization with Stochastic Approximation using Common Random Numbers. *Management Science*, 45(11):1570–1578, Nov 1999.

7. L. Kocsis. *Learning Search Decisions*. PhD thesis, Universiteit Maastricht, Maastricht, The Netherlands, 2003.
8. L. Kocsis, H. J. van den Herik, and J. W. H. M. Uiterwijk. Two Learning Algorithms for Forward Pruning. *ICGA Journal*, 26(3):165–181, 2003.
9. P. L'Ecuyer and G. Yin. Budget-dependent Convergence Rate of Stochastic Approximation. *SIAM J. on Optimization*, 8(1):217–247, 1998.
10. D. Levy. Some Comments on Realization Probabilities and the Sex Algorithm. *ICGA Journal*, 25(3):167, 2002.
11. M. Riedmiller and H. Braun. A Direct Adaptive Method for Faster Backpropagation Learning: The RPROP Algorithm. In E.H. Ruspini, editor, *Proceedings of the IEEE International Conference on Neural Networks*, pages 586–591, 1993.
12. J. C. Spall. Multivariate Stochastic Approximation Using a Simultaneous Perturbation Gradient Approximation. *IEEE Transactions on Automatic Control*, 37:332–341, 1992.
13. J. C. Spall. Adaptive Stochastic Approximation by the Simultaneous Perturbation Method. *IEEE Transactions on Automatic Control*, 45:1839–1853, 2000.
14. G. Tesauro. Practical Issues in Temporal Difference Learning. *Machine Learning*, 8:257–277, 1992.
15. J. Theiler and J. Alper. On the Choice of Random Directions for Stochastic Approximation Algorithms. *IEEE Transactions on Automatic Control*, 51:476-481, 2006.
16. Y. Tsuruoka, D. Yokoyama, and T. Chikayama. Game-tree Search Algorithm based on Realization Probability. *ICGA Journal*, 25(3):132–144, 2002.
17. M. H. M. Winands. *Informed Search in Complex Games*. PhD thesis, Universiteit Maastricht, Maastricht, The Netherlands, 2004.
18. M. H. M. Winands, L. Kocsis, J. W. H. M. Uiterwijk, and H. J. van den Herik. Learning in Lines of Action. In *Proceedings of BNAIC 2002*, pages 99–103, 2002.

Similarity Pruning in PrOM Search

H. (Jeroen) H.L.M. Donkers, H. Jaap van den Herik, and Jos W.H.M. Uiterwijk

Institute for Knowledge and Agent Technology,
MICC, Universiteit Maastricht, Maastricht, The Netherlands
{donkers, herik, uiterwijk}@micc.unimaas.nl

Abstract. In this paper we introduce a new pruning mechanism, called *Similarity Pruning* for Probabilistic Opponent-Model (PrOM) Search. It is based on imposing a bound on the differences between two or more evaluation functions. Assuming such a bound exists, we are able to prove two theoretical properties, viz., the bound-conservation property and the bounded-gain property. Using these properties we develop a Similarity-Pruning algorithm. Subsequently we conduct a series of experiments on random game trees to measure the efficiency of the new algorithm. The results show that Similarity Pruning increases the efficiency of PrOM search considerably.

1 Introduction to PrOM Search

PrOM search [3,4] is an extension of Opponent-Model (OM) search [1,3,6]. In OM search and in PrOM Search we call the player under consideration simply the player, and the other one the opponent. The opponent is assumed to use an evaluation function, known to the player. Moreover, the opponent's evaluation function is assumed to be weaker than the player's own evaluation function. OM search tries to exploit the opponent's weaknesses. PrOM search tries to exploit the weaknesses of the opponent too, but uses a more sophisticated model. In PrOM search the opponent is modeled by a *mixed strategy* of N known opponent types ω_i ($i = 0, 1, \ldots, N - 1$) in which (1) each opponent type is represented separately by an evaluation function $V_i(\cdot)$ and (2) each opponent type has a probability $\Pr(\omega_i)$. The true opponent is assumed to adopt at every move one of the opponent types ω_i according to probability $\Pr(\omega_i)$, and to act like this opponent type at the move under discussion.

To determine the best reply against such an opponent, the player has three tasks: (1) to determine the probability of each move for all opponent types at all min nodes, (2) to distill for each move the probability that it is played by the true opponent, and (3) to determine the expected outcome of each move.

The first task is executed as follows. At a min node h, each opponent type ω_i is assumed to play Minimax and therefore to select a move $m_j \in m(h)$ ($m(h)$ is the set of moves available at h) with a minimal Minimax value $v_{\omega_i}(h + m_k)^1$, using the evaluation function V_i of opponent type ω_i. The values for $v_{\omega_i}(\cdot)$ are thus computed by:

---

[1] With $h + m$ we indicate the node that is the result of performing move m at node h.

H.J. van den Herik et al. (Eds.): ACG11, LNCS 4250, pp. 57–72, 2006.
© Springer-Verlag Berlin Heidelberg 2006

$$v_{\omega_i}(h) = \begin{cases} \max\limits_{m_j \in m(h)} v_{\omega_i}(h + m_j) & h \text{ is a max node,} \\ \min\limits_{m_j \in m(h)} v_{\omega_i}(h + m_j) & h \text{ is a min node,} \\ V_i(h) & h \text{ is a leaf node.} \end{cases}$$

Now the Minimax values $v_{\omega_i}(h+m_j)$ of the nodes $h+m_j$ are used to determine the probabilities $\Pr(m_j | \omega_i, h)$. Here we have to inspect the moves with the lowest value. If only one move has a minimal value, then the probability of that move is 1 and all other moves have probability 0. If two or more moves (say K moves) have the same minimal value for $v_{\omega_i}(h+m_j)$, then there are two possible behaviors of the opponent: (1) the opponent always selects the first one and (2) the opponent selects one move out of the set of moves with minimal value at random. (This behavior is also part of the opponent model.) If the opponent always selects the first minimal move, only that move has probability 1 and all other moves have probability 0. If the opponent selects randomly between equipotent minimal moves, the probability of all K minimal moves is $1/K$ and the probability of all other moves is 0.

The second task is to distill for each move m_j at a min node h the probability that the move is actually played by the true opponent. These move probabilities $\Pr(m_j | h)$ depend on the probabilities $\Pr(\omega_i)$ of the opponent types and the probability $\Pr(m_j | \omega_i, h)$ that the move m_j is selected by the opponent type:

$$\Pr(m_j | h) = \sum_i \Pr(\omega_i) \Pr(m_j | \omega_i, h).$$

The third task is to determine the expected outcome of each move, i.e., to compute the value for the player for all nodes so that the player can pick the best move to play next. For min nodes h, the PrOM-search value $v_0(h)$ is the *expected* value of the child nodes' values $v_0(h+m_j)$, given the move probabilities. For max nodes, $v_0(h)$ is the maximum of the child nodes' values. At leaf nodes the player's own evaluation function V_0 is used directly. So the complete PrOM-search value is given by:

$$v_0(h) = \begin{cases} \max\limits_{m_j \in m(h)} v_0(h + m_j) & h \text{ is a max node,} \\ \sum_j \Pr(m_j | h) \, v_0(h + m_j) & h \text{ is a min node,} \\ V_0(h) & h \text{ is a leaf node.} \end{cases}$$

In [5] we showed that PrOM search as described above, can in some cases outperform both OM search and Minimax search (or equivalents), provided fixed-depth search trees are used and there are no time restrictions. A problem is that it is hard to compute the PrOM-search values efficiently, which makes the practical application of PrOM search difficult. Although pruning mechanisms for PrOM search have been described earlier [3,4] (see also Section 3), we feel that additional pruning mechanisms are needed to further practical application. Inspired by good results in OM search [2] and multi-player algorithms [12], we investigate in this paper whether putting a bound on the differences between the evaluation functions can increase the efficiency of PrOM search.

The remainder of this paper is organized as follows. In Section 2 we will assume that such a bound exists and use it to prove the existence of two properties of PrOM-search game trees, viz. the bound-conservation property and the bounded-gain property. In Section 3 we briefly explain an existing pruning algorithm for PrOM search and present the new Similarity-Pruning algorithm. The vehicle for our experiments, random game trees, is discussed in Section 4. The experimental set-up is given in Section 5 and the results in Section 6. We complete the paper with conclusions in Section 7.

2   Similarity

The algorithm proposed in this paper is inspired by the $\alpha\beta^*$ algorithm as suggested by Carmel and Markovitch [2]. In turn, they are inspired by the pruning algorithms proposed for multi-player games by Korf and Sturtevant [7,12] (see Subsection 3.2). The basic observation in both algorithms is that when a bound can be put on the sum of the evaluation functions applied, this bound can be used to prune the search tree.

In the case of PrOM search it would mean that when the evaluation functions $V_i(\cdot)$ within the probabilistic opponent model are rather similar to $V_0(\cdot)$, then this similarity can be used to prune the search tree of PrOM search. We define the similarity of the evaluation functions in PrOM search as follows:

$$\forall_{h\in E}, \forall_{i>0} \quad |V_0(h) - V_i(h)| \leq B,$$

where E is the set of leaf nodes (which is a subset of the set H of all nodes h in the game tree) and where bound B is a non-negative number. Below we prove two properties of PrOM-search game trees: the *bound-conservation property* and the *bounded-gain property*, that hold if the similarity condition of the evaluation functions is met. Both properties are fundamental and as such they form the basis for Similarity Pruning in PrOM search.

2.1   Bound-Conservation Property

The first property of PrOM-search game trees that is fundamental for Similarity Pruning is the *bound-conservation* property.

Property 1 (Bound Conservation). If the absolute difference between two evaluation functions $V_a(h)$ and $V_b(h)$ is bounded by some positive limit B for all leaf nodes $h \in E$ of a PrOM-search game tree, then the absolute difference between the Minimax values $v_a(h)$ and $v_b(h)$ is also bounded by B for all nodes $h \in H$.

Proof. The truth of the property is easily proven by induction. The property is clearly true for all $h \in E$. Assume that for a given h, $|v_a(h+m) - v_b(h+m)| \leq B$ for all $m \in m(h)$. Further, assume that $v_a(h) = v_a(h+m_i)$ and $v_b(h) = v_b(h+m_j)$ (Player a selects m_i and Player b selects m_j). If $i = j$ then $v_a(h)$ and $v_b(h)$ were propagated from the same node and therefore their difference is bounded by B. If $i \neq j$ at a max node, it must hold that $v_a(h + m_i) \geq v_a(h + m_j)$ and $v_b(h + m_i) \leq v_b(h + m_j)$; otherwise moves m_i and m_j would not have been selected. There are six different ways in which the four values for player a and b can be arranged under these constraints (see Fig. 1). It is noted that we do not treat cases with equal values separately. The pairs A and B, C and D, and E and F are symmetrical cases (by swapping a with b and i with j), so we only consider cases A, C, and E. In cases A and E, it is clear that the difference between $v_a(h + m_i)$ and $v_b(h + m_j)$ is bounded by B, since $v_b(h + m_j)$ falls inside the interval $[v_a(h + m_i), v_b(h + m_i)]$, which size is bounded by B. In case C, $v_b(h + m_j)$ cannot be larger than $v_a(h + m_i) + B$ since $v_a(h + m_j)$ cannot exceed $v_a(h + m_i)$ and the difference between $v_a(h + m_j)$ and $v_b(h + m_j)$ is bounded by B.

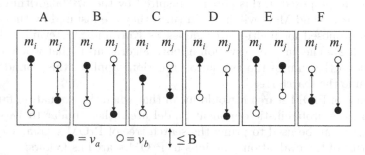

Fig. 1. Six possible configurations of $v_a(h + m_i)$, $v_b(h + m_i)$, $v_a(h + m_j)$, and $v_b(h + m_j)$. The heights of the black and white dots indicate the relative value of v_a and v_b, respectively.

A similar but mirrored reasoning is valid for min nodes. All together, it implies that the difference between $v_a(h)$ and $v_b(h)$ is bounded by B for all $h \in H$, which concludes the proof. $\qquad \square$

The consequence of the bound-conservation property for PrOM-search game trees is as follows. If the opponent types are constructed in such way that the differences between $V_{\omega_0}(\cdot)$ and $V_{\omega_i}(\cdot)$, $(i > 0)$ are bounded by B for any pair of opponent types ω_0 and ω_i, then the differences between values $v_{\omega_0}(h)$ and $v_{\omega_i}(h)$ are bounded by B too for all $h \in H$. It means that in a pruning procedure, the value $v_{\omega_0}(h)$ can be used at max nodes to predict the values of $v_{\omega_i}(h)$.

2.2   Bounded-Gain Property

The second property of PrOM-search game trees that is fundamental for Similarity Pruning is the *bounded-gain* property. Since the evaluation function $V_{\omega_0}(\cdot)$ is not only used by opponent type ω_0, but also by the player to determine the values of $v_0(\cdot)$, it is possible to derive a bound on the difference between $v_0(h)$ and $v_{\omega_0}(h)$ for all $h \in H$. Since in [3] it is already proven that $\forall h \in H$:

$v_0(h) \geq v_{\omega_0}(h)$, such a bound on the difference between $v_0(h)$ and $v_{\omega_0}(h)$ puts a bound on the gain of PrOM search.

Property 2 (Bounded Gain). If for all evaluation functions $V_a(\cdot)$, the difference between $V_0(h)$ and $V_a(h)$ is bounded by some positive limit B for all leaf nodes $h \in E$ of a PrOM-search game tree, then the difference between $v_0(h)$ and $v_{\omega_0}(h)$ is bounded by $\lfloor d(1-p)B \rfloor$ for all $h \in H$, where d is the search depth at h and $p = \Pr(\omega_0)$.

Proof. Again, the proof is constructed by induction. First we prove the property in the case when only 2 opponent types, ω_0 and ω_1, are present.

Since for all $h \in E$, $v_0(h) = v_{\omega_0}(h) = V_0(h)$, the induction hypothesis is clearly valid in set E. Now take h to be a min node. Assume that for all $m \in m(h)$ $|v_{\omega_0}(h+m) - v_0(h+m)| \leq xB$, for some $x \geq 0$. Assume further that ω_0 selects move m_i and ω_1 selects move m_j. We know that the following four pairs of values are each bounded:

$$|v_0(h+m_i) - v_{\omega_0}(h+m_i)| \leq xB \quad \text{(induction assumption)},$$
$$|v_{\omega_0}(h+m_i) - v_{\omega_1}(h+m_j)| \leq B \quad \begin{array}{l} \text{(since } v_{\omega_0}(h+m_i) = v_{\omega_0}(h) \\ \text{and } v_{\omega_1}(h+m_j) = v_{\omega_1}(h)), \end{array}$$
$$|v_{\omega_1}(h+m_j) - v_{\omega_0}(h+m_j)| \leq B \quad \text{(bound conservation)},$$
$$|v_{\omega_0}(h+m_j) - v_0(h+m_j)| \leq xB \quad \text{(induction assumption)}.$$

By the definition of PrOM search, $v_0(h) = pv_0(h+m_i) + (1-p)v_0(h+m_j)$. Now assume the worst case for the (positive) difference between $v_0(h)$ and $v_{\omega_0}(h)$. This happens when:

$$v_0(h+m_i) = v_{\omega_0}(h+m_i) + xB, \quad v_{\omega_1}(h+m_j) = v_{\omega_0}(h+m_i) + B,$$
$$v_{\omega_0}(h+m_j) = v_{\omega_1}(h+m_j) + B, \quad \text{and } v_0(h+m_j) = v_{\omega_0}(h+m_j) + xB.$$

Since $v_{\omega_0}(h) = v_{\omega_0}(h+m_i)$, the value of $v_0(h)$ can be expressed in $v_{\omega_0}(h)$:

$$v_0(h) = p(v_{\omega_0}(h) + xB) + (1-p)(v_{\omega_0}(h) + (2+x)B) = v_{\omega_0}(h) + (x+2(1-p))B$$

This means that the bound on the difference between $v_0(h)$ and $v_{\omega_0}(h)$ has grown with a factor $2(1-p)B$ with respect to the difference between $v_0(h+m)$ and $v_{\omega_0}(h+m)$. At max nodes, the difference between $v_0(h)$ and $v_{\omega_0}(h)$ does not increase. The proof of this statement is similar to the proof in Subsection 2.1 and is omitted here. Together it means that the difference between $v_0(h)$ and $v_{\omega_0}(h)$ increases maximally with a factor $2(1-p)B$ at every second (min) ply of search. When d is the depth of the subtree at max node h, then the difference between $v_0(h)$ and $v_{\omega_0}(h)$ will thus be bounded by $\lfloor d(1-p)B \rfloor$.

When more than two opponent types are involved, the bounded-gain property still holds. The worst case in this situation appears when all values $v_{\omega_i}(h)$ $(i > 0)$ are equal and differ B from $v_{\omega_0}(h)$ in the same way as above. However, this means that all opponent types (except ω_0) select the same move (say j) and the formula for $v_0(h)$ remains $pv_0(h+m_i) + (1-p)v_0(h+m_j)$. □

The bounded-gain property implies that the value of $v_{\omega_0}(h)$, which actually is the standard Minimax game value, can be used to predict the PrOM-search value $v_{\omega_0}(h)$.

3 Similarity Pruning

The two properties are a direct help when implementing Similarity Pruning. In Subsection 3.1 we describe the implementation. Related work is reviewed in Subsection 3.2.

3.1 Implementing Similarity Pruning

Below we describe the components of the Similarity-Pruning algorithm. The algorithm is an extension of the probing version of β-pruning PrOM search [3]. For convenience, this is reproduced in Fig. 2. In the algorithm, β pruning takes place in line 10 where α-β search is called with an appropriate upper bound ($\bar{\beta}_i$) and evaluation function ($V_i(\cdot)$). The upper bounds are determined in line 14: when a node already has been probed, the value found can be used as an upper bound for the next level of PrOM search in the subtree of the node.

PromSearchBeta$(h, \bar{\beta}, d)$
1   **if** $(d = 0)$ **then return** $(V_0(h), \mathbf{null})$
2   **if** $h = $ max node **then**
3        $L \leftarrow m(h)$; $v^* \leftarrow -\infty$; $m^* \leftarrow \mathbf{null}$
4   **for all** $m \in L$
5            $(v, \cdot) \leftarrow \mathbf{PromSearchBeta}(h + m, \bar{\beta}, d - 1)$
6   **if** $(v > v^*)$ **then** $v^* \leftarrow v$; $m^* \leftarrow m$
7   **if** $h = $ min node **then**
8        $L \leftarrow \varnothing$
9   **for** $i \in \{0, \ldots, n - 1\}$
10           $(\bar{v}_i^*, \bar{m}_i^*) \leftarrow \alpha\beta$ **search**$(h, -\infty, \bar{\beta}_i, V_i(\cdot), d)$
11           $L \leftarrow L \cup \{\bar{m}_i^*\}$
12       $v^* \leftarrow 0; \ m^* \leftarrow \mathbf{null}$
13   **for all** $m \in L$
14   **for** $i \in \{0, \ldots, n - 1\}$ **if** $(m = \bar{m}_i^*)$ $\bar{\beta}_i \leftarrow \bar{v}_i^* + 1$ **else** $\bar{\beta}_i \leftarrow \infty$
15           $(v, \cdot) \leftarrow \mathbf{PromSearchBeta}(h + m, \bar{\beta}, d - 1)$
16   **for** $i \in \{0, \ldots, n - 1\}$ **if** $(m = \bar{m}_i^*)$ $v^* \leftarrow v^* + \Pr(\omega_i) v$
17   **return** (v^*, m^*)

Fig. 2. Probing version of β-pruning PrOM search

Figure 3 lists the implementation of Similarity-Pruning PrOM search. In contrast to the algorithm of Fig. 2, this algorithm treats two plies at once: a max node and all its min child nodes. It means that in the code h always is a max node and that $h + m$ are the child min nodes.

The algorithm starts using the bounded-gain property to prune all min nodes $h + m$ that cannot contribute to the root value. Since we know that $v_0(h)$ can not be smaller than $v_{\omega_0}(h) - b$, where $b = \lfloor B(1 - \Pr(\omega_0))(d - 1) \rfloor$, all min nodes for which $v_{\omega_0}(h + m) < v_{\omega_0}(h) - b$ cannot contribute and can be pruned. So when determining the $v_{\omega_0}(h + m)$ using α-β search, we can use value $v^* - b$ for α when we scan the child nodes, where v^* is the maximum value of v_{ω_0} so far

PromSearchSim$(h, \bar{\beta}, d)$

```
1     if (d = 0) then return (V₀(h), null)
```
1   if $(d = 0)$ then return $(V_0(h), \textbf{null})$
2      $b \leftarrow \lfloor B(1 - \Pr(\omega_0))(d-1) \rfloor$
3      $L \leftarrow m(h)$; $m^* \leftarrow \textbf{null}$
4      $v^* \leftarrow -\infty$; $R \leftarrow \varnothing$
5   **for all** $m \in L$
6          $(v, mm) \leftarrow \alpha\beta$ **search**$(h + m, v^* - b, \bar{\beta}_0, V_0(\cdot), d-1)$
7          $R \leftarrow R \cup \{(v, m, mm)\}$
8   **if** $(v > v^*)$ **then** $v^* \leftarrow v$; $m^* \leftarrow m$
9   **if** $(d = 1)$ **then return** (v^*, m^*)
10   **for all** $(w, m, mm) \in R$: **if** $(w < w^* - b)$ $R \leftarrow R - \{(w, m, mm)\}$
11     $v^* \leftarrow -\infty$; $m^* \leftarrow \textbf{null}$
12   **for all** $(\bar{v}_0^*, m, mm) \in R$
13         $w^* \leftarrow -\infty$; $mm^* \leftarrow \textbf{null}$
14   **if** $(\bar{v}_0^* + b < v^*)$ **then** $w^* \leftarrow \bar{v}_0^*$
15   **else**
16             $L \leftarrow \{mm\}$
17   **for** $i \in \{1, \ldots, n-1\}$
18                 $(\bar{v}_i^*, \bar{m}\bar{m}_i^*) \leftarrow \alpha\beta$ **search**$(h + m, \bar{v}_0^* - B, \min(\bar{\beta}_i, \bar{v}_0^* + B + 1), V_i(\cdot), d-1)$
19                 $L \leftarrow L \cup \{\bar{m}\bar{m}_i^*\}$
20             $w^* \leftarrow 0$
21   **for all** $mm \in L$
22   **for** $i \in \{0, \ldots, n-1\}$ **if** $(mm = \bar{m}\bar{m}_i^*)$ $\bar{\beta}_i \leftarrow \bar{v}_i^* + 1$ **else** $\bar{\beta}_i \leftarrow \infty$
23                 $(v, \cdot) \leftarrow$ **PromSearchSim**$(h + m + mm, \bar{\beta}, d-2)$
24   **for** $i \in \{0, \ldots, n-1\}$ **if** $(mm = \bar{m}\bar{m}_i^*)$ $w^* \leftarrow w^* + \Pr(\omega_i) v$
25   **if** $(w^* > v^*)$ **then** $v^* \leftarrow w^*$; $m^* \leftarrow m$
26   **return** (v^*, m^*)

Fig. 3. Similarity-Pruning PrOM search

(see line 6). This provides the first case of Similarity Pruning. In list R we store the value of $v_{\omega_0}(h+m)$ (called v), the child node m, and its best move mm. The second case of Similarity Pruning occurs in line 10 where list R is inspected and all child nodes are removed for which $v_{\omega_0}(h+m) < v_{\omega_0}(h) - b$.

For all remaining child nodes at list R, the algorithm computes the PrOM-search value $v_0(h+m)$ and the maximum of these values produces $v_0(h)$. However, when the best value of $v_0(h)$ so far is larger than $v_{\omega_0}(h+m) + b$ for the current child node $h + m$, then that child node can be pruned (line 14), which is the third case of Similarity Pruning.

The computation of the PrOM search values for the remaining min nodes is not fundamentally different from the procedure in lines 8 to 16 of Fig. 2. Since the first opponent type (ω_0) has already be handled, list L can start with its best move mm. The only real difference occurs in line 18. Here, the bound-conservation property is used to control the window of the α-β search: the value of α and β are confined to an interval of $2B$ around $v_{\omega_0}(h+m)$ (in the algorithm indicated by $\bar{v}_0^*$). This constitutes the final, fourth case of Similarity Pruning.

Without a formal proof, we claim that this implementation of Similarity Pruning is safe: it produces the same root value of v_0 as PrOM search does.

3.2   Related Work

The idea of using restrictions on the values of several evaluation functions at the same leaf node for pruning a search tree is first described by Korf [7]. He introduces an α-β-like pruning algorithm for the Max^n approach of multi-player games [8]. In this approach, every player has an own evaluation function and every player tries to maximize its own results, independent of the other players (i.e., there is no co-operation between players). It means that at leaf nodes, there is a separate node value for every player. Luckhardt and Irani [8] show that without any restrictions on the evaluation functions, no safe pruning is possible. The two restrictions that Korf puts on the values to allow pruning are that (1) at all leaf nodes, the *sum* of the values must be equal to or less than a given upper bound, and (2) that all leaf-node values must be larger than a given lower bound (e.g., zero). We note that these restrictions are not equivalent with the similarity restriction as proposed above since the latter does not restrict the absolute values of the leaf nodes.

Korf proofs that two types of pruning are possible in the Max^n algorithm if the two restrictions hold. First, *immediate* pruning is possible if the value for the player to move of one child node is equal to the upper bound: in this case the other child nodes can be pruned immediately. The second pruning opportunity is *shallow* pruning, where the value of a previously investigated sibling node can be used to prune the children of a node under consideration. In contrast to alpha-beta pruning in two-player zero-sum games and to similarity-pruning as described above, *deep* pruning appears not to be possible in Korf's approach.

Additional pruning can be achieved in the Max^n approach when an extra restriction is put on the separate evaluation functions. Sturtevant and Korf [12] show that when the functions are *monotonic*, a branch-and-bound approach can be combined with the α-β-like pruning in Max^n. Monotonicity means that the (heuristic) score for a player cannot decrease during the course of a game, a rather severe restriction. Sturtevant [11] later presented two supplementary pruning techniques for Max^n that are based on the two first restrictions on evaluation functions: Last-Branch pruning and, an extension of the same, Speculative Max^n pruning. These techniques allow extra pruning for the last player based on the sum of lower bounds of values of all other players.

Carmel and Markovitch [2] present a pruning algorithm for their recursive (multi-model) version of (two-player) opponent-model search: M^* search. In this version, the opponent model that the first player uses of the second player (M^i) contains the second player's model of the first player (M^{i-1}), which can contain a subsequent model of the second player (M^{i-2}), *et cetera*. Each model M^i is connected to a separate evaluation function, say $V_i(\cdot)$. Even-numbered models are connected to the first player, odd models to the second player. Pruning in

M^* search is not possible, except for applying α-β search at the M^0 level. At higher model levels, a similar situation arises as in Max^n.

Carmel and Markovitch [2] therefore define the following constraint on the evaluation functions. For every leaf node h, and for every model M^i, $i > 0$, bounds B_i must exist such that $|V_i(h) + V_{i-1}(h)| \leq B_i$, where $B_i \geq 0$. This *bounded-sum constraint* is fundamentally different from the constraints in the Max^n approach. In Max^n all players are maximizing and therefore, all evaluation functions have the same sign. In M^* search, the models alternate between a maximizing and a minimizing player, so V^i and V^{i-1} have opposite signs. When the bounded-sum property holds, it can be proven that a bound exists on the sums of values in all internal nodes of the search trees. Namely, if for all children h' of h and for all models M^i this constraint holds, then it holds that $|V_i(h) + V_{i-1}(h)| \leq B_i + 2B_{i-1}$. Using this property, Carmel and Markovitch constructed an α-β-like pruning version for the multi-pass ($\alpha\beta^*$) and one-pass ($\alpha\beta_{1p}^*$) versions of M^* search.

Although our similarity constraint is defined as a bound on the *difference* of two evaluation functions, it is nevertheless equivalent to the bound on the *sum* in M^* search. In PrOM search all opponent types are representing the *second* player, and therefore they all have the same sign and, as already mentioned, V^i and V^{i-1} have opposite signs in M^* search. However, there are three differences. First, in PrOM search, the bound must hold for all pairs (V_0, V_i), whereas in M^* search it must hold for pairs (V_i, V_{i-1}) of subsequent models. The latter means that the differences between models can grow larger, the farther they are apart, whereas in PrOM search all opponent types have the same distance to V_0. This is intuitive since M^* and PrOM search can be seen as two orthogonal approaches that even can be combined into a single multi-model probabilistic algorithm [3].

The second difference is that in the pruning $M*$ approach, separate bounds are used *per model*, whereas in Similarity-Pruning PrOM search, we use only a single bound B. Although it is possible to specify a bound B_i per evaluation function V_i in line 18 of Fig. 3 to increase pruning efficiency, in line 2, still the maximum of all bounds B_i is needed. For clarity, we used only a single bound B in this paper.

The third difference is that in the pruning versions of $M*$, the bounds increase in two ways: first, with increasing level of modeling, and second, with increasing search depth. This leads to less pruning with larger trees and more complex models. In the case of Similarity Pruning PrOM search, the number of opponent types does not influence the bound. The bounded-gain property, however, dictates that with increasing search depth, the bound on the differences increases with a factor $(1 - p)B$ at every ply of search depth.

4   Random Game Trees

In the experiments we applied PrOM search with Similarity Pruning on random game trees. All leaf nodes in a tree had the same depth and all internal nodes had the same number of children (branching factor or width). The evaluation function in our game trees was generated analogously to the approach by Newborn [10]. In his approach, every node in the tree receives a random number. The value of a leaf node is the average of all random numbers at the nodes on the path from the leaf node to the root of the tree. This assured some correlation between the values of nodes in neighboring parts of the tree, partly simulating real game trees.

For our experiments with Similarity Pruning, we had to adapt Newborn's approach at two points: (1) we needed one evaluation value $V_{\omega_i}(h)$ for each of the N opponent types ω_i per leaf node h and (2) the difference between $V_{\omega_0}(h)$ and $V_{\omega_i}(h)$ $(i > 0)$ should be bounded by a given B. To achieve this, we assigned a series of N random numbers $x_i(\cdot)$ to every node in the tree. These values were drawn uniformly from $[0,1]$. For each leave node h we first computed the average $\bar{x}_0(h)$ of the $x_0(\cdot)$ values on the path from the root to h. For ω_0 we used the formula $V_{\omega_0}(h) = \text{Round}(Z \cdot \bar{x}_0(h))$ ($Z = 1,000$, resulting in a precision of 3 decimals), and for ω_i we used the formula $V_{\omega_i}(h) = \text{Round}(Z \cdot \bar{x}_0(h) + 2.B.(x_i(h) - 0.5))$. The $x_i(\cdot)$ values in the nodes were of float precision, the evaluation values were integer values. For generating the pseudo-random numbers we used the Mersenne Twister [9], also called MT19937. This pseudorandom-number generator has a period of 2^{19937} which makes it suitable for the generation of many random game trees with a large number of nodes. In order to allow paired comparisons of the search methods, we needed to apply the different methods to exactly the same random game tree. Instead of storing a complete game tree in memory, we developed a method to (re)construct the random game tree efficiently during search (see [3]).

The random game trees as produced by this method have a random ordering of moves with respect to the move quality. It is well understood that α-β pruning is influenced heavily by the move ordering in the tree and we expect the ordering to influence our search methods as well. Therefore we developed a new mechanism to allow for a gradual increase of ordering in the random game tree. To this end, we added an *ordering bias* to the value of $x_0(\cdot)$, depending on the move number. For max nodes h: $x_0(h_i) = x_0(h_i) - F.i$, for min nodes h: $x_0(h_i) = x_0(h_i) + F.i$, where h_i are the child nodes of h. The effect of the ordering bias depends on the size of the *ordering factor* F. If $F = 0$, no ordering takes place. When F increases, its effect will first overcome the variation in $x_0(\cdot)$, eventually causing a perfect ordering of the tree for V_{ω_0}, but not necessarily for other V_{ω_i}. When F increases further, bound B will be overcome too, and the tree will be ordered perfectly for all evaluation functions V_{ω_i}.

5   Experimental Set-Up

In the following experiments, we aimed to measure the efficiency of our implementation of Similarity Pruning on random game trees as described above. The base case in all experiments is a game tree of depth 8, branching factor 5, two opponent types having probabilities 0.3 and 0.7, a bound of 10, and ordering factor 0. We conducted six experiments. In each one we varied one setting and kept the other settings as stated in the base case.

- Experiment 1: vary the bound B;
- Experiment 2: vary the probability of opponent type ω_0;
- Experiment 3: vary the search depth d;
- Experiment 4: vary the branching factor b;
- Experiment 5: vary the number of opponent types N; and
- Experiment 6: vary the ordering factor F.

For every setting, we generated 100 random game trees and applied to each of them α-β search, β-Pruning PrOM search, and Similarity-Pruning PrOM search. We collected the search value and number of leaf-node evaluations for the three algorithms on all 100 trees. We present the results as follows: for every tree we determine the number of evaluations that β pruning and Similarity-Pruning PrOM search need *divided* by the number of evaluations that α-β search needs on the same tree. Since the distribution of these quotients is heavily skewed, we do not provide the means, but plot the median (solid line), the upper and lower quartile (dashed lines) and the 5th and 95th percentile (dotted lines) of the samples.

6   Results

Figure 4 shows the results of experiment 1. We varied the bound B from 0 to 1, 000. At value 1, 000 the similarity bound is the least restrictive; at value 0, all evaluation functions are equal. Due to the nature of the results, we present a log-plot. The results for bound 0 are therefore not presented, but they are similar to the results at 1. The bounded-gain property predicts that the most pruning can take place at low values of B, which is confirmed by the results. It appears that also β pruning profits from a smaller bound. This can be explained by the fact that opponent types will select the same move more often when their evaluation functions are similar than when they are non-similar. The results show, however, that Similarity Pruning needs significantly less evaluations than β pruning when the bound drops below 10. When the bound is near to 1, Similarity-Pruning PrOM search needs only less than 2 times the number of evaluations that α-β search needs.

The results of experiment 2 are presented in Fig.5. Here we varied the probability $\Pr(\omega_0)$ of opponent type ω_0 from 0.001 to 0.999. We did not include the border values 0 and 1 since at those values, PrOM search is equivalent to OM search and α-β search, respectively, and PrOM search is of no use. The results show that β pruning is insensitive to changes in $\Pr(\omega_0)$. As predicted by the bounded-gain property, the amount of evaluations needed by Similarity Pruning decreases when $\Pr(\omega_0)$ increases. At values larger than 0.9, the amount of pruning drops more sharply from 2.0 times to only 1.2 times the number of evaluations of α-β search, where β pruning still needs 7 times that amount.

The results of experiment 3 are presented in Fig. 6. In this experiment we varied the search depth (i.e., the height of the random game trees) from 2 to 14 (both even and odd depths). With increasing search depth, both β pruning

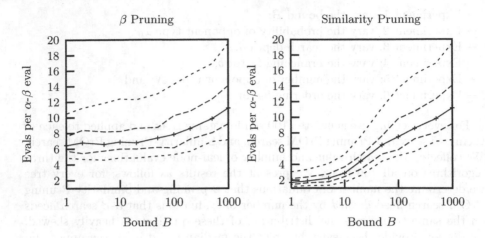

Fig. 4. Relative number of evaluations needed by β pruning and Similarity Pruning as a function of the bound B

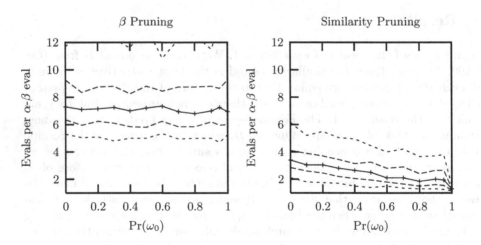

Fig. 5. Relative number of evaluations needed by β pruning and Similarity Pruning as a function of the probability $\Pr(\omega_0)$

and Similarity Pruning need, as expected, more evaluations in comparison to α-β search. However, the growth rate of Similarity Pruning is half of that of β pruning.

Figure 7 shows the results of experiment 4. Here, the branching factor was varied from 2 to 16. In contrast to experiment 3, the results of experiment 4 are unexpected. While β pruning needs more evaluations in comparison to α-β search with increasing numbers of moves per node, the amount of evaluations needed by Similarity Pruning remains about 3 times that of α-β search. The reason for this result might be that with increasing branching factor, the effect of the pruning based on the bounded-gain property might become larger since increasingly more min child nodes will have a value smaller than the $v_{\omega_0} - B$ bound.

The results of experiment 5 are presented in Fig. 8. In this experiment, we varied the number of opponent types from 2 to 10. The probability for opponent type ω_0 remained 0.3, the rest of the probability mass was spread evenly over the remaining opponent types. As expected, the number of evaluations needed increased with the number of opponent types, both in β pruning and in Similarity Pruning. However, in β pruning, each additional opponent type causes

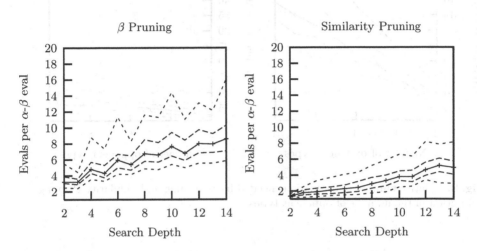

Fig. 6. Relative number of evaluations needed by β pruning and Similarity Pruning as a function of the search depth

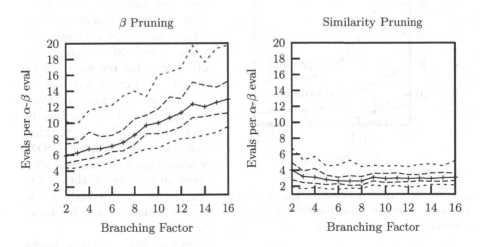

Fig. 7. Relative number of evaluations needed by β pruning and Similarity Pruning as a function of the branching factor

an increase of about 4 times, while in Similarity Pruning, each additional opponent type causes only an increase of roughly 1 times the number of evaluations that α-β search needs. It means that Similarity Pruning in this setting performs the equivalent of one time α-β search for every opponent type.

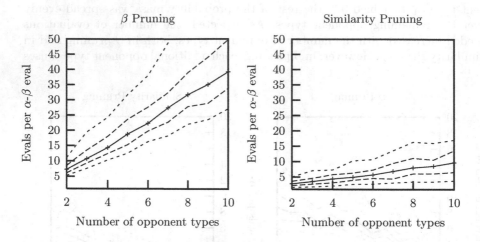

Fig. 8. Relative number of evaluations needed by β pruning and Similarity Pruning as a function of the number of opponent types

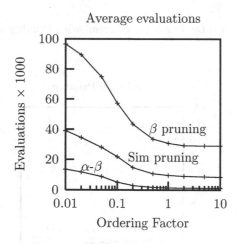

Fig. 9. Average number of evaluations needed by α-β search, β pruning, and Similarity Pruning as a function of the ordering factor

Experiment 6 concerned varying the ordering factor from 0 (no ordering) to 10 (perfect ordering) for all evaluation functions simultaneously. The results are presented in Figs. 9 and 10. To show the effect of the ordering mechanism, in Fig. 9 we plotted the absolute number of evaluations needed on average in α-β search and in β pruning and Similarity-Pruning PrOM search. Fig. 9 shows that with increasing value of the ordering factor, α-β search needs fewer evaluations, which was expected.

For ordering factors larger than 1, the tree is almost perfectly ordered for V_{ω_0}. α-β-search needs the minimum amount of $1,249$ evaluations at factor 5 instead of $13,939$ at factor 0

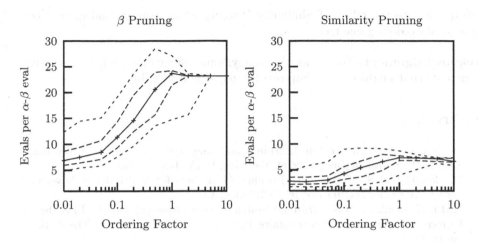

Fig. 10. Relative number of evaluations needed by β pruning and Similarity Pruning as a function of the ordering factor

when no ordering is present. (The game tree in the base case has depth 8 and branching factor 5; the full tree has $8^5 = 32,768$ leaf nodes.) Also the efficiencies of β-Pruning and Similarity-Pruning PrOM search increase with larger ordering factors, but Fig. 9 obfuscates the true relation with α-β search. Therefore, in Fig. 10 we present the relative number of evaluations needed, similar to the results of the previous five experiments.

The results show that for both β pruning and Similarity Pruning, the relative number of evaluations needed increases with growing ordering factors. However, the effect is much larger for β pruning than for Similarity Pruning. In perfectly ordered game trees, β pruning needs 23 times the number of evaluations in α-β search, where Similarity Pruning never needs more than 7 times that amount.

7 Conclusions

In this paper we introduced a new pruning mechanism for PrOM search, based on a bound imposed on the differences between two or more evaluation functions. We derived two theoretical properties of PrOM-search game trees and performed six experiments. In some cases, in particular with larger branching factors, the costs of applying PrOM search with Similarity Pruning are less than one additional ply of search with α-β search. Previous research has revealed that when PrOM search is provided with sufficient resources, it can outperform α-β search [3]. From the results of this paper we may conclude that when Similarity Pruning is applicable, PrOM search might become a feasible alternative for α-β search. Of course, future research is needed to investigate the effect and usability of a range of search enhancements on Similarity Pruning. In our future research we will investigate the effect of Similarity Pruning when using actual game trees instead of random game trees.

Acknowledgement. We thank the anonymous referees for their constructive comments that enabled us to improve the paper considerably.

References

1. D. Carmel and S. Markovitch. Learning and Using Opponent Models in Adversary Search. *Technical Report CIS9609*, Technion, Haifa, Israel, 1996.
2. D. Carmel and S. Markovitch. Pruning Algorithms for Multi-Model Adversary Search. *Artificial Intelligence*, 99(2):325–355, 1998.
3. H.H.L.M. Donkers. *Nosce Hostem: Searching with Opponent Models*. PhD thesis, Universiteit Maastricht. Universitaire Pers Maastricht, Maastricht, The Netherlands, 2003.
4. H.H.L.M. Donkers, J.W.H.M. Uiterwijk, and H.J. van den Herik. Probabilistic Opponent-Model Search. *Information Sciences*, 135(3–4):123–149, 2001.
5. H.H.L.M. Donkers, H.J. van den Herik, and J.W.H.M. Uiterwijk. Probabilistic Opponent-Model Search in Bao. In *Proceedings International Conference on Entertainment Computing - ICEC 2004* (ed. M. Rauterberg), LNCS 3166, pages 409–419, Springer-Verlag, Berlin, 2004.
6. H. Iida, J.W.H.M. Uiterwijk, H.J. van den Herik, and I.S. Herschberg. Potential Applications of Opponent-Model Search. Part 1: the Domain of Applicability. *ICCA Journal*, 16(4):201–208, 1993.
7. R.E. Korf. Multi-Player Alpha-Beta Pruning. *Artificial Intelligence*, 48(1):99–111, 1991.
8. C.A. Luckhardt and K.B. Irani. An Algorithmic Solution of n-Person Games. In *Proceedings AAAI-86*, pages 158–162, 1986.
9. M. Matsumoto and T. Nishimura. Mersenne Twister: A 623-Dimensionally Equidistributed Uniform Pseudorandom Number Generator. *ACM Transactions on Modeling and Computer Simulation*, 7(1):3–30, 1998.
10. M.M. Newborn. The Efficiency of the Alpha-Beta Search on Trees with Branch-Dependent Terminal Node Scores. *Artificial Intelligence*, 87(1-2):225–293, 1977.
11. N.R. Sturtevant. Last-Branch and Speculative Pruning Algorithms for Max^n. In *Proceedings IJCAI 2003*, pages 669–678, 2003.
12. N.R. Sturtevant and R.E. Korf. On Pruning Techniques for Multi-Player Games. In *Proceedings of AAAI 2000*, pages 201–208, 2000.

Enhancing Search Efficiency by Using Move Categorization Based on Game Progress in Amazons

Yoshinori Higashiuchi[1] and Reijer Grimbergen[2]

[1] Department of Computer Science,
Saga University, Saga, Japan
hi_yoshi@fu.is.saga-u.ac.jp
[2] Department of Informatics,
Yamagata University, Yonezawa, Japan
grim@yz.yamagata-u.ac.jp

Abstract. Amazons is a two-player perfect information game with a high branching factor, particularly in the opening. Therefore, improving the efficiency of the search is important for improving the playing strength of an Amazons program. In this paper we propose a new method for improving search in Amazons by using move categories to order moves. The move order is decided by the likelihood of the move actually being selected as the best move. Furthermore, it will be shown that the likelihood of move selection strongly depends upon the stage of the game. Therefore, our method is further refined by adjusting the likelihood of moves according to the progress of the game. Self-play experiments show that using move categories significantly improves the strength of an Amazons program and that combining move categories with game progress is better than using only move categories.

1 Introduction

Amazons is a two-player perfect information game with very simple rules [12]. From a computational point of view, its main feature is the large number of legal moves, particularly in the early stages of the game (in the initial position there are 2,176 possible moves). Even though the number of legal moves decreases as the game progresses, the average number of moves in an Amazons' position (479) [9] is considerably larger than chess (35), shogi (80) or Go (250) [8]. Because of the large average number of moves, the well-known search techniques that have been so successful in other games cannot be easily applied to Amazons. Therefore, Amazons has recently attracted some attention as a topic of research. The research efforts have focused on building an evaluation function that can evaluate positions accurately without deep search [5,7] and on how to do selective search in Amazons [1]. In this paper, we will present a new search method for Amazons which is partly a method for selective search but mainly aims at improving search efficiency.

H.J. van den Herik et al. (Eds.): ACG11, LNCS 4250, pp. 73–87, 2006.
© Springer-Verlag Berlin Heidelberg 2006

Amazons is a relatively new game, so there are no known strategies on how to play the opening and there is no expert feedback available to decide which moves are good in the middle game. In other game programs (e.g., chess, Go, or shogi), years of experience have led to heuristics for moves that are likely to lead to an advantageous position. These moves can then be given priority during the search. In chess, for example, moves that capture material or moves that cover the center are searched early, while sacrifices are searched last. By using these heuristics, the efficiency of α-β search can be improved, increasing search speed.

The research presented in this paper presents two methods to improve the search efficiency of an Amazons program. First, we will propose a set of move categories and use a calculation method from Realization Probability Search [11] to order moves based on these move categories. Second, we will show that the importance of move categories changes as the game develops. Therefore, when deciding the move ordering it is important to take the progress of the game into account.

In Sect. 2 we will start with an explanation of the properties of Amazons and how these properties lead to heuristics for good moves. These heuristics will be used to group moves into different categories. In Sect. 3 each category is assigned a priority based on statistical data from game records. In Sect. 4 a simple method to measure progress in Amazons is presented and the relation between game progress and the move categories is given. In Sect. 5 the results of a number of experiments comparing the performance of programs without move categories, with move categories, and with move categories based on game progress are presented. Finally, in Sect. 6 we provide conclusions and suggestions for future work.

2   Move Categories in Amazons

Because of the large number of possible moves in Amazons, under tournament conditions it is impossible to do a deep full-width search (the full-width search version of our program can only search to a depth of 2 or 3 ply in the opening, and to 6 or 7 ply near the end of the game). Therefore, using selective search is the only way to do a reasonable look-ahead. The most common domain-independent methods for selective search, like the null-move heuristic [2], ProbCut [3] and Multi-ProbCut [4] use a shallow search to estimate the result of deep searches. There are two reasons why these methods face problems in Amazons. One reason is that there is no deep search in Amazons until the endgame. Predicting the results of shallow searches by even shallower searches is risky. The second problem is that Amazons programs suffer from the even-odd iteration instability. There are no known methods to do quiescence search in Amazons, so there can be important changes in the evaluation function value after playing a move. A great deal of effort into building an evaluation function for Amazons goes into minimizing this effect, but there are still significant differences between the evaluation of even and odd iterations, especially in the opening. This makes it hard to predict the result of a d ply search with a $d - 1$ ply search. Also, in the case of shallow searches, the differences between a d ply search and a $d-2$ ply search are usually too large to be useful for a prediction.

Consequently, domain-dependent methods for selective search are needed in Amazons. One method, proposed by Avetisyan and Lorentz [1], is to use the evaluation function to evaluate each move after it has been played and discard moves for which the evaluation is below a certain threshold. This method was refined by making a difference between evaluation after the Amazon move and shooting the arrow.

Rather than eliminating a certain number of possible moves from the search, we will propose a method to use knowledge about good Amazons moves to improve the efficiency of α-β search. A common method for improving search efficiency is to use information that has become available during search. The best move of the previous iteration, killer moves, and the history heuristic are examples of such methods. By trying these moves first, the probability of a cut is increased, and search speed is improved. However, when a new position is encountered, this information is unavailable but searching good moves first will still improve search speed. In this case heuristic, game-specific information is needed to suggest which moves to search first.

In Amazons, with its short history, heuristics for selecting good moves are unknown. However, as pointed out by Lieberum [7] and confirmed by our own experience in playing and programming Amazons, confining one or more opponent Amazons to a small space is an important strategic theme. From this, a number of straightforward ways to limit the opponent Amazons' moving ability and improving the moving ability of one's own Amazons come to mind. We have categorized these heuristics into 10 basic categories that are given in Table 1.

As pointed out, in Amazons it is important to confine the opponent Amazons to a small space, in particular in the opening. Blocking the movement of the opponent Amazons, either by the move or by the arrow, is therefore a candidate for a good move. These are categories 1 and 2 in Table 1. An example of a move that blocks three opponent Amazons with both the Amazon and the arrow is given in Fig. 1. The white move blocked three black Amazons with the move (A7,

Table 1. Basic move categories in Amazons

No.	Category	Type	#
1	Move blocks opponent Amazons	0, 1, 2, 3 or more	4
2	Arrow blocks opponent Amazons	0, 1, 2, 3 or more	4
3	Move adjacent to opponent Amazon	true, false	2
4	Arrow adjacent to opponent Amazon	true, false	2
5	Blocking a single Amazon in multiple ways	true, false	2
6	Move previously blocked Amazon	0, 1, 2	3
7	Move Amazon to which Amazon moved adjacently	true, false	2
8	Move Amazon to which arrow was shot adjacently	true, false	2
9	Block Amazon that moved on previous move	true, false	2
10	Move Amazon not blocking any opponent Amazon	true, false	2

D10 and G10) and also three black Amazons with the arrow (D10, G10 and J7). This is a move that is often played as the first move in a game of Amazons. We make a difference between blocking 0, 1, 2, and 3 or more opponent Amazons, so the total number of categories is 4. Note that we are also making a difference between Amazons and arrows blocking the opponent. While an arrow is a simple block, the Amazon that blocked is not only blocking the opponent Amazon, but at the same time also blocked by the opponent Amazon. We feel that this difference between Amazon and arrow should be reflected in the move categories. This is an important difference with work by Soeda [10], a proposal for move categories for Amazons that made no difference between Amazon and arrow.

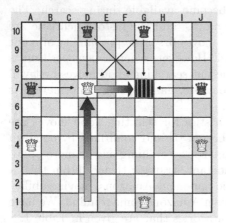

Fig. 1. D1-D7(G7): Amazon and arrow block 3 opponent Amazons

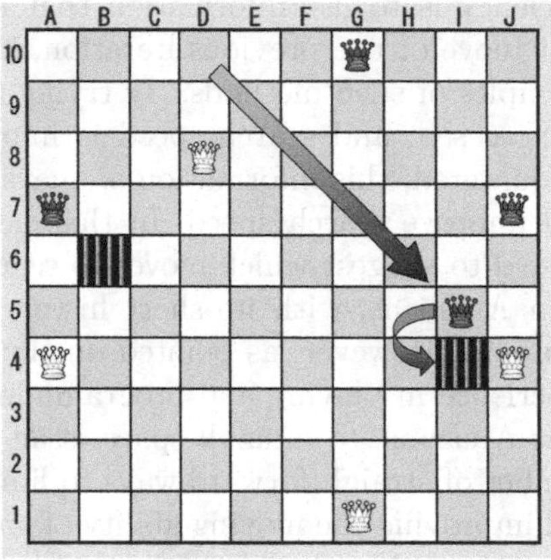

Fig. 2. D10-I5(I4): Amazon and arrow adjacent to opponent (J4), also blocking J4 twice

Even more aggressive is moving or shooting an arrow to the square adjacent to an Amazon (category 3 and 4). This is often a threat to trap the Amazon within the next few moves. In Fig. 2, the move D10-I5(I4) puts an Amazon and an arrow adjacent to the Amazon on J4. This Amazon on J4 now has very little space and to avoid being trapped White might have to move it next.

If a single Amazon can be blocked in more than one way, this is also a threat to trap this Amazon (category 5). If an opponent Amazon can be trapped early, the game becomes a fight of four Amazons against three, which is often a winning advantage. In Fig. 1, the black Amazons on D10 and G10 are blocked twice and so is the white Amazon on J4 in Fig. 2.

Moving the Amazon that is in danger of being trapped is the idea behind categories 6, 7, and 8. In Fig. 2, moving the Amazon on J4 (which was blocked twice) leads to the position of Fig. 3. Moving an Amazon that was blocked on the previous move helps avoiding a trap (category 6). Note that we make a difference between the number of times the Amazon was blocked on the previous move (0, 1 or 2 times), giving a total of 3 different categories. Moving the Amazon to which

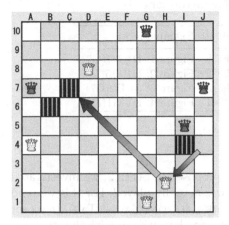
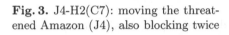

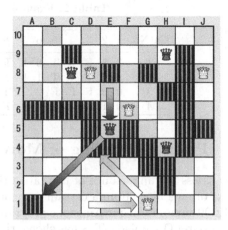

Fig. 3. J4-H2(C7): moving the threatened Amazon (J4), also blocking twice

Fig. 4. D1-G1(D4): blocking the Amazon that just moved (E5) with a free Amazon

an Amazon moved adjacently or an arrow was shot adjacently are categories 7 and 8, which mirror categories 3 and 4.

Blocking the Amazon that moved on the previous move (category 9) is based on the assumption that the previous move had meaning. An opponent Amazon tried to avoid being trapped or tried to claim or attack some territory. In Fig. 4, the previous move E7-E5(A1) is an attempt to enter the white territory at the bottom left. To keep the black Amazon from entering, white blocks this Amazon with D1-G1(D4).

Finally, category 10 is a move where an Amazon is not blocking any opponent Amazon. Often, this means that the Amazon is idle and should be moved or can be moved freely. In Fig. 4, the Amazons on G1 and J8 are Amazons that are not blocked by any opponent Amazon. J8 is an Amazon that is almost trapped and moving it to try and escape might be good. The Amazon on G1 is defending the white territory at the bottom left, so this Amazon will often move inside this territory to make sure that the opponent cannot enter.

The 10 basic categories are reflecting different aspects of moves, and we already gave examples of moves belonging to a combination of categories. The total number of combinations is $4^2 \times 3^1 \times 2^7 = 6,144$. Of these 6,144 combinations there are 4,724 combinations that are theoretically impossible. For example, when shooting an arrow adjacent to an opponent Amazon, this also automatically blocks an opponent Amazon at least once, so it never happens that category 4 is true and category 2 is 0. In our experiments, we have only used the 1,420 theoretically possible categories.

Using combinations of categories is important, because moves with multiple meanings are expected to be better than moves that belong to only a single category. Examples of move categorization for the moves in the figures are given in Table 2.

Table 2. Examples of move categories

CatNo.	1	2	3	4	5	6	7	8	9	10	Example move
1	3	3	F	F	T	0	F	F	F	F	Figure 1: D1-D7(G7)
2	1	2	T	T	T	1	F	F	F	F	Figure 2: D10-I5(I4)
3	0	2	F	F	F	2	T	T	F	F	Figure 3: J4-H2(C7)
4	0	1	F	T	F	1	F	F	T	T	Figure 4: D1-G1(D4)

3   Priority Ordering of Moves

To assess the importance of the categories proposed in Sect. 2, we investigated how often a move from a certain category was actually played. For this we used 11,000 games from our Amazons program THE AMAZONS SAGA (TAS). In recent Computer Olympiads, TAS has shown that it can play on a par with the strongest Amazons programs.

The method we used is the same as for determining the realization probabilities of move categories in Realization Probability Search (RPS), which has been a very successful approach in shogi [11]. The realization probability of a category is calculated with the following formula:

$$P_i = \frac{A_i}{B_i}$$

P_i: The realization probability of category i.
A_i: Number of times a move from category i was played.
B_i: Number of positions where a move from category i was possible.

The realization probability of a category is the ratio of positions where a move from a certain category was possible and the number of times that this move was actually played. The calculation has only been done until the stage where all territory is fixed. By fixed territory is meant that all Amazons have their own (i.e., non-overlapping) territory and the rest of the game is only about filling the territory with the maximum number of moves. Because this is a simple counting problem, usually a game of Amazons is stopped in a position with fixed territory. Both players then count the number of moves needed and the winner is agreed upon.

Although in RPS the game records of expert players are used, we believe it is important to use the game records of the same program to calculate the realization probability values. These values strongly depend upon the evaluation function, so improvement of search speed can only be expected if the moves are ordered in the way the program 'likes them', i.e. that have a high probability of leading to a good evaluation. Ideally, realization probabilities should be recalculated with every change in the evaluation function. However, we feel that the evaluation function of TAS is sufficiently stable to give reliable results.

A similar method for calculating realization probabilities in the absence of expert moves was proposed by Hashimoto et al. [6]. Their method calculates the probabilities by having a Lines of Action program play itself, recording the categories of the moves that were possible in each position and the category of the move that was selected after searching the position. If these positions and the selected moves are stored for future recalculation of the realization probabilities, the method is identical to ours. If not, our method has the advantage that if categories are changed, the recalculation of the realization probabilities can be done without search, using the standard set of games. If the evaluation function is changed (i.e., the program might select different moves), a new set of games is necessary for both methods.

The realization probabilities played by TAS (see above) of the 10 basic categories of Table 1 are given in Table 3.

Table 3. The realization probabilities of the basic categories

Category	RP (%)
Move blocked 0 Amazons	33.1
Move blocked 1 Amazon	56.9
Move blocked 2 Amazons	13.1
Move blocked 3 or more Amazons	6.6
Arrow blocked 0 Amazons	14.9
Arrow blocked 1 Amazon	71.4
Arrow blocked 2 Amazons	17.2
Arrow blocked 3 or more Amazons	5.3
Move adjacent to opponent Amazon	52.3
Arrow adjacent to opponent Amazon	72.3
Multiple opponent block	36.2
Blocking the previously moved Amazon	50.6
Moving the Amazon that was blocked once by the previous move	49.0
Moving the Amazon that was blocked twice by the previous move	33.9
Moving Amazon to which Amazon moved adjacently	32.8
Moving Amazon to which arrow was shot adjacently	34.5
Moving a non-blocking Amazon	45.6

The realization probabilities of Table 3 show that our basic categories in general have a high probability of being played and are therefore valid candidates for good moves. There are two notable exceptions: the categories "Move blocked 3 or more Amazons" and "Arrow blocked 3 or more Amazons" have a low realization probability which seems counter-intuitive. However, our experience with Amazons shows that the high mobility of Amazons makes it very rare to trap more than one Amazon. Rather than trying to confine multiple Amazons, it is better to try and trap a single Amazon and then use the advantage of having four Amazons fighting three.

In general RPS, the next step is to use the realization probabilities of move categories to decide the depth of the search. Moves with a high realization probability are searched more deeply than moves with a low realization probability.

Unfortunately, this method can currently not be used in Amazons programs because of the aforementioned even-odd iteration effect. Searching moves with a high probability one ply deeper makes the search unstable. General RPS is therefore not feasible until there is a solution to the even-odd iteration effect in Amazons.

Instead of using realization probabilities to decide the depth of the search, we propose to use the realization probabilities of Table 3 to order moves of new positions. Moves with a high realization probability have a higher probability of being played, so a correlation between realization probability and good moves can be expected.

As pointed out earlier, in our program the 10 basic categories are not used directly. Instead, the 1,420 theoretically possible combinations of basic categories are used. The calculation method for these categories is the same as those for the basic categories. Examples of the categories used in TAS and their realization probability are given in Table 4.

In Table 4, the realization probability of the moves in Table 2 are given. The table shows that D1-D7(G7) is a move belonging to a category with an extremely high probability (second highest among the 1,420 categories). The reason for this is that such moves are only possible in the opening. Actually, the category with the largest realization probability was only possible in 10 positions. In contrast, the probability of J4-H2(C7) is very low, even though this can be considered a good move. One way of improving the move ordering is to adjust the category priorities according to game progress, which we will explain next.

Table 4. Realization probabilities of move categories

CatNo.	Example move	RP (%)	Order
1	Figure 1: D1-D7(G7)	32.615	2
2	Figure 2: D10-I5(I4)	0.902	371
3	Figure 3: J4-H2(C7)	0.158	876
4	Figure 4: D1-G1(D4)	7.250	25

4   Adjusting Category Priorities Using Game Progress

The strategic features of Amazons shift as the game progresses. In the opening, the *mobility* and *balance* of the Amazons are most important. As the game progresses, it becomes more important to *secure territory*. Large territories in which Amazons can move freely will lead to more available moves at the end of the game, so the opponent will run out of moves first.

Because of this, the realization probabilities of Table 3 are likely to change as the game progresses. To reflect this shift in realization probabilities, the progress of the game has to be used to adjust the realization probabilities of the proposed move categories. There are a number of ways to measure progress in the game of Amazons, for example move number or using territory measurements. Here we will restrict ourselves to the most basic progress measurement: move number. In Amazons, the number of arrows on the board grows with each move. Therefore, the game is over after a maximum of 92 moves. Using the number of arrows (i.e., the number of moves) is therefore a natural choice for measuring progress in Amazons.

For this analysis the same 11,000 games as for calculating the realization probability were used. Despite this large number of games, near the end of the game the moves in certain categories are almost never played. For example, positions where it is possible to block three Amazons or more become very rare near the end of the game. Although this might be a problem of our method, the number of possible moves near the end of the game is relatively small, so deep search can be conducted even without using move categories. Therefore, we do not expect that this problem will influence playing strength much.

Because a sudden change in realization probability is undesirable, we grouped data for the change in realization probability into groups of 8 moves: 0-7 (note

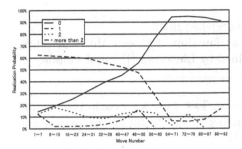

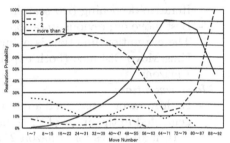

Fig. 5. Changes in realization probability for blocking by an Amazon

Fig. 6. Changes in realization probability for blocking by an arrow

Table 5. Game progress related realization probabilities (%) for the moves in Table 4

Mvno	Category number			
	1	2	3	4
1 - 7	32.65	0.51	0	0.52
8 - 15	0.01	0.99	0.18	1.20
16 - 23	0.01	0.88	0.21	4.20
24 - 31	0.01	2.22	0.20	8.32
32 - 39	0.01	2.41	0.80	18.39
40 - 47	0.01	1.76	2.53	17.75
48 - 55	0.01	17.65	0.01	30.86
56 - 63	0.01	0.01	0.01	50.44
64 - 71	0.01	0.01	0.01	37.50
72 - 79	0.01	0.01	0.01	75.00
80 - 87	0.01	0.01	0.01	0.01
88 - 92	0.01	0.01	0.01	0.01

that there is no data for move 0), 8-15, 16-23, etc. The results of this analysis for blocking by an Amazon and blocking by an arrow (categories 1 and 2) are given in Fig. 5 and Fig. 6 respectively.

From these graphs it is clear that the realization probability of categories can change drastically as the game progresses, so rather than taking the average realization probability over the whole game, it seems more promising to use realization probabilities based on game progress. Examples of the actual realization probabilities used in TAS for the four moves in Table 4 are given in Table 5.

5   Experimental Results

To investigate the significance of using game progress combined with move ordering based on our move category proposal, we have used our program TAS. TAS has the following properties.

General features:
  - Iterative deepening α-β search.
  - Best move of previous iteration is searched first.
  - The evaluation function is a linear function of Queen Move Distance, King Move Distance (both explained by Lieberum [7]), and mobility of Amazons.

Move generation:
  - Each Amazon is assigned a number in TAS. Moves for each Amazon are generated in the order of these numbers, i.e., first all moves for Amazon 1, then all moves for Amazon 2, etc.
  - The moves of each Amazon and arrow are generated in clockwise order: moving up, moving right up, moving right, etc.

Priority ordering:
  - Moves are ordered based on realization probability adjusted by game progress.
  - Moves that are not in any category (i.e., their realization probability is 0%) are pruned.

Our experiments have been conducted with three versions of the same program. The first version is searching without the use of move categories (called *NMC*), generating moves in the order explained above. The second version uses the realization probabilities, but no game progress (called *NGP*). The third version uses realization probabilities adjusted by game progress (called *GP*).

There are 263 move categories that have a realization probability of 0% (i.e., possible but not played) when game progress is not taken into account. When using realization probabilities based on progress, the number of categories with a probability of 0% changes from 590 categories in move 1-7 to 2 categories or less after move 48. These numbers should be seen relative to the number of possible categories. In move 1-7 there are 1324 categories that are possible, while in move 88-92 only 6 categories are possible. Details are given in Table 6 (NP means number of categories not played).

Table 6. Number of categories with a realization probability of 0%

Mvno	NP	(Pos)	MoveNo	NP	(Pos)	MoveNo	NP	(Pos)
1 - 7	590	(1324)	32 - 39	45	(1218)	64 - 71	2	(127)
8 - 15	214	(1404)	40 - 47	9	(944)	72 - 79	1	(90)
16 - 23	140	(1395)	48 - 55	1	(615)	80 - 87	0	(45)
24 - 31	100	(1335)	56 - 63	1	(276)	88 - 92	0	(6)

We have conducted two experiments. First, we selected a number of positions and let each of the test programs search to depth 3, comparing the search time. The results of this experiment give an indication of the speed-ups that can be expected. The second experiment is a self-play experiment between the different versions. By using categories, certain moves will be discarded without any search (the moves that do not fit into a category), so there is a risk of missing good moves that are not in our categories. The self-play experiments will show how playing strength is influenced by using move categories.

5.1   Comparing Search Times to Depth 3

To investigate the savings in search effort, we compared the search times of searching to depth 3 for the three program versions in a number of example positions. Because the number of legal moves in an average Amazons position is high, searching to depth 3 for a large number of positions is infeasible. We limited ourselves to a set of 1,521 positions from 30 Amazons games. The data for 30 games was quite similar to the data for 25 games, so we do not believe that using more positions will lead to many new insights.

In Fig. 7 the search times of searching to depth 3 in the positions of 30 games are given for NMC, NGP and GP. From this figure it is clear that without using move categories, there are many positions that require a long time to finish the search to depth 3. Also, without move categories there seems to be no strong relation between the number of legal moves and the search time. Using move categories improves the search times considerably and there is a much stronger relation between search times and the number of legal moves. Finally, using move categories and game progress further improves the search speed. Also, there is a very strong relation between search times and the number of legal moves when game progress is taken into account.

From the graph showing the average search times against the number of legal moves, it is clear that in case of a high number of legal moves searching without move categories becomes quickly infeasible. Using move categories is more promising; using move categories based on game progress is even more promising than using only move categories.

There were a number of positions where NMC returned a move different from the ones returned by NGP or GP. NGP played a different move in 95 positions, while GP played a different move in 98 positions. The reason for this difference was that a move in the principal variation of NMC was deleted because it

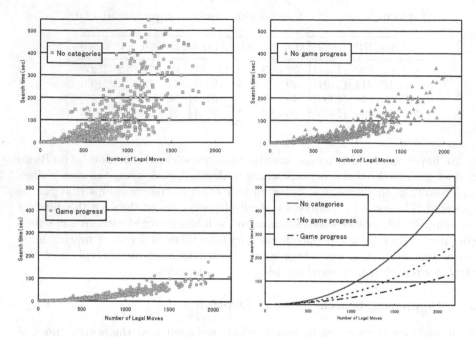

Fig. 7. Search times to depth 3 against number of legal moves for *NMC* (upper left), *NGP* (upper right), *GP* (lower left) and the average search times for each program version (lower right)

had a realization probability of 0%. This seems to indicate that there is a risk that good moves are being discarded, but investigating the positions in question showed that none of the discarded moves were particularly good or necessary. From these results, we concluded that the current categories are good enough to lead to an improvement in search speed without losing playing strength. To confirm this, we conducted a number of self-play experiments that we will explain next.

5.2   Results of Self-play Experiments

In Amazons, there is the problem of generating positions where the chances of winning can be considered equal, because there are no experts and there is almost no opening theory. We generated the initial positions of our self-play experiments by using 50 different positions that were randomly selected from the opening book (after the fourth move from the starting position of the game). These positions were then played twice by each program version, once as White and once as Black. It is possible that the initial positions are better for one side, so we also collected data of the number of squares that the winning side could freely move to at the end of the game. Even if both versions win a game from the same starting position, the difference in free squares at the end of the game gives an indication of difference in playing strength.

We played a total of 9 matches with 100 games each, giving each program 10 seconds, 30 seconds, and 60 seconds per move. The results are given in Table 7. The results of the matches show that our concerns about ending up in uneven positions were unfounded. Even without considering the differences in free squares at the end of the game, each match result except one gives a statistical probability of more than 95% that the winning version is stronger than the losing version (the 58-42 result between *NGP* and *NMC* gives a probability of 94.5% that *NGP* is stronger than *NMC*). Furthermore, the results are not influenced by the amount of time given per move. The results of the matches for 10 seconds and 60 seconds are very close and it seems that the only difference is that more time can reduce the margin of defeat, as can be seen in the drop of the total number of free squares.

Table 7. Self-play results for 100 games with 10, 30, and 60 seconds per move

Match	10 seconds		30 seconds		60 seconds	
	Result	SqDif	Result	SqDif	Result	SqDif
NGP - NMC	58 - 42	+146	60 - 40	+92	59 - 41	+87
GP - NMC	79 - 21	+428	72 - 28	+439	75 - 25	+388
GP - NGP	62 - 38	+214	70 - 30	+259	63 - 37	+165

In Table 8 the results for each program version are summarized. The self-play experiments show that using move categories for move ordering significantly improves the playing strength of an Amazons program and that playing strength can be further improved significantly by using game progress to adjust the realization probabilities of the move categories.

Table 8. Total self-play results

No	Version	10 seconds	30 seconds	60 seconds
1	GP	141 - 59	142 - 58	138 - 62
2	NGP	96 - 104	90 - 110	96 - 104
3	NMC	63 - 137	68 - 132	66 - 134

6   Conclusions and Future Work

In this paper, we have presented a method for improving the search efficiency of an Amazons program by ordering moves based on move categories. We have also investigated the influence of game progress on the move ordering. Experiments showed that a program using move categories is stronger than a program without move categories and that a program using move categories based on game progress is stronger than a program not taking game progress into account.

One area of future work concerns the realization probabilities of the categories. As explained, there are a number of categories that are so special that they occur in only a small fraction of the positions. As a result, these categories often get a very high or very low probability. To address this problem, different ways to decide realization probabilities need to be investigated.

A second, different problem is that the different program versions all have the same evaluation function, leading to mutual oversights. Improvement of general playing strength needs to be further assessed by playing other strong Amazons programs like AMAZONG or INVADER.

A third focus of future work concerns the representation of game progress. The game progress we have used in our experiments is based upon the number of moves played. This is not a perfect solution, because the moment when territory gets fixed differs from game to game and is only loosely related to the number of moves. The number of moves seems to be a good way of measuring progress when games develop in a normal way, but in case of abnormal development (very early or very late fixed territory) the proposed category probabilities can lead the program astray. As a future work, we intend to investigate different methods for measuring progress and compare these with our current findings.

Finally, using game progress to change realization probabilities is a fourth idea for future research. It can be applied to other games than Amazons as well. RPS in shogi might also benefit from dynamically updating realization probabilities using game progress. We are planning to investigate the feasibility of our method in other games than Amazons.

References

1. H. Avetisyan and R.J. Lorentz. Selective Search in an Amazons Program. In *Computers and Games: Proceedings CG2002* (eds. J. Schaeffer, M. Müller, and Y. Björnsson), LNCS 2883, pages 123–141. Springer-Verlag, Berlin, 2002.
2. D. Beal. A Generalised Quiescence Search Algorithm. *Artificial Intelligence*, 43:85–98, 1990.
3. M. Buro. ProbCut: An Effective Selective Extension of the Alpha-Beta Algorithm. *ICCA Journal*, 18(2):71–76, 1995.
4. M. Buro. Experiments with Multi-probcut and a New High-quality Evaluation Function for Othello. In *Games in AI Research* (eds. H.J. van den Herik and H. Iida), pages 77–96. Van Spijk, Venlo, The Netherlands, 2000.
5. T. Hashimoto, Y. Kajihara, N. Sasaki, H. Iida, and J. Yoshimura. An Evaluation Function for Amazons. In *9th Advances in Computer Games (ACG9)* (eds. H.J. van den Herik and B. Monien), pages 191–201. Van Spijk, Venlo, The Netherlands, 2001.
6. T. Hashimoto, J. Nagashima, M. Sakuta, J. Uiterwijk, and H. Iida. Application of Realization Probability Search for Any Games - a Case Study Using Lines of Action. In *Game Programming Workshop in Japan '02*, pages 81–86, Kanagawa, Japan, 2002. (In Japanese).
7. J. Lieberum. An Evaluation Function for the Game of Amazons. In , *10th Advances in Computer Games (ACG10), Many Games, Many Challenges* (eds. H.J. van den Herik, H. Iida, and E.A. Heinz), pages 299–308. Kluwer Academic Publishers, Boston, USA, 2004.
8. H. Matsubara, H. Iida, and R. Grimbergen. Natural Developments in Game Research: From Chess to Shogi to Go. *ICCA Journal*, 19(2):103–112, 1996.

9. N. Sasaki and H. Iida. Report on the First Open Computer-Amazon Championship. *ICCA Journal*, 22(1):41–44, 1999.
10. S. Soeda and T. Tanaka. Categories for Amazons Moves. In *Game Programming Workshop in Japan '03*, pages 118–121, Kanagawa, Japan, 2003. (In Japanese).
11. Y. Tsuruoka, D. Yokoyama, and T. Chikayama. Game-tree Search Algorithm Based on Realization Probability. *ICGA Journal*, 25(3):145–152, 2002.
12. Wikipedia. http://en.wikipedia.org/wiki/The_Game_of_the_Amazons, 2005.

Recognizing Seki in Computer Go

Xiaozhen Niu[1], Akihiro Kishimoto[2], and Martin Müller[1]

[1] Department of Computing Science,
University of Alberta, Edmonton, Canada
{xiaozhen, mmueller}@cs.ualberta.ca
[2] Department of Media Architecture,
Future University-Hakodate, Hakodate, Hokkaido, Japan
kishi@fun.ac.jp

Abstract. Seki is a situation of coexistence in the game of Go, where neither player can profitably capture the opponent's stones. This paper presents a new method for deciding whether an enclosed area is or can become a seki. The method combines local search with global-level static analysis. Local search is used to identify possible seki, and reasoning on the global level is applied to determine which stones are safe with territory, which coexist in a seki and which are dead. Experimental results show that a safety-of-territory solver enhanced by this method can successfully recognize a large variety of local and global scale test positions related to seki. In contrast, the well-known program GNU GO can only solve easier problems from a test collection.

1 Introduction

In the game of Go, the player who has the larger territory wins the game. It is therefore a fundamental requirement for programs to assess continuously the safety of territories correctly. Current computer Go programs use a combination of exact and heuristic techniques, including eye-space analysis [4], search, and heuristic rules based on influence. Heuristic approaches do not guarantee correctness. In order to improve the accuracy of the territory evaluation, they need to be replaced by exact techniques, using search.

An exact, state of the art search-based safety-of-territory solver is described in [12]. However, this solver could not recognize safe stones in seki. The new solver described in this paper extends the previous one and is able to recognize many, even complex seki situations.

Seki is a position where neither player can capture the opponent's stones, so coexistence is the best result. There are two major issues about how to recognize seki. First, a pass is often the only good move for both players. Consecutive passes can cause the Graph History Interaction (GHI) problem [13], which leads to incorrect search results. As a second issue, even if a seki position is recognized locally, the recognition is done in a specific context. The most pessimistic context for one player assumes that the outside is completely filled by safe opponent stones. The result of a local search depends on such assumptions and might need

H.J. van den Herik et al. (Eds.): ACG11, LNCS 4250, pp. 88–103, 2006.
© Springer-Verlag Berlin Heidelberg 2006

to be updated once its surrounding information has been changed by searches in other regions.

This paper presents search-based methods that correctly deal with enclosed areas involving seki. The contributions of the paper are summarized as follows. (1) The status of local seki is correctly determined by using the DF-PN(r) algorithm [5]. (2) An efficient re-search method distinguishes a seki from a win for one player by capturing the opponent. (3) An algorithm for solving global seki problems combines the outcome of local seki searches.

The paper is organized as follows. Section 2 describes the terminology. Section 3 briefly reviews related work. Section 4 explains recognizing seki and its importance to territory evaluation. Section 5 introduces the safety solver and the search algorithms for solving the local and global seki problems. Section 6 presents and describes experimental results and Section 7 provides conclusions and discusses future work.

2 Terminology

Terminology follows [12]. A *block* is a maximal connected set of stones with the same color on the Go board. The adjacent empty points of a block are called *liberties*. A block that loses all its liberties is *captured* and removed from the board. Stones of one color divide the rest of the board into *basic regions*. A *merged region* is the union of two or more basic regions of the same color. The term *region* refers to either a basic or a merged region. If a block is adjacent to only one region of its color, it is called an *interior block*, otherwise it is called a *boundary block* of the region.

A liberty of a boundary block of a region r is called *internal liberty* if it is in r, and *external liberty* otherwise. The color of boundary blocks is called the *defender*, the opponent is called the *attacker*. A *shared liberty* is an empty point that is adjacent to blocks of attacker and defender. A defender's region is called *safe* if all its boundary blocks can be proved safe and the attacker cannot live inside the region.

In a region, simple seki are positions that falls into either of following two types.

1. The defender cannot capture the attacker's stones, neither can the attacker. They both do not have two clear eyes. The best result for both players is to pass.
2. The defender and the attacker are allowed to capture each other's stones. However it will lead to repetitions of the board.

In simple cases, a region can be recognized as a seki by using static rules.[1] However, verifying complicated seki may require deep search. A seki is often related to a *semeai*, a race to capture between two adjacent groups that cannot

---

[1] For a discussion of complex, strange cases, see http://senseis.xmp. net/?StrangeSekis.

both live. For a detailed discussion of semeai, and static methods for evaluation of semeai classes including many seki, see [10].

Figure 1 shows two seki examples. The left black region is a *static seki*. Both the defender and the attacker must pass locally, to ensure that their blocks ● and ⊘ are not captured. They share two liberties in A and B. In the right black region, after 3 moves, with White as the attacker playing first, the black region becomes a seki. Even though the region is not safe, the boundary block ● is safe by seki. This kind of seki that must be discovered by search will be called *dynamic seki*.

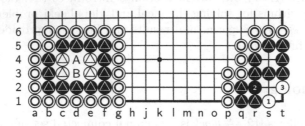

Fig. 1. Left: a static seki. Right: a dynamic seki

3   Related Work

Many approaches have been proposed for recognizing safe stones and territories, including both static and search-based methods.

Unconditionally alive blocks [2] are safe even when the attacker can play an unlimited number of moves and the defender always passes. Related work on seki analysis also includes Müller's static analysis and search-based semeai algorithms [10], and Vilà and Cazenave's static classification rules to recognize safe blocks containing regions up to a size of 7 points [16], including some seki. Among these static methods, only [10,16] can detect seki.

Search-based methods are used in life and death (tsume Go) and safety-of-territory solvers. Müller introduces local search methods for identifying the safety of regions by *alternating play*, where the defender is allowed to reply to each attacker move [9]. Van der Werf *et al.* extend Müller's static rules and use them in the program CSA* (Cascaded Scoring Architecture) to score Go positions [15]. The methods in [9,12] prove territories safe by using static rules and search.

Although some of the static approaches above can handle seki, they do not provide a general solution for recognizing seki in enclosed areas. A practically successful heuristic way of dealing with seki is based on tactical search. Most strong Go programs contain tactical search engines, typically with a threshold of 5 liberties, to recognize blocks that seem safe from capture. This approach does not work for complicated seki cases. In addition, most programs fail to detect vulnerable territories which do not contain any attacker stones yet, where a seki can be reached after an invasion sequence. Even in the tiny example in Fig. 1 on the right, most current programs will pass as both Black and White in the starting position.

4   Recognizing Seki

Recognizing seki is strongly related to both the safety of territory and life and death problems. When evaluating the safety of a territory, seki can be viewed as a win for the attacker, since the defender's territory is destroyed. However when viewed as a life and death problem, the roles are switched since the defender can survive by achieving coexistence in seki.

The search methods to recognize seki in this paper use the most pessimistic assumptions about unproven outside properties for recognizing local seki. The worst-case assumption is that for a given region, all its boundary blocks are completely surrounded by safe attacker stones and have no external liberties. However, the algorithm can take external eyes that have been detected before into account. Any seki recognized under worst-case assumptions will be called as a *local seki*.

The result of local seki can be modified when information about the surroundings changes. For example, external liberties of boundary blocks of a region might affect the result of a local seki. An example is shown in Fig. 2. The Black region is a local seki. However, since the boundary block of Black has one external liberty at A, the status of this black region is unsettled. If White plays first at A, then the black region is a seki. Black playing first can capture four white stones inside the region and the black region becomes territory.

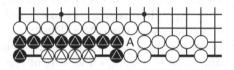

Fig. 2. External liberties can change seki status

Recognizing a global seki is much more complicated because the result of a local seki might be affected when its surrounding conditions are changed. For example in Fig. 1, if the surrounding white blocks ◎ , which were assumed safe, are proved to be dead later by a search on the outside, then in both examples the black region changes from seki to safe, since the local seki collapses.

This paper describes a more general region-based approach to recognize local and global seki positions efficiently and correctly. A detailed discussion of the algorithms is provided in the following section.

5   Safety Solver and Algorithms

5.1   Safety Solver

The safety solver described in this paper is based on the ones in [8,12]. It has
been integrated into the Go program EXPLORER [8]. The search-based solver
in [12] sequentially processes single regions and tries to prove their safety. To
prove a region safe, all boundary blocks must remain safe - none of them can
be captured by the attacker, and the attacker may not live inside the region
surrounded by safe blocks. The solver uses Alpha-Beta search with enhancements
including iterative deepening, a transposition table, move ordering, and heuristic
evaluation functions.

The safety solver used in this paper utilizes a more powerful DF-PN(r) search
algorithm [6,5]. It includes the techniques of [12], such as the solution of strongly
and weakly dependent regions, as well as static and heuristic region evaluation
functions [12]. The following new features will be described in detail in later
sections.

- Solutions to local and global seki.
- Solutions to basic ko situations and the GHI problem.
- The ability to switch goals between attacker and defender. The solver can be
  used both to find successful invasions, and to defend against them. For seki
  detection, the solver is used to prove that stones are safe while territories
  are not.

5.2   The DF-PN(r) Search Algorithm

DF-PN(r) [6,5] is an extension of Nagai's depth-first proof-number (DF-PN)
search algorithm [11]. DF-PN modifies Allis *et al.*'s best-first proof-number search
(PNS) [1] algorithm to use depth-first search. It can expand fewer interior nodes
and use a smaller amount of memory than PNS. DF-PN utilizes local thresholds
for both proof and disproof numbers, selects the most promising node, and performs iterative deepening until exceeding either one of the thresholds. Because
DF-PN is an iterative deepening method that expands interior nodes again and
again, the heart of the algorithm is its use of the transposition table. Whenever
a node is explored, the transposition table is used to cache previous search efforts (i.e., proof and disproof numbers). DF-PN(r) solves a problem of DF-PN
on computing proof and disproof numbers in the presence of repetitions, while
inheriting the good properties of DF-PN. DF-PN(r) contributed to the strength
of a one-eye solver and the currently best tsume-Go solver [5].

5.3   The Local Search Algorithm

The local search algorithm is region-based. It takes a region r as an input and
generates all legal moves in r as well as a pass move. It searches the region
until either the result is decided as safe/safe by seki, or unsafe, or a time limit

is exceeded. r is evaluated as safe if and only if all boundary blocks of r are proved to be safe, and no attacker's stones can live inside. For details, see [12]. r is evaluated as unsafe if and only if any boundary block b of r is captured, or b has only one liberty left and it is the attacker's turn to make a move, or the attacker lives inside the region.

Since the outcome of the search sometimes depends on ko, the local search algorithm needs to model ko threats. The current model assumes that the attacker has the unlimited number of ko threats to win the ko. Therefore, the attacker is always allowed to recapture ko immediately. To deal with a more complicated ko such as double or triple ko, the algorithm uses the *situational super-ko (SSK) rule*. Under the SSK rule, any move that repeats a previous board position, with the same color to play, is illegal. However, assuming an unlimited number of ko threats is often unrealistic, and in some special cases the safety solver fails in proving the safety. See Subsection 6.1 for an example and further discussion.

5.4   Seki and the Graph History Interaction Problem

Let P be a position where player p_1 is to play and Q be the position after p_1 passes in P. If p_1 passes in P and the opponent p_2 passes in Q, the second pass leads back to P. In the local search algorithm, this repetition is allowed, and p_2's pass leads to a local seki. The two passes mean that neither player can win this position. Because of the unlimited ko threat model, ko is not affected by this.

Since DF-PN(r) uses a transposition table, in the presence of repetitions the Graph History Interaction (GHI) problem must be addressed [13,3]. GHI is a notorious problem that may cause search algorithms to solve positions incorrectly. A typical transposition table implementation ignores paths that DF-PN(r) takes to cache search results. However, if a search result that depends on a path is saved in the transposition table, an incorrect cached result may be retrieved from the transposition table.

Figure 3, adapted from [7], shows an example of the GHI problem. Assume that G is a win for the attacker. Let $E \rightarrow H$ and $H \rightarrow E$ be pass moves. If H is searched via $A \rightarrow B \rightarrow E \rightarrow H$, seki is saved in the transposition table entry for H. Two consecutive passes lead back to E. However, this is incorrect if H is reached via $A \rightarrow C \rightarrow F \rightarrow H$, since $A \rightarrow C \rightarrow F \rightarrow H \rightarrow E \rightarrow G$ leads to an attacker win at H.

The GHI problem related to seki is solved by the techniques in [7]. This approach uses an additional field in each transposition table entry to hash the path leading to a position.

5.5   Algorithm for Recognizing Local Seki

Since DF-PN(r) can only answer binary questions, the algorithm for recognizing local seki may have to perform two searches to determine the outcome as win, seki, or loss. Assume that the attacker plays first, and the first search evaluates terminal seki positions as a win for the defender. If the root position r is a loss for the defender, the result is established. However if the result of this search is a

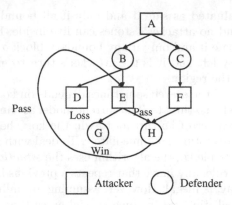

Fig. 3. The GHI problem related to seki

win, it could be either a defender win or a seki. In this case, for the second search all seki terminal positions are considered attacker wins. If the second search is still a defender win, then r is a defender win. However, if the second result is an attacker win, then r is a seki since neither player can win without using seki terminal positions.

The two searches with different seki winners are often quite similar. A method similar to speeding up re-search for ko in [6] decreases the overhead of the second search as follows: each transposition table entry contains a *seki flag* f. In the first search, f is set if and only if a position's disproof tree contains seki.

Consider the second search, when DF-PN(r) looks up the transposition table entry for a position n. If n's table entry contains either a proof, or the proof and disproof numbers of an unproven node, the information can be used safely. However, if n's table entry contains a disproof from the first search, DF-PN(r) checks n's seki flag. If the flag is not set, the position is an unconditional win for the defender and this result can be reused. However, if the seki flag is set, the position is unsolved for this search. The proof and disproof numbers are initialized to 1 to perform a re-search.

The second search often has a low overhead because of the reuse of previous search results in the transposition table. If n's disproof tree does not involve seki, no search is performed. In the extreme case, the second search consists only of a single table lookup, if disproving the root did not involve seki in the first search. Even if a transposition table entry does not contain a disproof, using the proof and disproof numbers in the transposition table results in better move ordering. However, if the solution changes drastically by seki, it might need a high overhead. One example is test position 15 from Test Set 1 (see Sect. 6). The first search returns a loss for the attacker in 0.02 seconds, and the second search returns a seki win for the attacker in 0.93 seconds. In this example, the loss in the first search is obtained quickly by static rules. There is not much useful information in the transposition table that the second search could utilize.

5.6   Algorithm for Recognizing Global Seki

The local result of seki for a region can be proved by using the algorithm discussed in Subsection 5.5. However, this result might need to be updated once the assumptions about surrounding conditions are modified. Figure 4 provides an example. In the beginning, white region A is considered to be unsafe due to the following two reasons: (1) the external liberties for block ⊖ are not used during a local search, and (2) the block also has no internal liberties. White region B is a local seki because 1. the white block ⊗ is already proved to be safe with two eyes elsewhere, and 2. White's boundary block ⊖ and Black's block ▲ share two liberties. The White block ⊖ can be marked as "at least seki", because the *worst* result for this block is safe-by-seki. It might be proved to be unconditionally safe in the future. The *best* result for the black block ▲ is safe-by-seki, therefore it is marked as "at most seki". The information about safe "at least seki" blocks is used in a re-search for white region A, which will become safe. The black block ■ is now dead. The white boundary block ⊖ changes from "at least seki" to safe. Finally, by using this updated information region B is proved to be not a seki, but safe for White, and the status of the Black block ▲ changes from "at most seki" to dead.

Fig. 4. A local seki that collapses on the global level

In Fig. 4, the observation that region A affects region B is trivial because they are adjacent. However, in a real board position the situation can be much more complex. For example, there might be multiple regions that are local seki and form a chain of seki. If the regions in one end of the chain are proved to be not a seki later, then most likely the local result of every region in the chain has to be updated. In addition, care should be taken for using the information about "at least seki" blocks. Clearly any "at least seki" block can only be used to prove one of its adjacent regions that has the same color with the block. A quite important question is: given an "at least seki" block, when it is safe to use it to prove its adjacent regions and when is it not? As described in the previous example, region B can be proved to be a local seki by search in the first round. If the "at least seki" block ⊖ is used to perform the second search to B itself right away, then B will be proved to be safe instantly. However, this result is not correct. Region B can only be confirmed as safe once region A is confirmed as safe (so block ⊖ becomes safe indeed). The solution used in this implementation uses the following condition. For an "at least seki" boundary block b and a region r are to be proved the following.

– If b has no liberties inside r, then it is safe to be used as "safe" to prove r because it will not affect the liberties of any opponent's blocks that are inside r.
– If b does have internal liberties in r, then it should not be used as "safe" to prove r because it will affect the liberties of opponent's blocks that are in r.

By using this condition, it is obvious that in the previous example in Fig. 4 the "at least seki" block ⊖ should never be used to prove white region B directly because ⊖ does has two internal liberties in B.

The algorithm for global recognition first statically recognizes safe regions using the static rules in [2,9]. Then it calls the local search algorithm for proving the safety of the remaining unsolved regions. When processing regions globally, the information about "at least seki" blocks achieved by local searches is stored and possibly used for updating the status of other regions. Eventually, if no further updates can be made, because the status of every region becomes stable, either safe/safe by seki, unsafe, or unknown due to time out, then the global search terminates. At this moment all "at least seki" and "at most seki" blocks are marked to safe, and the results for all local regions are guaranteed to be correct. The pseudo code in Fig. 5 gives the global region processing algorithm for recognizing seki.

The two full-board positions from Test Set 2 shown in Fig. 6 and Fig. 7 are used to illustrate details of the global processing algorithm. In Fig. 6, the solver processes unsafe regions one at a time for both colors in several rounds. At the beginning, the three White regions in the left side of the board have already been proven safe by static rules. When processing a region, the attacker is the opponent and plays first. Assume that all unsafe black regions are processed first, followed by all unsafe white regions in the order A, B, C, and D. The order of regions and colors does not affect the final result, but it may influence the efficiency. In the example, A, B, C, and D are used to refer to different regions for Black and for White. For example, the white region A contains the black block to the left, while the black region A contains the white block to the right but not the black block.

1. Following the worst-case assumption for the outside, black Region A is unsafe because one of its boundary blocks does not have any internal liberties. Similarly black regions B and C are proved to be unsafe. Blocks $j8$ and $g1$ both have an outside eye, but it only provides one liberty. Region D is proved to be a local seki. Therefore its two boundary blocks at $b1$ and $g1$ are added to the "at least seki" list L. For White, region A is proved a local seki, and block $e9$ is added to L. Using this information, white region B is also a local seki and block $h9$ is added to L. Similarly, white region C is a local seki and block $h1$ is added to L, and then the merged region at $b1$ and $e1$ is also a local seki and blocks $c2$ and $d1$ are added to L. Techniques for merging strongly related regions from [12] are used here. After the first round of processing for both colors, L contains 2 black blocks at $b1, g1$ and 5 white blocks at $d9, h9, h1, c2, d1$.

List *at-least-seki* = ∅;
List *at-most-seki* = ∅;
R = all remaining unsafe regions of both colors;

bool updated = true;
while (updated) // The main loop terminates when there is no update can be made
    updated = false;
    for each region r in R perform a first local search
        if (the first search result is proved safe)
            updated = true;
            mark all points in region r and its boundary blocks as safe;
        else if (the first result is a local seki)
            Add attacker's blocks inside r to *at-most-seki*;
            Add unsafe defender's boundary blocks of r to *at-least-seki*;
            if (any new boundary block was added to *at-least-seki*)
                updated = true;
        else if (the first result is proved loss)
            Perform a re-search in region r by using information in *at-least-seki*;
            if (the second search result is proved safe)
                updated = true;
                Mark all points in region r and its boundary blocks as safe;
            if (the second search result is a local seki)
                Add attacker's blocks inside r to *at-most-seki*;
                Add unsafe defender's boundary blocks of r to *at-least-seki*;
                if (any new boundary block was added to *at-least-seki*)
                    updated = true;

Mark all blocks in both *at-least-seki* and *at-most-seki* as safe;

Fig. 5. Global region processing algorithm for determining seki

2. Re-process unsolved regions in the same order. Black regions A and B are still unsafe. Using the "at least seki" black block at $g1$, now black region C can be proved to be a local seki, and block $j8$ is added to L. For White, no progress is made.
3. Black region B is proved to be seki by using the "at least seki" block at $j8$, and block at $f9$ is added to L. Again, no progress for White can be made.
4. Black region A is proven seki by using the "at least seki" block at $f9$ and block $c9$ is added to L.
5. No update can be made in the fifth round. The main loop terminates and the remaining "at least seki" and "at most seki" blocks are marked as safe.

In this example, 10 "at least seki" blocks are marked as safe, and no "at most seki" blocks. In the end, all blocks are proved to be safe. A total of 70 points are marked as safe. The remaining empty points are neutral points in seki in Japanese rules. In Chinese rules, the points surrounded by a single player such as $a1$ are counted for that player.

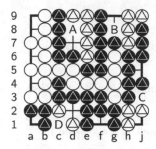

Fig. 6. Global search confirms local regions to be seki

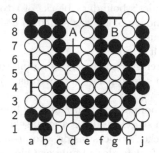

Fig. 7. Local seki collapse in global search

Figure 7 illustrates how local seki can collapse globally. The computation proceeds as follows.

1. In the first round, as in the previous example, assume black regions are processed first in the order A, B, C and D. Region D is proved a local seki and blocks $b1, g1$ are added to L. For White, region A is not a local seki because Black can capture White by playing at $d8$. Therefore white $d9$ is not added to L. Regions B, C and the merged region at $b1, e1$ are also not local seki. After the first round, L contains only the two black blocks $b1$ and $e1$.

2. Using the "at least seki" block at $g1$ black region C is a local seki, and $j8$ is added to L. There is no progress for White.

3. Similarly, using the "at least seki" block $j8$, black region B becomes a local seki, $f9$ is added to L, and there is no progress for White.

4. Using the "at least seki" block at $f9$, black region A is proved to be safe! Region A and block $f9$ are updated to safe and $f9$ is removed from L. Similarly, using safe block $f9$ and "at least seki" block at $j8$ region B is proven safe, and block $j8$ becomes safe and is removed from L. In the same way, black regions C and D are proved to be safe, blocks $e1$ and $b1$ are safe, and they are removed from L. All black regions A, B, C, D are safe and L is empty. Since no further progress can be made for either Black or White, the

main loop terminates. All 81 points on this board are proved to be safe for one of the players.

6   Experimental Results

Two seki test sets were created by combining examples from several resources, including [14,17], and positions from professional games. Test Set 1 is used for local seki testing and Test Set 2 for global seki testing. Both sets contain a mix of easy, moderate, and hard problems. Their difficulty levels are estimated by the CPU time used.[2] All experiments were performed on a Pentium IV/1.6GHz machine with 512 Mb memory.

6.1   Experiment 1: Local Seki Tests

Test Set 1 contains 45 positions, some of them seki and some not when external liberties are considered. Among them, 23 positions are classified as easy, 16 positions as moderate, and 6 positions as hard. The seki-enhanced safety solver solves 42 positions within a time limit of 120 seconds per position.

Table 1 summarizes the cost of seki re-search on all 45 positions. The total execution time for phase two search is 14.5% of phase one, and the total number of nodes expanded in phase two is 17.7% of phase one. The cost is relatively small compared to the version that does not detect seki.

Table 1. Overhead of seki re-search

Phase One Search		Phase Two Search	
Total nodes expanded	Total time (sec)	Total nodes expanded	Total time (sec)
1,101,733	403.12	195,039	58.47

Figure 1 showed two easy positions from Test Set 1. Figure 8 provides another four positions with different difficulty levels. The black regions in the top corners are two moderate cases, solved in 16 seconds (left) and 31 seconds (right) total execution time for the two phases. The bottom right corner is a hard case, solved in 116 seconds. In these examples, seki is reached dynamically through a sequence of moves from the starting position. The black region in the bottom left corner is safe with best defense. The attacker cannot achieve a seki. This hard case is solved in 44 seconds. Strongest move sequences for both players are shown in Fig. 8.

Sometimes seki depends on ko. It is important to model ko in the search of seki. The main technical limitation of the current solver is a class of ko positions called *Moonshine Life*.[3] The current solver defines the ko winner to be the attacker, and allows the ko winner get unlimited ko-threats to win the ko. In these situations,

---

[2] The test sets are available at: http://games.cs.ualberta.ca/go/seki.

[3] Described at http://senseis.xmp.net/?MoonshineLife.

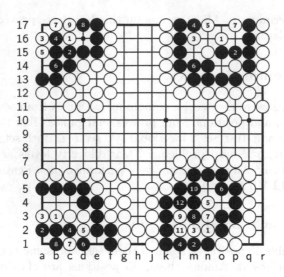

Fig. 8. Four examples from Test Set 1. White to play.

Table 2. Comparison with GNU Go in Test Set 1

Program	Easy (total 23)	Moderate (total 16)	Hard (total 6)	Solved / Total
EXPLORER	20	16	6	93.3%
GNU Go	19	7	0	57.8%

of which there are three in the test set, the solver produces a doubtful result. Figure 9 shows an example. Since White is considered to have infinite ko threats, there is no way for Black to capture White. So block ● is identified as "at least seki" and ⊖ is identified as "at most seki" in the local search. If no update can be made after the global search, both of them will be identified as safe by seki. This result is dubious because in a real game ko threats are only unlimited when there is a multiple ko elsewhere on the board. In this example, static rules can be added to solve the problem. However, a more general solution that takes a finite number of available ko threats into account remains as future work.

The correctness of global search is based on the correctness of local searches. Still, even if Moonshine Life occurs in a local search, it will not invalidate the result of the global search. The reason is that only "at least seki" blocks are used to update other regions in the global search. In the example, only block ● is marked as "at least seki" and it can safely be used when searching other regions.

A comparison is made between the safety solver in the program EXPLORER and one of the strongest programs, GNU Go 3.6.[4] Table 2 shows the number of test positions that are solved correctly by each program.

---

[4] Available at http://www.gnu.org/software/gnugo/.

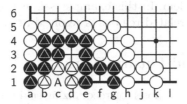

Fig. 9. An unsolved position

The seki-enhanced safety solver solved 42 positions with correct results while GNU GO solved 26. Both EXPLORER and GNU GO can solve most easy positions. The safety solver in EXPLORER does not use any static rules to recognize seki, while it seems that GNU GO use mainly static rules. GNU GO solves less than half of the moderate positions and none of the hard positions.

6.2   Experiment 2: Global Seki Tests

Test Set 2 contains 20 global seki problems. In each of these problems, the result of local seki search needs to be resolved on the global level. 19 of them are solved within a time limit of 200 seconds. GNU GO was not tested on these global positions because the correctness of global search is based on the correctness of local search. Two full-board positions shown in Fig. 6 and Fig. 7 have been used in Subsection 5.6 to illustrate the global processing steps. The total execution time for these examples was 192 and 121 seconds respectively.

Figure 10 shows the only unsolved global position in Test Set 2. Currently there is no semeai search used to compute how many liberties the white blocks can get from the big eyes in regions A and C, and how many liberties the black block can get from region B. By using the most pessimistic assumption for the defender (White), the solver believes that the blocks $e5$ and $m5$ have only 2 internal liberties and one external liberty each (one liberty per eye). Therefore

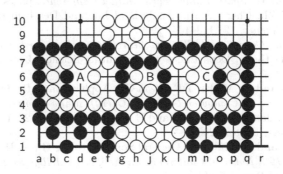

Fig. 10. Unsolved seki example

it cannot solve this problem. In this example, seki can be confirmed only if the semeai status of several related regions is resolved first.

7   Conclusions and Future Work

This paper presents two methods for recognizing seki positions locally and globally. Although we may conclude that the results are quite encouraging, there are still numerous ideas to improve the performance. The limitations of the current solver and other possible topics for future work include the following.

- Handling of ko situations. Instead of allowing the ko winner unlimited ko threats, handling a finite number of threats must be implemented.
- The current solver is purely search-based. Static rules could improve the efficiency in many simple cases.
- The solver is region-based, and does not work for more open-ended areas that occur in games.
- Integrate the functionality of tactical solvers and semeai search into a safety solver.
- In the current global method, there is no priority for selecting which region to process in the main loop of the global search. Ordering regions by considering their adjacency should increase the efficiency of the global search.
- Seki spanning multiple regions such as in Fig. 10 need a different approach.
- When to call the safety solver to recognize seki during a game?

References

1. L.V. Allis, M. Meulen, and H.J. van den Herik. Proof-Number Search. *Artificial Intelligence*, 66(1):91–124, 1994.
2. D.B. Benson. Life in the Game of Go. *Information Sciences*, 10:17–29, 1976. Reprinted in Computer Games, (ed. D.N.L. Levy), Vol. II, pp. 203-213, Springer-Verlag, New York, 1988.
3. M. Campbell. The Graph-History Interaction: On Ignoring Position History. In *Association for Computing Machinery Annual Conference*, pages 278–280, 1985.
4. K. Chen and Zh. Chen. Static Analysis of Life and Death in the Game of Go. *Information Science*, 121:113–134, 1999.
5. A. Kishimoto. *Correct and Efficient Search Algorithms in the Presence of Repetitions*. PhD thesis, Department of Computing Science, University of Alberta, 2005.
6. A. Kishimoto and M. Müller. DF-PN in Go: Application to the One-Eye Problem. In *10th Advances in Computer Games (ACG10), Many Games, Many Challenges* (eds. H.J. van den Herik, H. Iida, and E. A. Heinz), pages 125–141, Kluwer Academic Publishers, Boston, 2004.
7. A. Kishimoto and M. Müller. A General Solution to the Graph History Interaction Problem. In *19th National Conference on Artificial Intelligence (AAAI'04)*, pages 644–649. AAAI Press, 2004.
8. M. Müller. *Computer Go as a Sum of Local Games: An Application of Combinatorial Game Theory*. PhD thesis, Diss. ETH Nr. 11.006, ETH Zürich, 1995.

9. M. Müller. Playing it Safe: Recognizing Secure Territories in Computer Go by Using Static Rules and Search. In *Game Programming Workshop in Japan '97* (ed. H. Matsubara), pages 80–86, Computer Shogi Association, Tokyo, Japan, 1997.

10. M. Müller. Race to Capture: Analyzing Semeai in Go. In *Game Programming Workshop in Japan, Vol.99(14) of IPSJ Symposium Series*, pages 61–68, 1999.

11. A. Nagai. *DF-PN Algorithm for Searching AND/OR Trees and Its Applications.* PhD thesis, Department of Information Science, University of Tokyo, 2002.

12. X. Niu and M. Müller. An Improved Safety Solver for Computer Go. In *4th Computers and Games Conference (CG 2004)* (eds. H. J. van den Herik, Y. Björnsson, and N. S. Nethanyahu), LNCS 3846 , pages 97–112, Springer-Verlag, Berlin, 2006.

13. A.J. Palay. *Searching with Probabilities.* PhD thesis, Carnegie Mellon University, 1983.

14. Sh. Tao. *Guan Zi Pu.* 1689. Reprinted in Shu Rong Qi Yi Chu Ban She (Eds. Wei Qi Ji Qiao Da Quan, Jiang MingJiu, and Jiang ZhuJiu), Cheng Du, China, 1996.

15. E. van der Werf, H.J. van den Herik, and J.W.H.M. Uiterwijk. Learning to Score Final Positions in the Game of Go. In *10th Advances in Computer Games (ACG10), Many Games, Many Challenges* (eds. H.J. van den Herik, H. Iida, and E.A. Heinz), pages 143–158, Kluwer Academic Publishers, Boston, 2004.

16. R. Vilà and T. Cazenave. When One Eye is Sufficient: a Static Classification. In *10th Advances in Computer Games (ACG10), Many Games, Many Challenges* (eds. H.J. van den Herik, H. Iida, and E.A. Heinz), pages 109–124, Kluwer Academic Publishers, Boston, 2004.

17. E. van der Werf. *AI Techniques for the Game of Go.* PhD thesis, University of Maastricht, 2005.

Move-Pruning Techniques for Monte-Carlo Go

Bruno Bouzy

UFR de mathematiques et d'informatique,
Université René Descartes, Paris, France
bouzy@math-info.univ-paris5.fr

Abstract. Progressive Pruning (PP) is employed in the Monte-Carlo Go-playing program INDIGO. For each candidate move, PP launches random games starting with this move. The goal of PP is: (1) to gather statistics on moves, and (2) to prune moves statistically inferior to the best one [7]. This papers yields two new pruning techniques: Miai Pruning (MP) and Set Pruning (SP). In MP the second move of the random games is selected at random among the set of candidate moves. SP consists in gathering statistics about two sets of moves, GOOD and BAD, and it prunes the latter when statistically inferior to the former. Both enhancements clearly speed up the process of selecting a move on 9×9 boards, and MP improves slightly the playing level. Scaling up MP to 19×19 boards results in a 30% speed-up enhancement and in a four-point improvement on average.

1   Introduction

Computer Go remains a difficult task for computer science [14,12] mainly for two reasons. First, the branching factor of the game tree and the game length prohibit global tree search. Second, evaluating non-terminal Go positions is hard [13]. Meanwhile, computer Go has been used as an appropriate testbed for AI methods [6] during the last decade. I started twelve years ago, with the development of the Go-playing program, INDIGO [5]. Since 2002, INDIGO includes a Monte-Carlo approach that enriches the knowledge-based approach developed previously. Our Monte-Carlo approach is inspired by usual experiments [7] reproducing the original approach of Monte-Carlo Go [8]. Subsequently, these experiments introduced different enhancements to the basic Monte-Carlo algorithm. Currently, Progressive Pruning (PP) is the umbrella enhancement to be used in INDIGO. In 2003, we combined our Monte-Carlo Go approach with a knowledge-based approach [3] and with a global tree-search approach [4]. The result was successful because INDIGO won the bronze medal at the 2004 Olympiad on 19×19 Go [10]. Yet this successful combination is not the topic of the current paper. The innovative question that motivates the work presented here is: how can we improve Progressive Pruning (a) in the game of Go, and (b) in general? To this purpose we assess two pruning techniques intended to improve PP: Miai Pruning (MP) and Set Pruning (SP). We will introduce these two new pruning techniques, and provide an experimental assessment.

H.J. van den Herik et al. (Eds.): ACG11, LNCS 4250, pp. 104–119, 2006.
© Springer-Verlag Berlin Heidelberg 2006

Section 2 discusses related work that deals with Monte-Carlo games, and it recalls the underlying idea of PP. Section 3 defines the two pruning techniques, MP and SP. Then, Section 4 yields the results of the experiments assessing these two techniques in isolation, and in combination. Some remarks are discussed in Sect. 5. Section 6 provides a conclusion and some prospects.

2   Related Work and Motivations

Below we discuss three topics, viz. Monte Carlo in computer games (2.1), progressive pruning (2.2), and motivations (2.3).

2.1   Monte Carlo in Computer Games

Monte-Carlo methods were designed in order to simulate physical models. Because they used random number generation such as games in the casino, the name Monte Carlo was adopted. Then Monte-Carlo methods were embraced by computer games, and so, to some extent, a loop has been closed. In games such as Poker and Scrabble, hidden information is sampled with the help of random distributions that are plausible according to past actions performed in the game. In such games, random generation can also be used to perform random simulations of games, which is done by POKI at Poker [2] and by MAVEN at Scrabble [15]. In games containing randomness in their rules, such as Backgammon, random simulations are used quite naturally [17]. In complete information games not containing any chance, such as Go, Chess, and Othello, the idea of simulating games at random is less natural. Nevertheless, this is not the first time that Monte-Carlo methods have been tried in complete information games.

In 1990, Abramson [1] gave a seminal description of evaluating a position of a two-person complete information game with statistics. He proposed the *expected-outcome model*, in which the evaluation of a game-tree node is the expected value of the game's outcome given random play from that node on. The author showed that the expected outcome is a powerful heuristic. He concluded that the expected-outcome model of two-player games is "precise, accurate, easily estimable, efficiently calculable, and domain-independent". In 1990, he tried the expected-outcome model on the game of 6 × 6 Othello.

Brügmann [8] was the first to develop a Go program based on random games. The architecture of the program, GOBBLE, was remarkably simple. In order to choose a move in a given position, GOBBLE played a large number of random games from this position to the end, and scored them. Then, he evaluated a move by computing the average of the scores of the random games in which it had been played.

We believe that Abramson's approach (or Brügmann's) are quite appropriate for the game of Go because they enable the program to reach terminal positions that are easy to evaluate and particularly representative of the current position. By computing a mean on terminal positions reached at random, the program obtains a first-rate evaluation of the current position. We admit that computing a Monte-Carlo evaluation costs much more time than computing a conceptual evaluation using domain-dependent knowledge, but we believe that the cost is worthwhile. This is why we follow the Monte-Carlo approach in INDIGO. The next subsection recalls how PP is used in INDIGO.

2.2   Progressive Pruning

The aim of PP is to be able to choose the best move. The current description is based on Bouzy and Helmstetter [7]. As contained in the basic idea of Abramson, each move has a mean value m, a standard deviation σ, a left expected outcome m_l and a right expected outcome m_r. For a move, $m_l = m - \sigma r_d$ and $m_r = m + \sigma r_d$. r_d is a ratio fixed by practical experiments. Currently, $1.5 \leq r_d \leq 2.0$ is for us a good tradeoff between playing level and time. A move M_1 is said to be statistically inferior to another move M_2 if $M_1.m_r < M_2.m_l$. Two moves M_1 and M_2 are statistically equal when $M_1.\sigma < \sigma_e$ and $M_2.\sigma < \sigma_e$ and no move is statistically inferior to the other. σ_e is called the standard deviation for equality, and its value is determined by experiments.

In PP, after a minimal number $N_{\min}$ of random games (currently 50 per move), a move is pruned as soon as it is statistically inferior to another move. N_{rg} is the current number of random games performed; the standard deviation of the mean value computed after N_{rg} random games is $\sigma / \sqrt{N_{\mathrm{rg}}}$. Therefore, moves are pruned as N_{rg} increases, and the number of candidate moves decreases while the process is running. The process stops if one of three conditions is fulfilled: (1) when there is only one move left, (2) when the moves left are statistically equal, or (3) when a maximal threshold of iterations N_{total} is reached. In all cases, the move with the highest expected outcome is chosen. This progressive pruning algorithm is similar to the one described in [2].

Due to the increasing precision of mean evaluations while the process is running, the mean value of the current best move is decreasing. Consequently, a move can be statistically inferior to the best one at a given time and not later. Thus, the pruning process can be either hard (a pruned move cannot be a candidate later on) or soft (a move pruned at a given time can be a candidate later on). Of course, soft PP is more precise than hard PP. Nevertheless, in the experiments shown here, we use hard PP.

2.3   Motivations

Using PP or any move-pruning scheme is debatable. For example, Sheppard [16] uses a clever scheme to drive the choice of which simulation to perform on which move, and this scheme does not prune any move. In contrast, we assume that PP is used, and attempt to improve it. We do not debate on the use or non-use of move pruning.

The background of this work is the architecture of INDIGO. It is made up of a pre-selection module and a Monte-Carlo module. N_{select} is the number of moves; it is the output of the pre-selection module, and the input of the Monte-Carlo module. As long as PP is running, the number of possible moves is decreasing, namely from N_{select} down to 1 at which point the process stops. Two remarks can be made. First, PP spends most of its time when two possible moves are left. Frequently, the evaluations of the two moves left are almost equal, and a great deal of iterations are necessary to separate them. Thus, the first goal is to reduce the time spent by PP when two moves are left. Second, INDIGO's playing level highly depends on N_{select}. INDIGO's playing level roughly increases with N_{select}. (Admittedly, this is not completely correct. Actually, INDIGO's playing level reaches an optimal value with $N_{select} = 8$, 16, or 32, depending on the size of the board 9×9, 13×13, or 19×19, and then, it decreases.) However, the optimal value of N_{select} can be quite high and therefore the second goal is to reduce the time spent by PP to eliminate moves at the beginning of the process. To sum up, we need (1) a pruning technique that lowers the time spent when two or a few moves are left at the end of the process, and (2) another technique to eliminate quickly most of the moves at the beginning of the process; both techniques should operate under the same statistical confidence when pruning. With this aim, we designed two pruning techniques.

3 Two Pruning Techniques

This section describes the two pruning techniques. Miai Pruning (MP) is the technique that speed-up the end of PP process when a few moves are left (in particular two), and Set Pruning (SP) is the technique that speeds up the beginning of the PP process when many moves are involved.

3.1 Miai Pruning

Without any loss of generality, we assume that two moves are left, A and B, and that Black is to move. PP aims at finding the move with the best mean. Therefore, PP launches many games (1) starting with Black A and with the following moves randomly chosen, and (2) starting with Black B, and with the following moves also randomly chosen. Unfortunately, in the half of the games starting with move A, move B is also played by Black. In addition, in the half of the games starting with move B, move A is also played by Black. Thus, in the half of the random games played out to separate the moves A and B, A and B are played by the same player. When the order of the moves of a sequence is not important to reach a position (which is not rare in Go) the half of the random games does not help much to discriminate A and B. So, the idea of MP is (1) to launch games starting with Black A and White B, and (2) to launch games starting with Black B and White A to separate A and B. Now we should make MP working when the number of possible moves is arbitrary small (say three or more). MP works as follows. For each possible move A, B being another possible move different from A chosen at random, MP launches games starting with Black A in the first move, and White B in the second move. The term "miai" emerged because it is used by human Go players for the same concept: "miai" means equivalent. When two moves are miai, then it happens that if a player plays one of them, the other player plays the other one. Finally, we remark that MP is designed to separate moves which are not miai by imitating the actual way of human playing. When played after move A, B can be an illegal move. In such case, MP is simply not used.

3.2   Set Pruning

Let us assume that $N_{possible}$ moves are left. SP applies two stages. First, to be cautious and avoid rather bad pruning due to bad chance, N_{min} is devised by PP to forbid any pruning before N_{min} random games per move are arrived at. Second, at the beginning of the PP process, the σ of a move is high for each move. Thus, no move has a good chance to be statistically inferior to the best one with a given statistical confidence. The idea underlying SP is to associate a mean value and a σ not only to the possible moves but also to all the possible *sets* of moves of size $N_{possible}/2$. N_{rgpm} is the number of random games performed per candidate move, then the mean value of the random games performed given that the first move belongs to a given set of size $N_{possible}/2$ is known with a σ that is in $1/\sqrt{N_{rgpm} \times N_{possible}/2}$. Consequently, the width of the confidence interval around the mean value associated to sets of moves is $\sqrt{N_{possible}/2}$ times smaller than the width of the confidence interval around the mean value associated to moves. Thus, it is possible to prune a set of moves at once, with a statistical confidence which equals the statistical confidence at which PP prunes moves one by one. This way, the number of possible moves is divided by two, each time a set of moves is pruned. The idea is quite attractive because in practice we do not need to consider all of the possible sets of size $N_{possible}/2$ (a big effect) but only two sets. Because the moves are ranked from the best move down to the worst move, they can be grouped into two sets, the set of the $N_{possible}/2$ best moves, called GOOD, and the set of the $N_{possible}/2$ worst moves, called BAD. The mean associated to GOOD is the highest mean associated to any other set of size $N_{possible}/2$, and the mean associated to BAD is the lowest mean associated to any other set of size $N_{possible}/2$. Therefore, the first set to be pruned is the pruning of BAD. In practice, SP works as follows. In addition to the mean and σ computed by PP for each move, SP builds the two sets GOOD and BAD, and computes their means and their σ. When BAD is found to be statistically inferior to GOOD, it is pruned with a statistical confidence identical to the statistical confidence at which moves are pruned by PP.

4   Experiments

Starting from PP we determine empirically the size of N_{select} (4.1). Thereafter, this section evaluates the relative merits of MP (4.2) and SP (4.3), of their direct combination (M+SP) (4.4), their strong combination (M+gbSP) (4.5), and of a special combination of the two (M+gbP) (4.6), all of this regarding time and playing level. We end up this section with an all-against-all tournament (4.7) gathering the best programs of the experiments.

Since we explore move-pruning abilities of a Monte-Carlo Go program, we wish first to observe the move-pruning effect isolated from deep-tree search effects. Thus, we have performed experiments with depth-one search only (4.8). Furthermore, because (1) we need a large amount of game results to obtain a sufficient statistical significance, and (2) 19×19 games are too long, we have used 9×9 boards. When a pruning technique has been demonstrated as performing well on 9×9 boards at depth-one, it is assessed in a second stage either at depth-n on 9×9 boards (4.9) or at depth-one on 19×19 boards (4.10).

For each technique, we set up experiments to assess its effect on the time level and on the playing level. An experiment consists of a match of 200 games between the program to be assessed and the experiment reference program, each program playing 100 games with Black. The result of an experiment is generally a set of relative scores assuming that the assessed program is the max player. Given that the standard deviation of 9×9 games played by our programs is roughly 15 points, 200 games enable our experiments to lower σ down to 1 point and to obtain a 95% confidence interval of which the radius equals 2σ, i.e., 2 points. We have used 2.8 GHz computers. Furthermore, all programs in these experiments do not use any conservative or aggressive style depending on who is ahead in a game, they only try to maximize their own score. The score of a game is more significant than the winning percentage. N_{select} is a power of 2 between 2 and 64. r_d is set to 2.0.

PRUNE is the name of the program to assess. In its basic version, PRUNE uses PP only. The notation to name a program to be assessed is simple: for example, PRUNE($MP = true$) is the program that uses additionally the MP technique, and so on for PRUNE($SP = true$) or PRUNE($N_{\text{select}} = N$).

4.1 N_{select} Versus $N_{\text{select}}/2$

We start attempting to obtain the best playing level and the minimal response time, knowing that the effect of increasing the value of N_{select} is worth remembering. Table 1 shows the effects of simply doubling N_{select}. Each number corresponds to a confrontation between PRUNE(N_{select}) and PRUNE($N_{\text{select}}/2$).

Table 1 shows an increase of the playing level in N_{select}. However, the returns diminish as N_{select} increases. Although being not statistically significant, Table 1 shows that PRUNE($N_{\text{select}} = 32$) is slightly superior to PRUNE($N_{\text{select}} = 16$), and even that PRUNE($N_{\text{select}} = 64$) is almost equal to PRUNE($N_{\text{select}} = 32$). Regarding the relative time between the two programs, for low values of N_{select},

Table 1. Result of doubling N_{select}

	N_{select}				
	4 vs. 2	8 vs. 4	16 vs. 8	32 vs. 16	64 vs. 32
Mean score	+8.0	+6.0	+4.3	+0.4	−0.2
Winning percentage	72	65	63	54	52
Mean relative time	2.0	1.75	1.58	1.30	1.12

PRUNE(N_{select}) is about twice slower than PRUNE($N_{\text{select}}/2$), but for high values of N_{select}, PRUNE(N_{select}) is almost as fast than PRUNE($N_{\text{select}}/2$).

The results of this introductory experiment show that increasing N_{select} from 2 up to 16 is worthwhile considering in terms of playing level on a 9×9 board. The explanation is straightforward for PRUNE. We assume that the pre-selection module being based on hand-crafted domain-dependent knowledge still contains errors, and that the Monte-Carlo module is quite adequate at selecting the right move. With this assumption, the larger N_{select}, the larger the probability of selecting a good move input of Monte Carlo. However, one thing cannot be omitted: the pre-selection module gives a penalty to tactically bad moves but it does not eliminate them. Thus, when N_{select} is sufficiently large, the tactically bad moves are also input of Monte Carlo. Monte Carlo is bad at recognizing tactically bad moves, which explains that PRUNE($N_{\text{select}} = 64$) is worse than PRUNE($N_{\text{select}} = 32$). Meanwhile, the time for obtaining the overall playing-level jump is multiplied by a factor 6.

4.2   Miai Pruning Versus Progressive Pruning

This subsection first compares MP with PP. Then it shows the effects of doubling N_{select} while using MP.

Table 2. Result of MP vs. PP

	N_{select}					
	2	4	8	16	32	64
Mean score	+2.9	+1.3	+2.1	−1.3	−0.6	−0.7
Winning percentage	53	50	51	49	50	47
Mean relative speed	1.50	1.45	1.37	1.33	1.31	1.27

Table 2 shows that MP is worth considering for low values of N_{select}. Regarding the motivations of this paper (see 2.3), the result of PRUNE($MP = true$, $N_{\text{select}} = 2$) is crucial to comment upon. First, PRUNE($MP = true$, $N_{\text{select}} = 2$) has a non-negative result against PRUNE($MP = false$, $N_{\text{select}} = 2$): +3 points and 53% wins. The mean score is statistically significant because 3 points is superior to the radius of the confidence interval which equals 2 points. Second, the speed is enhanced significantly, multiplied by 1.5. Thus, the first column of Table 2 experimentally proves the relevance of MP, and it adheres the goals set in Section 3. The non-negative mean score and the speed enhancement of two next columns ($N_{\text{select}} = 4, 8$) of Table 2 confirm the effectiveness of MP. The right part of the table then shows slightly negative mean scores. MP appears to be less adapted to situations in which many moves are candidate than to situations with a few candidate moves.

As already shown in Table 1, Table 3 shows that with MP the playing level also increases in N_{select}. Between $N_{\text{select}} = 2$ and $N_{\text{select}} = 4$, the return improves faster with MP than without MP. However, for high values of N_{select} the return diminishes faster with MP than without MP. Moreover, PRUNE($N_{\text{select}} = 64$)

Table 3. Result of doubling N_{select} while using MP

	N_{select}				
	4 vs. 2	8 vs. 4	16 vs. 8	32 vs. 16	64 vs. 32
Mean score	+15.4	+3.7	+1.0	−1.2	−4.3
Winning percentage	67	58	52	49	43
Mean relative speed	0.47	0.52	0.61	0.60	0.62

looks like inferior to PRUNE($N_{select} = 32$) with some statistical significance; this is remarkable. It confirms the experimental fact that MP is less adapted to situations in which many moves are candidate than to situations with a few candidate moves. Our current explanation is the following. Without MP, the second move of a random game is selected pseudo-randomly with domain-dependent knowledge: one-liberty string or 3×3 pattern urgencies [3]. With MP, the second move is selected at random with uniform probability among the set of candidate moves. When N_{select} is small, the candidate moves are all approximately good, thus the second move selected by MP has a good chance to be better than the move generated pseudo-randomly with domain-dependent knowledge: one-liberty string or 3×3 pattern urgencies. When N_{select} is high, the candidate moves are approximately average, thus the second move selected by MP has a good chance to be worse than the move selected pseudo-randomly with domain-dependent knowledge. In conclusion, selecting the second move of the random game is a question of superiority between MP and the pseudo-random generator based on domain dependent knowledge. To be effective, MP must be better than the current pseudo-random move generator. If random games based on uniform probability were used, then MP would have no difficulty to be superior. In the background of the pseudo-random generator using domain-dependent knowledge, employing MP when N_{select} is high is consequently a bad idea. Subsection 4.8 will show a remedy to this problem.

4.3   Set Pruning Versus Progressive Pruning

This subsection compares SP with PP. Table 4 shows the results.

For low values of N_{select}, PRUNE($SP = true$) plays at the same level as PRUNE($SP = false$) and the increase in speed is not high. For high values of N_{select}, the relative speed of the two programs is significantly superior to 1, which was expected, because the SP technique is designed for high

Table 4. Result of SP vs. PP

	N_{select}					
	2	4	8	16	32	64
Mean score	+0.1	+0.5	+0.2	−1.3	−2.4	−3.5
Winning percentage	50	49	52	48	44	43
Mean relative speed	1.00	1.05	1.08	1.12	1.14	1.17

values. However, while the relative speed increases, the playing level decreases significantly, PRUNE($SP = true$, $N_{select} = 64$) being significantly inferior to PRUNE($SP = false$, $N_{select} = 64$). This result confirms the fact that using SP is debatable.

4.4 M+SP Versus Progressive Pruning

This subsection directly combines MP and SP, and compares this combination with PP. The direct combination means that SP is used in addition to MP, and that no other enhancement is used, as will be shown in the next subsection.

Table 5. Result of M+SP vs. PP

| | N_{select} | | | | | |
	2	4	8	16	32	64
Mean score	−0.5	+0.4	−3.9	−8.5	−11.4	−12.5
Winning percentage	49	51	42	33	28	29
Mean relative speed	1.5	1.5	1.7	2.2	2.6	2.8

Table 5 shows the results of MP+SP versus PP. The results are bad. Losing by eleven or twelve points on average on 9×9 boards is huge in Go standards. This result is disappointing. While SP did not give good results, it was risky to combine them so directly. The next experiment aims at combining MP in a sophisticated way.

4.5 M+gbSP Versus Progressive Pruning

Since the direct combination M+SP did not work well, we tried a more sophisticated combination of MP and SP, called M+gbSP. In addition to MP and SP, the two sets, GOOD and BAD updated by SP, were used by M+gbSP to launch the random games: for each possible move A, MP+gbS launches games starting with Black A and White B, B being picked up at random among BAD if A is in GOOD, and picked up in GOOD otherwise. Thus, when launching the random games, the idea underlying M+gbSP is to apply the miai principle on the two sets GOOD and BAD instead of applying them on moves.

Table 6 shows the results of M+gbSP versus PP. The results are still bad. Losing by ten or fifteen points on average on 9×9 boards is huge in Go standards. Our current explanation is that the strong combination reinforces the pruning

Table 6. Result of M+gbSP vs. PP

| | N_{select} | | | | | |
	2	4	8	16	32	64
Mean score	+0.3	+1.7	−8.3	−10.2	−11.3	−15.4
Winning percentage	52	52	35	33	32	26
Mean relative speed	1.5	1.6	1.9	2.8	3.5	3.8

strategy. Here, games are launched in order to enhance move pruning instead of neutrally finding the mean value of moves. When the first part of a run badly ranks a move - a "good" move M is put into BAD - the next part of the run tries to reinforce the current finding: when it launches games starting by M, the second move of the game (played by the opponent) is picked up from GOOD, thus the mean value of M is penalized. Symmetrically, when a "bad" move M is put into GOOD in the beginning of the run, the next part of the run launches games starting by M, then the second move of the game (played by the opponent) is picked up from BAD, thus the mean value of M is optimistic, and M remains in GOOD.

4.6 M+gbP Versus Progressive Pruning

Since M+gbSP does not work, probably due to its complexity, this subsection tries a simplification. It combines MP with the sole use of the two sets, GOOD and BAD, but not with SP. We call this combination M+gbP. GOOD and BAD are used in the same way as they are used in the strong combination M+SP: for each possible move A, M+gbP launches games starting with Black A and White B, B being picked up at random among BAD if A is in GOOD, and picked up in GOOD otherwise.

Table 7 shows the results obtained by M+gbP versus PP. The relative speed is higher than it was in Table 2. For $N_{select} = 2, 4, 8, 16$, the playing level of M+gbP seems identical to the playing level of MP. However, for $N_{select} = 32$ or 64, the results are still bad. The arguments highlighted by the previous subsection could still explain them.

4.7 All-Against-All Tournament

In the previous subsections, we have made relative assessments of the pruning techniques against PP with constant N_{select}, and relative assessments of doubling N_{select} with a fixed pruning technique (either MP or PP). In this subsection, we look for the best programs, in term of time and playing level. Thus, based on the previous experiments' results, we have built two tables approximating the values of the programs against PRUNE(PP, $N_{select} = 2$). Table 8 yields the average time used by the programs PRUNE to play one game, and Table 9 the relative playing level of PRUNE estimated with the previous results.

Table 7. Results of M+gbP vs. PP

	N_{select}					
	2	4	8	16	32	64
Mean score	+0.7	+2.1	+0.0	−1.1	−3.5	−8.1
Winning percentage	51	55	50	44	43	35
Mean relative speed	1.5	1.5	1.5	1.5	1.6	1.7

Table 8. Average time (in minutes) spent by PRUNE(N_{select}, P) to play out one game

N_{select}	PP	MP	SP	M+SP	M+gbSP	M+gbP
2	1.0	0.7	1.0	0.7	0.7	0.7
4	2.0	1.4	2.0	1.3	1.2	1.3
8	3.5	2.7	3.5	2.1	1.8	2.5
16	5.5	4.3	5.5	2.5	1.9	3.7
32	7.0	5.2	6.2	2.7	2.0	4.5
64	7.5	6.0	6.5	2.7	2.0	4.5

Table 9. Relative playing level of PRUNE(N_{select}, P) estimated by the previous subsections

N_{select}	PP	MP	SP	M+SP	M+gbSP	M+gbP
2	0	+3	0	0	0	0
4	+8	+11	+8	+8	+8	+10
8	+14	+15	+13	+10	+8	+14
16	+18	+18	+17	+10	+8	+17
32	+18	+16	+16	+9	+7	+15
64	+18	+14	+15	+8	+4	+10

Table 9 clearly shows that PRUNE(M+SP OR M+GBSP) and PRUNE(N_{select} = 2 OR 4), are not worth considering. Table 8 shows that PRUNE(N_{select} = 64) is slower than PRUNE(N_{select} = 8, 16, 32). Meanwhile, it is slightly weaker, thus eliminated. Thus, we kept nine programs PRUNE(PP, MP, SP, N_{select} = 8, 16, 32) for an all-against-all tournament.

Table 10 gives the final rankings with the average score per game. The σ of each average result is about 1.2. The radius of the 95% confidence interval is 2.4. Consequently, clear conclusions can hardly be drawn from this tournament. Concerning the playing level, all the players are on a par. The best value of N_{select} seems to be 32, unfortunately lowering the relevance of the pruning techniques. SP seems to be a better enhancement than MP regarding both playing level and time. However this tournament is not fair for MP because of the high values of N_{select}. This leads to the perspective to apply MP only when the number of candidate moves is inferior to a threshold. Conclusions on the playing level are hard to draw; the time considerations may break the tie. Table 8 shows that MP8 is the fastest program among the players of the all-against-all tournament, enhancing the interest of MP.

Table 10. Final ranking of the all-against-all tournament

				Rank				
	1	2	3	4	5	6	7	8	9
Prune	P32	S32	S16	P8	S8	M8	M16	P16	M32
Mean score	+3.6	+1.7	+1.2	0.0	−0.3	−1.2	−1.5	−1.7	−2.0

4.8   Weak Miai Pruning (WMP)

Since the all-against-all tournament has shed the light on a weakness of MP when $N_{select} = 16$ or 32, this subsection considers a weak version of MP which consists in using the MP rule only when the number of candidate moves is strictly inferior to a threshold T.

Conversely to straightforward MP which is effective for low values of N_{select} only, Table 11 shows that WMP (with $T = 5$) is worth considering for any values of N_{select}. First, PRUNE($WMP = true$, $N_{select} = 2, 4$) keeps the positive result shown by Table 2. Second, PRUNE($WMP = true$, $N_{select} = 8, 16$) has quite a positive result against PRUNE($WMP = false$, $N_{select} = 8, 16$): +3 points and 60% wins. As mentioned in Subsection 4.2, the mean score is statistically significant. Third, the positive mean score of two next columns ($N_{select} = 32, 64$) of Table 11 confirms the effectiveness of WMP, and removes the negative mean scores of Table 2. Finally, the speed is enhanced significantly, multiplied by 1.5 for $N_{select} = 2$, and not lowered for high values of N_{select}. To sum up, WMP is experimentally demonstrated to be superior to PP on 9×9 boards for any value of N_{select}, both in time and in playing level. This experiment is a success.

4.9   Integrating WMP with Global Tree Search on 9×9 Boards

The result obtained by WMP within the basic MC framework on 9×9 boards, namely depth-one search, suggests using WMP in the framework combining MC and Tree Search [4]. Currently, INDIGO uses a depth-3 global tree search on 9×9 boards. Consequently, we set up an experiment assessing PRUNE($WMP = true$, $N_{select} = 8$, $Depth = 3$) against PRUNE($WMP = false$, $N_{select} = 8$, $Depth = 3$) on 9×9 boards. It turns out that, although playing 5% faster, PRUNE ($WMP = true$, $N_{select} = 8$, $Depth = 3$) is 1.7 point inferior to PRUNE ($WMP = false$, $N_{select} = 8$, $Depth = 3$) and wins 45% of games only. Thus, integrating WMP with global tree search on 9×9 boards is not a success.

We have the following explanation. First, the MP principle can be discussed in front of depth-2 search. Actually, since MP launches games beginning by two given moves, the mean values computed correspond to depth-2 nodes. However, the background in which WMP is used cannot be forgotten. WMP is used only when previous random games have pruned moves, and moreover, after move A, MP develops move B only, and not all the children of move A. Thus depth-2 search dominates MP. MP is a trick used because of time constraints, when

Table 11. Result of Weak MP (WMP) vs. PP for $T = 5$

	N_{select}					
	2	4	8	16	32	64
Mean score	+1.5	+1.0	+3.1	+4.5	+2.7	+3.0
Winning percentage	51	50	60	61	53	55
Mean relative speed	1.50	1.48	1.25	1.12	1.05	1.02

depth-2 search cannot be used. Second, [4] and MP both expand the child nodes of a parent node when the number of children decreases and reaches a threshold ($W-$ in [4] and T in MP). Therefore, the two techniques do not live well all together. Finally, [4] being a kind of iterative deepening algorithm, we have also tried to use WMP at the maximal depth only, and not at intermediate depths, but this attempt was not satisfactory.

4.10   Scaling WMP Up to 19 × 19 Boards

As explained in the introduction of Section 4, to speed up the validation process of our ideas, we have first performed our experiments on 9 × 9 boards. After such experiments, SP does not fulfil our initial expectation, but MP, and in particular WMP, is still worth considering. Therefore, in a second stage, WMP deserves a 19 × 19 assessment. To make the programs playing in adequate time on 19 × 19 boards, Monte-Carlo parameters are set differently. For example, r_d is set to 1.5 and not to 2.0. Moreover, when scaling up to 19 × 19 boards from 9 × 9 boards, the maximal number of random games is reduced in a 40% proportion. Obviously on 19 × 19 boards, the time constraints bring about a depth-one search. The value of the parameters being different, it was not certain that WMP behaves on 19 × 19 boards in the same way as it does on 9 × 9 boards. After 400 games, PRUNE($WMP = true$, $N_{\text{select}} = 8$) turns out to be +4 point superior to PRUNE($WMP = false$, $N_{\text{select}} = 8$) winning 51.6% of the games. The 95% confidence interval is [-3.4, +11.8] and the 68% confidence interval is [+0.4, +8.0]. Hopefully, this result shows that WMP scales well on 19 × 19 with a depth-one search. More interesting is the fact that PRUNE($WMP = true$) used 36 minutes on average to complete one 19 × 19 game. Meanwhile, PRUNE($WMP = false$) used 46 minutes on average. Thus PRUNE($WMP = true$) is 1.27 faster than PRUNE($WMP = false$). This positive result on 19 × 19 boards is explained by the fact that a depth-one search is mandatory. In this context, WMP appears to be a trick when depth-two search remains forbidden.

5   Discussion

To explain why MP works, we introduce the notion of the *incentive of a move* as the difference between the MC evaluation of the position reached by this move and the MC evaluation of the current position. Let a be the incentive of Black playing move A, and b the incentive of Black playing move B. PP launches games to assess a and b. It stops when the difference $a - b$ is statistically different from zero. Let us assume that the incentive of White playing move A (resp. B[1]) is $-a$ (resp. $-b$), which is plausible in most cases. When A and B are not dependent, the games launched by MP starting with Black A (resp. B) and White B (resp. A) contribute to assess $a - b$ (resp. $b - a$). MP stops when the difference between the two means, i.e., $2 \times (a - b)$, is statistically different from zero. Thus, when A

---

[1] The notation "(resp. B)" is shorthand for "(or B respectively)".

and B are not miai, the average number of games launched by MP to separate A and B is smaller than the average number of games launched by PP.

SP was designed to speed up the beginning of the PP process. The experiments show that SP is a failure. One could say that SP has no more theoretical foundation than PP itself. One could expect that pruning the set of moves would exhibit the same move quality versus CPU time tradeoffs obtained by pruning the individual moves. The experimental results are not inconsistent with this hypothesis.

Scared by the long lasting experiments on 19×19 boards, we have chosen to spend the CPU time to perform experiments on 9×9 games first. Considering that INDIGO uses depth-3 on 9×9 boards, and that MP does not work well with depth-n search, all this work does not result in profitable behavior of INDIGO on the 9×9 board. Hopefully, scaling up to 19×19 boards after assessing the quality of MP was a good surprise. MP works well with depth-one search and INDIGO uses depth-one on 19×19 boards. Considering INDIGO's development, the speed-up and move quality improvements are finally effective on 19×19 boards, and not on 9×9 boards. In Subsection 4.8, WMP experiments were presented. The difference between WMP and MP lies in the use of a threshold enabling the program to use the MP heuristic only when the number of candidate moves is lower than the threshold.

A more important reason of the success of WMP over MP as been revealed since the time of the experiments. In fact, as specified in 3.1, MP was designed to discriminate moves that are tied *after several iterations* of PP to speed up the end of the process. Thus, it is quite important not to start MP at the beginning of the PP process. If the MP technique is applied from the beginning of PP, then, on tactical positions, the program may select a very bad move. This kind of blunder occured recently in a game against CRAZYSTONE [9], the new Monte-Carlo Go program of Rémi Coulom. INDIGO overlooked the good move on a tactical position near the end of the game. This error did not change the outcome of the game, but it costed the loss of a large group of stones, which could have been avoided. As against this, we may state that its merit was to point out the problem. On this position which was quite near the end of the game, two moves only (and not eight as usual) were selected by the knowledge-based move selector, let us say A, the good one, and B, a bad one. Consequently, the PP process started with two moves, i.e., less than the threshold. Thus, the MP rule was applied from the start of PP. On the mentioned position, playing random games starting by B and A lead to positions in which INDIGO, in its view, expected to obtain everything (because move A did not work for CRAZYSTONE, but it still was obliged to be played as second move by the MP rule). Besides, playing random games starting by A and B led to stable positions in which half of the points was for INDIGO, and the other half for CRAZYSTONE. Thus, INDIGO assessed (wrongly) that B was superior to A. To debug this blunder, we observed what would have happened without MP. Then, we saw that playing random games starting by B lead to positions in which CRAZYSTONE obtained everything (because move A that did not work for CRAZYSTONE, was not forced as second move by the MP rule). Furthermore, playing random games starting by A led to stable positions in which half of the points was for INDIGO. Thus, without MP, PP quickly and correctly found out that A was superior to B. Consequently, to fix up the bug, we added another threshold forbidding to use MP before a minimal number of random games. This error underlines the fact that MP has the big downside of considering that a good move for the opponent is also a good move for itself, which is a famous Go proverb, but which is wrong in many tactical situations.

6   Conclusion and Perspectives

We presented two pruning heuristics: Miai Pruning and Set Pruning. They were intended to improve the existing Progressive Pruning technique. Miai Pruning actually simulates the miai principle used by human players which consists in playing the second move when the first is played. Set Pruning manages two sets of moves, GOOD and BAD, and tries to prune BAD when possible. These two pruning techniques have been assessed on 9×9 Go boards first. MP is domain-dependent and experimentally effective both in time and playing level, when N_{select} is low and when combined with a depth-one search. Weak MP has been shown effective both in time and playing level on 9×9 boards with depth-one search. However, MP and WMP are not effective within a depth-n search. Moreover, the recent history of INDIGO showed that MP must be used only after a minimal number of random games, when PP alone has shown that it cannot break the tie between the remaining moves. SP seems experimentally effective in time as well, but it does not offer a satisfactory compromise between time and move quality. Combinations of MP and SP has also been tested but they all failed. Finally, we scaled WMP up to 19×19 boards, and we obtained a significant speed-up (about 1.3). Besides, we gained 4 points in terms of playing level.

Our experiments assessed the effect of MP within the global tree search algorithm proposed in [4]. A further step consists in translating MP into the game-tree search framework forgetting the statistical framework presented here. Besides, SP is general, and to this extent, it should be tried on other games with a high branching factor, such as Amazons [11]. Finally, assessing the ideas of [16] within the Monte-Carlo Go landscape, and performing local statistical search are still on our to-do list.

References

1. B. Abramson. Expected-outcome: a General Model of Static Evaluation. *IEEE Transactions on PAMI*, 12:182–193, 1990.
2. D. Billings, A. Davidson, J. Schaeffer, and D. Szafron. The Challenge of Poker. *Artificial Intelligence*, 134:201–240, 2002.
3. B. Bouzy. Associating Domain-Dependent Knowledge and Monte-Carlo Approaches within a Go Program. *Information Sciences*, 175:247–257, 2005.

4. B. Bouzy. Associating Shallow and Selective Global Tree Search with Monte-Carlo for 9×9 Go. In *4th Computers and Games conference (CG 2004)* (eds. H.J. van den Herik, Y. Björnsson, and N.S. Netanyahu), LNCS 3846, pages 67–80, Springer-Verlag, Berlin, 2006.
5. B. Bouzy. INDIGO Home Page. www.math-info.univ-paris5.fr/ ~bouzy/INDIGO.html, 2005.
6. B. Bouzy and T. Cazenave. Computer Go: an AI-oriented Survey. *Artificial Intelligence*, 132:39–103, 2001.
7. B. Bouzy and B. Helmstetter. Monte-Carlo Go Developments. In *10th Advances in Computer Games (ACG10), Many Games, Many Challenges* (eds. H.J. van den Herik, H. Iida, and E.A. Heinz), pages 159–174, Kluwer Academic Publishers, Boston, 2004.
8. B. Brügmann. Monte-Carlo Go. www.joy.ne.jp/welcome/igs/Go/computer/-mcgo.tex.Z, 1993.
9. R. Coulom. CRAZYSTONE Home Page. remi.coulom.free.fr/CrazyStone/, 2005.
10. D. Fotland. GO INTELLECT Wins 19x19 Go Tournament. *ICGA Journal*, 27(3):169–170, 2004.
11. J. Lieberum. An Evaluation Function in the Game of Amazons. In *10th Advances in Computer Games (ACG10), Many Games, Many Challenges* (eds. H.J. van den Herik, H. Iida, and E.A. Heinz), pages 299–308, Kluwer Academic Publishers, Boston, 2004.
12. M. Müller. Computer Go. *Artificial Intelligence*, 134:145–179, 2002.
13. M. Müller. Position Evaluation in Computer Go. *ICGA Journal*, 25(4):219–228, 2002.
14. J. Schaeffer and H.J. van den Herik. Games, Computers, and Artificial Intelligence. *Artificial Intelligence*, 134:1–7, 2002.
15. B. Sheppard. World-Championship-Caliber Scrabble. *Artificial Intelligence*, 134:241–275, 2002.
16. B. Sheppard. Effective Control of Selective Simulation. *ICGA Journal*, 27(2):67–80, 2004.
17. G. Tesauro and G. Galperin. On-line Policy Improvement using Monte-Carlo Search. In *Advances in Neural Information Processing Systems*, pages 1068–1074. MIT Press, 1996.

A Phantom-Go Program

Tristan Cazenave

Labo IA,
Université Paris 8, St-Denis, France
cazenave@ai.univ-paris8.fr

Abstract. This paper discusses the intricacies of a Phantom-Go program. It is based on a Monte-Carlo approach. The program called ILLUSION plays Phantom Go at an intermediate level. The emphasis is on strategies, tactical search, and specialized knowledge. The paper provides a better understanding of the fundamentals of Monte-Carlo search in Go.

1 Introduction

Phantom Go is a variant of Go in which part of the information is hidden for the players. Since Phantom Go is therefore to be considered as a game with imperfect information[1], Monte-Carlo methods are expected to work well. This paper demonstrates that the expectation is true and shows the results of a Monte-Carlo based program that plays Phantom Go.

In Sect. 2 we present the game of Phantom Go. In Sect. 3 we recall previous work on Monte-Carlo Go. In Sect. 4 we detail how the Monte-Carlo method is adapted to Phantom Go. In Sect. 5 we give experimental results of our program ILLUSION. Section 6 provides a conclusion and outlines future work.

2 Phantom Go

Phantom Go is a two-player game. There are two players and a referee. It is played on three boards, one for each player and one for the referee. The board of the referee is called the reference board. It is usually played on 9×9 boards. The referee can see all three boards. Each player can only see his[2] own board. When it is a player's turn, he chooses a move and asks the referee if it is legal for him to play on the intersection intended by pointing at the intersection. The referee answers 'legal move' or 'illegal move' according to the reference board. If the move is illegal, the player chooses another move, and so on until he arrives at a legal move. When the player has indicated a legal move, it must be played on the reference board by the referee, and by the player on his own board. If

---

[1] Sometimes the term 'incomplete information' is also used for games with hidden information. In the case of Phantom Go, both players have complete information of the game rules, possible states, and possible outcomes.

[2] For brevity we use 'he' ('his') if 'he or she' ('his or her') is meant.

H.J. van den Herik et al. (Eds.): ACG11, LNCS 4250, pp. 120–125, 2006.

there is a capture, the referee announces the number of stones being captured and communicates to the other player which stones have been captured. After a move has been played, it is the other player's turn. The game ends when both players pass. Phantom Go (as a variant of Go) is the equivalent to Kriegspiel in Chess [2].

3   Monte-Carlo Go

Monte-Carlo methods compute statistics on a set of random games in order to find the best move. They have been used in games such as Bridge [7], Poker [1], Tarok [8], and Scrabble [9]. All these games have hidden information which make them particularly suited for Monte Carlo. However, Monte-Carlo methods have also proven to be useful in complete information games, and particularly in Go. Brügmann [5] was the first researcher to experiment with Monte Carlo in Go. Recently, other Go programs have started using it, and improved the method in many directions, among others by (1) simplifying the method and proposing basic improvements [4], (2) combining it with a knowledge-based program that selects a few number of moves that are evaluated by the Monte-Carlo method [3], and (3) by combining it with tactical search [6].

There are several slightly different ways to write a Monte-Carlo Go program [4]. In this paper, we use for Monte-Carlo Go the following algorithm: the program plays a large number (usually 1,000 to 10,000) of random games starting at the current position. The moves of the random games are chosen almost randomly among the legal moves, except that they must not fill the player's eyes. A player passes in a random game when his only legal moves are on his own eyes. The game ends when both players pass. In the end of each random game, the score of the game is computed using Chinese rules (in our case, it consists in counting one point for each stone and each eye of the player's color, and subtracting the opponent count from the player's count). The program computes for each intersection the mean results of (1) the random games in which the player starts with a move at that intersection, and (2) the random games in which the opponent starts at that intersection. The value of a move is the difference between the two means. The program plays the move with the highest value.

4   Monte-Carlo Phantom Go

Monte-Carlo Go is a game of complete information. Monte-Carlo Phantom Go has to deal with hidden information. In order to cope with this hidden information, the program has to guess where the stones of the opponent are, and the best move on average against different configurations of the opponent stones.

In essence, for Monte-Carlo Phantom Go the basic Monte-Carlo Go method is reused: the program plays many games randomly with the constraint of not filling its own eyes. However, all the random games do not start with the same position, as the program does not know exactly the real position. The program memorizes all the forbidden moves. It places an opponent stone on each of the forbidden moves. The program also knows the number of the opponent stones that are present on the reference board. Subtracting the number of forbidden moves from the number of opponent stones gives the number of stones with an unknown position.

From the beginning of each random game, the program places its own stones and places opponent stones on the forbidden intersections. Moreover, it randomly places the opponent stones on the empty intersections left. More precisely, it randomly places as many opponent stones as there are opponent stones with an unknown position. Once all the stones are placed, it plays a random game starting with a move of its color, and performs moves randomly on empty intersections, provided they are not a player's eye.

A player passes when his only moves left are his own eyes. When both players pass, the game is ended. At the end of a random game, the score is computed using Chinese rules. For each move played during the random game, the mean score of playing this move for all the random games is updated to take into account the result of the game. When 10,000 random games have been played, the program subtracts, for all the legal moves, the mean score of the move for the opponent's color from the mean score of the move for the player's color. The move that has the highest difference is tried.

If the move is announced to be illegal by the referee, the program memorizes it as a forbidden move, and starts its process again. It plays 10,000 new random games, taking into account the new information given by the referee on the forbidden move.

5   Experimental Results

The number of random games played at each move is set to 10,000. The program plays a move in a few seconds on a Pentium 3.0 GHz. Subsection 5.1 details an example game by ILLUSION. Subsection 5.2 gives results against different Go players.

5.1   An Example Game

The author is an European one-dan Go player. Playing games with the program ILLUSION usually results in a small win by the author. An example 9 × 9 game is given in Fig. 1. When a player tries an illegal move, it is reported in the game's notation, and therefore multiple moves by the same player follow each other. All but the last move by the same player are illegal, they are mentioned explicitly because they give information on the knowledge of the game by the player and are required to analyze and understand the game. The author is White and the program is Black.

In this game, the author followed the strategy of (1) dividing the board into two parts, and (2) trying to kill one side of the board after it has been divided. This strategy is also used by other experienced Phantom-Go players and admittedly it is quite efficient.

A Phantom-Go Program   123

1 B(D4), W(E5); 2 B(E3), W(E6); 3 B(F5), W(E4); 4 B(E6-F6), W(E7);
5 B(F7), W(E3-D4-F4); 6 B(E5-F4-E4-G4), W(F3); 7 B(F3-E7-E8), W(F2);
8 B(G3), W(F1); 9 B(F2-D8), W(E8-D7); 10 B(F8), W(D8-C7); 11 B(G2),
W(C8); 12 B(G5), W(C9); 13 B(D3), W(F7-D3-B7); 14 B(C7-D7-C8-F1-C2),
W(A7); 15 B(B4), W(G4-G3-G2-B6); 16 B(B6-B3), W(B5); 17 B(C9-B5-E2),
W(B4-A5); 18 B(C5), W(A4); 19 B(B8), W(A3); 20 B(B7-H6), W(B3-A2); 21
B(E1), W(B2); 22 B(A8), W(C2-B1); 23 B(G1), W(C1); 24 B(D5), W(D1);
25 B(A5-A7-H7), W(F6-F5-D9); 26 B(C3), W(E9); 27 B(B9), W(F8-F9);
28 B(G8), W(G9); 29 B(A4-A3-A2-C1-D1-B1-B2-D9-E9-F9-G9-H9), W(G8-
H9-H8); 30 B(H8-J8), W(J9-J8-J7); 31 B(J7-H5), W(H8-G1-H7-H6-Pass);
32 B(J6).

Fig. 1. Example game: ILLUSION (B) – Cazenave (W)

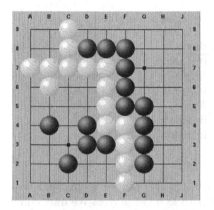

Fig. 2. The reference board after 15 moves of both sides

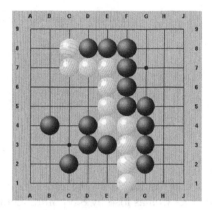

Fig. 3. The program board after 15 moves of both sides

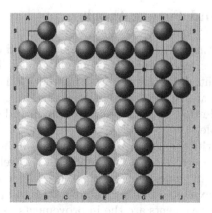

Fig. 4. The final position

Figure 2 gives the reference board after the first 15 moves (30 plies) of the game. Figure 3 gives the program board after the first 15 moves of the game. In the latter position, we see that ILLUSION has guessed most of the white stones, and is ahead of White. It has to close the borders of its right group, and to make two eyes with its left group. Figure 4 gives the reference board at the end of the game. The program failed to make its left group live and lost the game.

5.2 Results Against Go Players

In order to test the program ILLUSION adequately, it played 9 × 9 games against Go players of different levels. The results are in Table 1. For each game we give the level of the human player, the division of colors, and the result.

Table 1. Results of ILLUSION against Go players

Level	White	Black	Result
13 kyu	Nicolas	ILLUSION	W+8.5
13 kyu	ILLUSION	Nicolas	W+resign
5 dan	Bernard	ILLUSION	W+13.5
13 kyu	ILLUSION	Arpad	W+17.5
5 dan	ILLUSION	Bernard	W+27.5
5 dan	Bernard	ILLUSION	W+47.5

As the results show, the program can win by 27.5 points against a 5 dan Go player as well as lose by 8.5 points against a 13 kyu Go player. We may conclude that the program has the level of experienced Go players who only played a few number of Phantom-Go games. We do not know experienced (and even less ranked) Phantom-Go players to whom we could test the program.

6   Conclusion and Future Work

We presented our Phantom-Go program ILLUSION based on a Monte-Carlo approach. ILLUSION plays interesting Phantom-Go games. The peculiarity of the application of the Monte-Carlo method to Phantom Go is that unknown stones are placed at random at the beginning of each random game. From the results so far we may conclude that strategy, tactical search, and specialized knowledge plays an important role.

Below we suggest three essential improvements of the current program. A first improvement is to deal more accurately with the well-known Phantom-Go strategies in the random games. For example, the divide-and-kill strategy used by the author can be exploited in the random games to bias the move selection.

A second set of improvements are the improvements used in Monte-Carlo Go programs. For example, it is possible to combine the current search with a tactical search by computing the results of simple tactical searches at the beginning of the random games, so as to compute statistics in the random games on the goal to be achieved. Once the statistics on the goals are computed they give a better evaluation of the corresponding move than the basic statistics on arbitrary moves. Such an approach has worked in Monte-Carlo Go [6], and could well work too in Phantom Go.

A third improvement is to use patterns to bias the selection of moves in the random games so as to improve their quality as in [3]. Finally, we remark that the program also has problems dealing with semeais, and opponent eyes. This can be improved by specialized knowledge.

References

1. D. Billings, A. Davidson, J. Schaeffer, and D. Szafron. The Challenge of Poker. *Artificial Intelligence*, 134(1-2):210–240, 2002.
2. A. Bolognesi and P. Ciancarini. Computer Programming of Kriegspiel Endings: The Case of KR Versus K. In *Advances in Computer Games 10 (ACG10) Many Games, Many Challenges* (eds. H.J. van den Herik, H. Iida, and E.A. Heinz), pages 325–342, Kluwer Academic Publishers, Boston, 2004.
3. B. Bouzy. Associating Domain-Dependent Knowledge and Monte Carlo Approaches within a Go Program. In *Joint Conference on Information Sciences*, pages 505–508, Cary, 2003.
4. B. Bouzy and B. Helmstetter. Monte Carlo Go Developments. In *Advances in Computer Games 10 (ACG10) Many Games, Many Challenges* (eds. H.J. van den Herik, H. Iida, and E.A. Heinz), pages 159–174, Kluwer Academic Publishers, Boston, 2004.
5. B. Brügmann. Monte Carlo Go. ftp://ftp-igs.joyjoy.net/go/computer/mcgo.tex.z, 1993.
6. T. Cazenave and B. Helmstetter. Combining Tactical Search and Monte-Carlo in the Game of Go. In *IEEE Symposium on Computational Intelligence and Games (CIG'05)* (eds. G. Kendall and S. Lucas), Colchester, UK, 2005.
7. M.L. Ginsberg. GIB: Steps Toward an Expert-Level Bridge-Playing Program. In *IJCAI-99*, pages 584–589, Stockholm, Sweden, 1999.
8. M. Lustrek, M. Gams, and I. Bratko. A Program for Playing Tarok. *ICGA Journal*, 26(3):190–197, 2003.
9. B. Sheppard. Efficient Control of Selective Simulations. *ICGA Journal*, 27(2):67–80, 2004.

Dual Lambda Search and Shogi Endgames

Shunsuke Soeda[1], Tomoyuki Kaneko[1], and Tetsuro Tanaka[2]

[1] Computing System Research Group,
The University of Tokyo, Tokyo, Japan
{shnsk, kaneko}@graco.c.u-tokyo.ac.jp
[2] Information Technology Center,
The University of Tokyo, Tokyo, Japan
ktanaka@ecc.u-tokyo.ac.jp

Abstract. We propose a new threat-base search algorithm which takes into account threats by both players. In full-board Semeais in Go or Shogi endgames, making naive attack moves often result in losing the game. The reason is that a player must first make a defense move if the opponent has a better attack move than his[1] own. Some attack moves weaken the defense of the player who made the move. Thus, players must be aware of the threats by both players to avoid such naive attack moves. However, existing threat-based search algorithms are only aware of threats by one player, and cannot detect such naive attacks efficiently. We propose a solution to this problem, by applying λ-search mutually recursively so that it searches the best move by taking into account threats by both players. We call this search algorithm *dual λ-search*. Dual λ-search can handle inversions efficiently compared to previous algorithms by making passes for both players. We implemented dual λ-search with DF-PN as the driver, and made experiments with difficult Shogi-endgame problems. We showed the effectiveness of our algorithm by solving 32 problems out of 97. It includes solving problems that even one of the strongest Shogi program had not yet been able to solve correctly.

1 Introduction

In many games, a variety of strong game programs have been built using game-tree search algorithms with heuristic evaluation functions. The Shogi endgame is a particular research domain that has successfully employed this approach over many years. However, even top-level programs still make fatal mistakes that turn winning positions into losing ones [7]. One reason is that Shogi is a complex game. It has huge branching factors due to the great number of legal moves, even in endgames. Another reason is that, although threats and threat sequences play a large role in Shogi endgames, it is difficult to construct evaluation functions that take threats into account.

In such domains, tactical search algorithms have been effective. For example, λ-search [14] and Generalized Threats Search [2] have solved capture games in

---

[1] For brevity, we use 'he' ('his') if 'her or she' ('his or her') is meant.

H.J. van den Herik et al. (Eds.): ACG11, LNCS 4250, pp. 126–139, 2006.
© Springer-Verlag Berlin Heidelberg 2006

Go, and DF-PN can solve checkmate problems in Shogi [11]. These algorithms focus on a part of the game. Thus they are not sufficient to find a winning move in the full game. In a Shogi endgame, a move to attack an opponent King often threatens the King of the player who made the move. This is partly because the player must first counter his opponent's attack move provided that the move could be achieved faster than his own attack move. Moreover, an attack move itself can weaken the attacker's King. From combinatorial game theory [4] we know the theory on how to combine the results of searches in independent sub-games to yield the globally best move. However, this theory does not work well for games that cannot be divided into independent sub-games.

In order to solve this problem, we propose a new search algorithm based on the analysis of threats by both players. Therefore, we extended λ-search to be mutually recursive. As a consequence, our algorithm measures the threat of one player while also taking into account the threat of the other player in the form of inversions. By handling the attack and defense moves of one King, we can use heuristics for searching efficiently, while still achieving the global goal by taking into account the threats of both players. With our search algorithm, we solved difficult Shogi problems, including some which require tricky moves that influence threats by both players.

In Sect. 2, we give an explanation of Shogi and the Shogi endgame. In Sect. 3, we place our work in the framework of related works. In Sect. 4, we apply λ-search to Shogi. In Sect. 5, we propose our method, dual λ-search. In Sect. 6, we show our experimental results, and in Sect. 7 we make some concluding remarks.

2   The Game of Shogi

Shogi is a Japanese board game played by two players. It is believed to have the same origin as "Western" chess and Chinese chess. Compared with other Chess-like games, Shogi has a unique "drop rule" [7]. In brief, when a piece is captured it is not totally removed from the game, but the player who took the piece can play it back on the board later on. This means that the number of pieces involved in the game does not decrease towards the end of game.

Due to this rule, Shogi has a significantly complex endgame. For example, endgame databases [13] which work well for (western) Chess are not practical in Shogi endgames. Together with the larger search space, Shogi is considered to be a much harder game for computers than Chess.

2.1   Checkmate in Endgame

We start defining *checkmate*, which plays an important role in the Shogi endgame.

Definition 1. Checkmate (narrow sense). *A position is in checkmate (narrow sense), if any legal move[2] by the player is followed by a position where the player's King could be captured by the opponent.*

---

[2] Strictly speaking, a move that does not resolve a check is an illegal move in Shogi.

Definition 2. Checkmate Tree. *A checkmate tree is a game tree, where one player (the attacker) only plays check moves, and the other player (the defender) plays all possible moves, and all leaf nodes are positions in checkmate.*

Definition 3. Checkmate (wide sense). *A position is said to have a checkmate, if it is in a checkmate tree.*

A search to verify if a position is in a checkmate tree is more efficient than searching for a win in normal positions, as only check moves and moves to escape from checks need to be considered. The introduction of DF-PN [11] has enabled Shogi programs to solve complicated checkmate problems, and the ability of computers to prove a checkmate has surpassed that of human grand masters.

2.2   Threats of Higher Order

Although computers are able to solve complicated checkmate problems, it is still difficult for computers to find checkmates hidden two or three moves away from the root position in a normal search tree. Most Shogi programs use forward pruning based on heuristics, together with a hand-tuned evaluation function, but it is not rare for even the top-level programs to miss the correct move which is evident even for an intermediate human Shogi player.

Threatmate moves and brinkmates are concepts similar to check moves and checkmate, which also play an important role in the Shogi endgame [6]. A threatmate move is a move by the attacker, which if neglected by the defender, allows the attacker to establish a checkmate within next move. The defender loses unless either the defender can checkmate the attacker's King, or the defender can make a defense move that can prevent the checkmate by the attacker.

Definition 4. Threatmate move. *A threatmate move is a non-check move by a player (attacker), that is followed by a position in which, if the opponent (defender) passes, the player (attacker) has a checkmate.*

If the player has no move to resolve an checkmate by the opponent, the position is said to be in a brinkmate.

Definition 5. Brinkmate (narrow sense). *A position is in brinkmate, if any legal move by the player (defender) is followed by a position where the opponent (attacker) has a checkmate.*

Definition 6. Brinkmate Tree. *A brinkmate tree is a game tree, where one player (the attacker) only plays check moves or threatmate moves, and the other player (the defender) plays all possible moves, and all leaf nodes are positions in brinkmate.*

Definition 7. Brinkmate (wide sense). *A position is said to have a brinkmate, if it is in a brinkmate tree.*

Although some algorithms have been proposed to search brinkmates [6,1], their effectiveness is limited because of the large search costs caused by the difficulties in the identification of attack and defense moves.

3   Related Work

Game-tree search algorithms and their enhancements have been intensively studied for a long time now, including mini-max search, alpha-beta search, MTD(f),
null-move pruning, and transposition tables. Although they have made great
successes possible in many kinds of games, they do not work well in some games
with huge branching factors such as Go [10], in spite of all efforts. Our algorithm
searches more selectively by focusing on the threats, so that it can still work in
such games.

Tactical search algorithms are specialized in solving a part of the game, or a
partial goal. For example, trying to take a specific stone in the game of Go is a
partial goal. Trying to checkmate one player's King in Chess or Shogi is another
example of a partial goal. Tactical searches are efficient because they limit their
moves related to the specific goal they handle and only search a part of the game
tree. Thus, they are still effective in games where global search does not work well.

λ-search [14], Generalized Threats Search [2], and iterative widening [3] were
applied to solve capture games or life-or-death problems in Go. Proof number
search and DF-PN were applied to solve checkmate problems in Shogi [11].

3.1   Simulation

Simulation [8] was first proposed to solve effectively Shogi positions with useless
interposing piece drops in checkmate search. It is also shown to be effective in
a solver for the game of Go [9]. The idea is as follows. Assume that position P
is proven, and position Q is a position similar to P. Then simulation borrows
moves from the proof tree of P and tries to find a quick proof for Q.

3.2   Generalized Threats Search

Generalized Threats Search [2] is a search algorithm that could model existing
algorithms based on threat analysis. Generalized Threats Search is based on the
idea of generalized threats, which defines the depth of search and where players
could make consecutive moves. Generalized threats can represent a search tree
with a multiple level of threats at various depths. However, only passes by one
player are allowed in a given generalized threat. In contrast, dual λ-search uses
passes by both players.

An enhancement to Generalized Threats Search, called 'forced move for left',
is also explained in [2]. The idea is quite similar to dual λ-search in the sense that
it looks at the threats introduced not only by the attacker, but also the defender.
A similar idea was also proposed by Thomsen [14]. In dual λ-search, threats by
the defender are used to cut inversions by the defender, while in Generalized
Threats Search, it is used to cut moves by the attacker that could not block the
inversion.

3.3   Shogi Endgame

If we ignore inversions and consider only attack moves, it would be much easier to find a winning move sequences made up of threatmate moves. However,

Yamashita reported that it is dangerous to ignore inversions, as a check move by the defender to the attacker's King often prevents such naive winning sequences [15].

Although checkmate searchers are efficient, they are still too heavy to be called at every leaf node of the normal search. So most Shogi programs call checkmate searchers only at shallow depth of the search tree.

IS-SHOGI, one of the strongest Shogi program, uses simulation in two ways to to find checkmates deep in the search tree [12]. The first idea is to use simulation to verify if a possible checkmate by the opponent was prevented properly by IS-SHOGI. In the root node, IS-SHOGI first passes and sees if the opponent has a checkmate. If so, IS-SHOGI checks by simulation if the checkmate by the opponent still holds after IS-SHOGI has made a move. The second idea is to record a successful checkmate made by IS-SHOGI found in normal search, and use this to verify if the same checkmate holds in the descendant nodes.

4   λ-Search for Shogi

In this section, we start with a brief explanation of λ- search [14], then show how it could be applied to Shogi. We call the player to move at the begin position of the search *Black*, and the other player *White*. The aim of the search is (1) to prove that there is a win for Black, and (2) to obtain its proof tree.

We assume that either pass is allowed, or zugzwang is not a motive. Although this does not strictly hold in Shogi, it should cause no problem as there is practically always a harmless move to play which could substitute a pass.

4.1   λ-Search

λ-search is an algorithm for searching a binary-valued game tree. It uses passes together with different orders of threat sequences. A more direct threat has a lower *threat level* and a more indirect threat has a higher *threat level*. We denote with n the level of threat. The basic idea of λ-search is to reduce the number of positions searched, by reading along the positions where there is a threat. Formally, λ-search could be defined by λ^n-trees and λ^n-moves.

Definition 8. λ^n**-tree.** *A λ^n-tree is a search tree which consists solely of λ^n-moves; a λ^n_{attack}-tree is a λ^n-tree where Black moves first.*

The value of a λ^n-tree is the result of evaluating the tree as an AND-OR tree. For the leaf positions, with no λ^n-moves, it is disproven if it is Black to play, and it is proven if it is White to play.

Definition 9. λ^n**-move.** *A λ^n-move is a move with the following characteristics. If Black is to move, it is a move that implies – if White passes – that there exists at least one subsequent λ^i_{attack}-tree that is proven, where $0 \leq i \leq n - 1$. If White is to move, it is a move that implies that there does not exist any subsequent λ^i_{attack}-tree that is proven, where $0 \leq i \leq n - 1$.*

A check move is an example of a λ^1-move, as if neglected, the player who made the check move can capture the opponent King. A threatmate move is a λ^2-move. An example of a λ^2-move is shown in Fig. 1. In this position, the first player can drop the Silver at 7b. If the second player neglects this move and plays elsewhere, the first player can make another move to checkmate the second player.

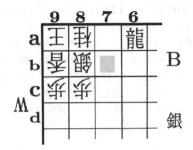

Fig. 1. Part of a Shogi endgame position, where Black has a λ^2 move (drop Silver at 7b)

The only move that prevents an immediate checkmate by the second player is to move his Silver to 7a, which would be responded by the first player by capturing it, either with Silver or promoted Rook. This move by the first player is another λ^2-move.

4.2   Lambda Search with One King

We introduce the notation ϕ, which represents a λ-search regarding only one King. In Shogi, the game ends if either King is captured. But for the search to determine the value of ϕ, we allow the search to continue beyond the capture of the attacker's King. When the attacker's King is captured, we simply remove it from the game. So, when we search for the value of ϕ with the goal of Black capturing the white King, we allow Black to leave its King under check, and White capturing the black King. In other words, White concentrates on defending its own King, and forgets about attacking the black King.

For a given position P, player p and threat level n, we define $\phi_p^n(P)$ as follows.

Definition 10. $\phi_p^n(P)$ *is the value of a λ-search with the goal p capturing the opponent King, starting from the position P. If p is to play in P, $\phi_p^n(P)$ is true if there is at least one move from P, which is followed by a position which $\phi_p^i(0 \le i \le n)$ is true. If the opponent of p is to play in P, $\phi_p^n(P)$ is true if the opponent of p passes, $\phi_p^i(0 \le i \le n-1)$ is true in the subsequent position, and all move from P is followed by a position for which $\phi_p^i(0 \le i \le n)$ is true.*

4.3   Lambda Search with Two Kings

Next, we represent a composite goal, in which one player can capture its opponent King without its own King being captured. This means that the player has a successful ϕ starting from that position, for every position in the sequence it is not in a successful ϕ sequence with a lower threat level for its opponent. We introduce a notation similar to the previous notation ϕ to represent the composite goals.

To represent that in position P, player p can safely capture the King of its opponent o with a sequence made up of moves with a threat level less than n, we use the following notation: $(\phi_p^n \wedge \neg \phi_o^{n-1})(P)$.

Definition 11. $(\phi_p^n \wedge \neg\phi_o^{n-1})(P)$ *is the value of a* λ*-search with the goal p cap-turing the opponent King, without its own King being captured, starting from the position P. If p is to play in P,* $(\phi_p^n \wedge \neg\phi_o^{n-1})(P)$ *is true if there is at least one move from P, which is followed by a position which* $(\phi_p^i \wedge \neg\phi_o^j)(P)(0 \leq i \leq n, 0 \leq j \leq i-1)$ *is true. If the opponent of p is to play in P,* $(\phi_p^n \wedge \neg\phi_o^{n-1})(P)$ *is true if the opponent of p passes,* $(\phi_p^i \wedge \neg\phi_o^j)(P)(0 \leq i \leq n-1, 0 \leq j \leq i-1)$ *is true in the subsequent position, and all moves from P are followed by a position for which* $(\phi_p^i \wedge \neg\phi_o^j)(P)(0 \leq i \leq n, 0 \leq j \leq i-1)$ *is true.*

Note that this composite goal is not the same as the conjunction of simple goals $\phi_p^n(P) \wedge \neg\phi_o^{n-1}(P)$, as the latter notation allows the move to support $\phi_p^n(P)$ and the move to support $\neg\phi_o^n(P)$ to be different. For example, take a position P, in which p is to play and which has two children, Q and R (Fig. 2). Let Q and R have the following values (see table in Fig. 2).

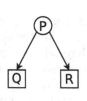

	P	Q	R
ϕ_p^0	false	false	false
ϕ_o^0	false	true	false
ϕ_p^1	true	true	false
ϕ_o^1	true	true	true
$\phi_p^1(X) \wedge \neg\phi_o^0(X)$	true	false	false
$(\phi_p^1 \wedge \neg\phi_o^0)(X)$	false	false	false

Fig. 2. A position P, p to move, with two child nodes Q and R

From $\phi_p^0(Q)$ = false and $\phi_p^0(R)$ = false follows $\phi_p^1(P)$ = false. The composite goal $(\phi_p^1 \wedge \neg\phi_o^0)(P)$ is false, as neither Q nor R can support it. However, $\phi_p^1(P) \wedge \neg\phi_o^0(P)$ is true, as $\phi_p^1(Q)$ supports $\phi_p^1(P)$ and $\neg\phi_o^0(R)$ supports $\neg\phi_o^0(P)$. Moreover, the result by the non-composite goal is incorrect, for this position is a o to win position, with ϕ_o^1 = true and ϕ_p^0 = false for both moves.

5   Dual Lambda Search

In this section, we show how to expand λ- search [14] into dual λ-search. λ-search is a rather efficient way of searching when threats and threat breaking moves are important. However, λ-search with a single goal can only be applied to search in local games, and does not always yield correct results for global searches of the game. λ-search with a composite goal could handle global games properly, but lacks the efficiency of the original λ-search, as passes of only one player are concerned. Below, we propose the dual λ-search algorithm where threats of both players are taken both into account.

5.1 Formalization

We formalize dual λ-search with μ_j^n-moves and μ_j^n-trees, where p denotes a player and n denotes the threat level of the search. We denote by o the opponent of player p.

Definition 12. μ^n-**tree.** *A μ_p^n-tree is a search tree which consists solely of μ_p^n-moves.*

Definition 13. μ^n-**move.** *A μ_p^n-move is a move with the following characteristics.*

If p is to move, it is a move that implies it is followed by a position where, if o passes, there exists at least one subsequent μ_p^i-tree that is proven, and there does not exist any subsequent μ_o^i-tree that is proven, where $0 \le i \le n-1$.

If o is to move, it is a move that implies that there does not exist any subsequent μ_p^i-tree that is proven, where $0 \le i \le n-1$.

Unlike previous algorithms, where attack by only one player is concerned, our algorithm takes account of the threats by both players. This could be well illustrated by the example shown in Fig. 3. This position is almost identical with Fig. 1, except that the black King and the surrounding pieces have been added.

In this position, Black dropping Silver to 7b is a naive attack move, and if played back by White moving promoted Rook to 6i. The only move by Black to avoid immediate checkmate is to move Silver to 7i. However, White can still checkmate Black by capturing the Silver with moving promoted Rook to 7i. Moreover, not making any attack move is better than dropping Silver to 7b, as even when Black passes, there is no immediate checkmate of the black King by White. This is an example where a λ-search with a single goal fails to yield a correct result.

This problem could be also handled by λ-search with composite goals, but it lacks efficiency. After Black chose to drop a Silver at

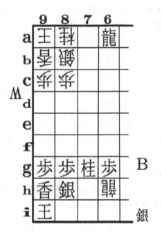

Fig. 3. Part of a Shogi endgame position. If the first player uses Silver to attack, the second player can move the promoted Rook to 6i, which the first player cannot defend.

7b, the question is: which move should White make? This is a easy question for dual λ-search, as 6a promoted Rook is a μ_W^1 move, and should be tried to see if this position is a win for White, before μ_B^2 moves should be considered. However, as λ-search with composite goals does not consider threats by White, it thus cannot limit candidates to be searched first.

5.2   Dual Lambda Search with Df-pn

DF-PN is an algorithm for searching AND-OR trees, that searches the thinner part of the search tree first [11]. For AND-nodes (attacker's positions), it starts searching from the node with the least number of leaves required to be expanded to prove the node. For OR-nodes (defender's positions), it starts searching from the node with the least number of leaves required to be expanded to disprove the node. DF-PN is efficient in searching non-uniform trees, and is the best known algorithm to solve Shogi checkmate problems. Below, we show how DF-PN could be used as the driver for dual λ-search.

We denote the player to move at the initial position as p, and his opponent o. We start from a lower threat level, and for each level we start from trying to prove a win for p. That is, we start from λ_p^0, then go on to $\lambda_o^0, \lambda_p^1, \lambda_o^1, \lambda_p^2...$ until a win is proven for one player.

For each position in the search, we do a similar iteration. If we are searching to prove λ_p^n, we start from proving λ_p^0, up to λ_p^n. The position is proven if λ_p^i is proven and λ_o^{i-1} is disproven for $i(i \le n)$. The position is disproven if λ_p^i is disproven for all $i(0 \le i \le n)$ or λ_o^{i-1} is proven.

The proof number and disproof number of each position is defined for each threat level and player. For position P, $\lambda_p^i.p(P)$ denotes the proof number for player p at threat level i, and $\lambda_p^i.d(P)$ denotes the disproof number for player p at threat level i.

If P is a position without any moves to play, it is a terminal position. If p is to play in P, $\lambda_p^i.p(P) = \infty$ and $\lambda_p^i.d(P) = 0$. If o is to play in P, $\lambda_p^i.p(P) = 0$ and $\lambda_p^i.d(P) = \infty$.

If P is a position with moves to play, it is an internal position. The proof number and disproof number are calculated from the children of P. If the threat level is lower than 0, both the proof number and the disproof number of the position is 0. The proof number and disproof number of an unexpanded position is 1.

If p is to play in P,

$$\lambda_p^i.p(P) = \min_{s \in childs(P)} (\lambda_p^i.p(s) + \lambda_o^{i-1}.d(s))$$

$$\lambda_p^i.d(P) = \min(\sum_{s \in childs(P)} (\lambda_p^i.d(s)), \sum_{s \in childs(P)} (\lambda_o^{i-1}.p(s))).$$

Let t denote the position after a pass from P. If o is to play in P,

$$\lambda_p^i.p(P) = \lambda_p^{i-1}.p(t) + \sum_{s \in childs(P)} (\lambda_p^i.p(s)) + \sum_{s \in childs(P)} (\lambda_o^{i-1}.d(s))$$

$$\lambda_p^i.d(P) = \min(\lambda_p^{i-1}.d(t), \min_{s \in childs(P)} (\lambda_p^i.d(s)), \min_{s \in childs(P)} (\lambda_o^{i-1}.p(s))).$$

Positions are expanded until there are no more positions to expand. The position to be expanded is chosen as follows. For a position p to play, the child position with the least proof number is chosen, and for a position o to play, the child position with the least disproof number is chosen, until an unexpanded position is found.

6   Experimental Results

To show the effectiveness of our algorithm, we solved some Shogi endgame positions with our search algorithm. In our experiments, we used full-board endgame positions where the first player can find a win by brinkmate if he played properly. We chose the problems from a test set provided by Grimbergen [5]. None of the problems in the test set was solved by the top Shogi programs on the market at the time when [5] was written. Thus, the test set represents the weakness of the current Shogi programs. The test set had 97 problems with endgame positions, none of them with an immediate checkmate.

6.1   Move Generator

We heuristically limited our move generation. In general, if we generate more moves, we obtain more accurate results under the sacrifice of the search space and the speed. However additional moves sometimes do reduce the search cost by rejecting naive attacks and defenses. So what we want is the smallest set of moves that contains "the right" moves. To decide what moves to generate, we started with the minimal set of moves and gradually added moves that were needed to solve some problems.

For the attack moves, we decided to generate three categories of moves: (1) moves that the pieces to be moved could move into the 25 squares surrounding the defender's King on its next move, (2) moves to open way for the big pieces, and (3) moves that captures any opponent piece.

For the defense moves, we came up with six categories of moves: (1) moves that moves the King, (2) moves that capture the piece that has been just played by the attacker, (3) moves that increases the liberty of the King by moving a piece occupying a square where the King can move, (4) drop moves to the 8 squares surrounding the King, (5) moves that capture the attacker's pieces that could move into the 8 squares surrounding the King on its next move, and (6) moves of the piece to be moved that can move into the 8 squares surrounding the King on its next move.

Note that whenever the player to move is under check, instead of generating the moves introduced above, all check escaping moves are generated.

6.2   Solving Shogi-Endgame Problems

Dual λ-search depends on other search techniques for searching the μ-trees. In our preliminary experiments, dual λ-search driven by iterative-deepening depth-first search was only able to solve quite simple problems. So we chose DF-PN as the driver for the dual λ-search in our experiments.

Implementing Dual λ-Search. We have incorporated an enhancement proposed by Kishimoto to handle loops efficiently with DF-PN [9]. However, we have not incorporated enhancements to handle GHI problems correctly. We used an estimator for the proof numbers and disproof numbers of the pre-expanded positions, based on the liberty of the defender's King.

We implemented both λ-search and dual λ-search with DF-PN as its driver. We used a dual Opteron 250 PC, with 12GB of memory. We did not give a explicit time limit or table size limit to the program, so the program stopped either when it found a winning move or when it exhausted the memory.

Comparison with λ-Search with One King. The results of the comparison with λ-search are shown in Table 1. If the program gave the correct first move, we counted it as "correct", and if it gave a wrong move, we counted it as "incor-

Table 1. Comparison with λ-search

	Correct	Unknown	Incorrect
dual λ-search	32	63	2
λ-search	20	46	31

rect". For most problems, the program ran out of memory which we counted "unknown". Dual λ-search gave only 2 incorrect answers, while λ-search with one King gave as much as 31 incorrect answers. This shows the importance of taking into account the threats by both players.

Dual λ-search was able to give a correct solution of some problems that even could not be correctly answered by the development version of the strongest Shogi program YSS.[3]

Comparison with General Search. We had our Shogi playing program GPS-SHOGI solve some of the problems, to compare dual λ-search with a general Shogi playing program.

GPS-SHOGI searches with MTD(f) for normal positions, and with DF-PN for a checkmate position. We had GPS-SHOGI solve the problems for which dual λ-search was able to give the correct answers. Let us see how many nodes GPS-SHOGI required to explore, to find the correct answer. We started with a small search limit, and increased the limit until either GPS-SHOGI was able to find the correct move, or the limit was surpassing a large threshold.

For most problems, GPS-SHOGI was not able to find the correct move although the program searched considerably more nodes compared to dual λ-search. This clearly shows the strength of dual λ-search in the Shogi endgame.

Incorrect Answers. As we limited the defense moves, our solver gave incorrect answers for two problems. The problems were 44 and 85.

Problem 44 is shown in Fig. 5. The correct sequence for this problem starts with 2a Gold. However, our program gives 2a Bishop (promote) as the answer. The sequence goes on 2a King (capture Promote Bishop) by the defender, 4b Rook (promote) by the attacker. The best move for the defender is 7i Bishop (drop) which is a check to the attacker's King. Although this move does not establish a checkmate, it is a brinkmate-breaking brinkmate move, which brings a win to the second player.

Problem 85 is shown in Fig. 6. The correct sequence for this problem starts with 2c Silver, but our program generates 2c Gold. The reason for this is that the

---

[3] A message from the Computer Shogi Association mailing list.

Table 2. Search space and search time of dual λ-search for the problems solved

Problem ID	Table size	Positions explored	Problem ID	Table size	Positions explored
1	5,077,274	13,625,141	40	2,687,773	5,341,140
3	23,127,440	51,547,045	47	6,322,382	16,688,986
5	3,324,052	9,071,688	53	653,649	1,694,982
6	12,693,730	28,224,664	55	985,227	2,333,563
13	332,886	927,355	56	41,194	119,889
14	2,779,596	7,424,350	67	1,777,506	4,121,735
18	819,600	1,837,131	69	128,301	283,346
19	3,096,938	7,930,800	70	410,784	1,176,575
25	21,068	66,176	74	2,552,585	4,914,454
28	9,847,634	26,819,205	76	158,928	325,432
29	9,544,140	22,514,140	79	1,561,958	2,538,576
31	2,789,624	7,958,037	80	2,958,425	6,172,481
32	4,035,965	10,369,349	81	3,359,032	8,814,780
34	5,256,028	14,586,670	89	10,327,673	15,145,559
36	3,386,594	8,628,844	94	1,019,703	2,586,687
38	14,709,681	35,263,407	99	13,041,836	32,832,194

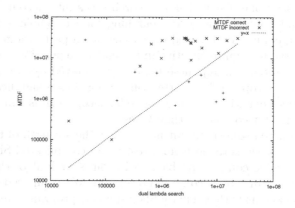

Fig. 4. Comparison of dual λ-search and GPS-SHOGI search

program does not generate the next move 3e Promoted Bishop by the defender with the current sets of move generating rules.

Both problems could be solved by generating more moves for the defender. Problem 85 could be solved by adding more defense moves, such as 3e Promoted Bishop. However, when we generate more defense moves, we had to give away many correct answers. That is why we are reluctant in adding the moves. Problem 44 could be also solved if more defense moves were generated. If check moves by the defender were generated, problem 44 could be answered correctly, but some other problems would then not be solved.

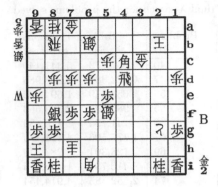

Fig. 5. Problem 44

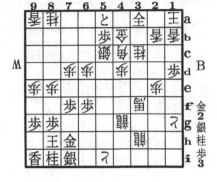

Fig. 6. Problem 85

7   Conclusion

We proposed a new search algorithm called dual λ-search. It is based on a search algorithm called λ-search. Although λ-search is an efficient algorithm, it is only aware of attacks by one player. Dual λ-search takes into account of attacks by both players and could be used for searching for the whole game, while the original λ-search could only be used in searches for a part of a game.

The new algorithm is quite effective in the endgame positions of many games, including Shogi. Shogi is a Japanese board game played by two players, which has a the unique "drop rule". For this rule, Shogi has a significantly complex endgame, and together with the larger search space, Shogi is considered to be a much harder game for computers than Chess.

We have conducted some experiments to show the strength of this algorithm, by solving some positions taken from a test set of complicated Shogi problems. We were able to solve complex problems with our search algorithm, some were solved for the first time by a computer program. We compared our algorithm with λ-search as well as with a general Shogi-playing program. We have achieved both high level of accuracy as well as competitive speed.

Based on these results, we may conclude that our algorithm has been successful in the Shogi endgame, and is expected to be also useful in other games that require a global search, provided that it is possible to decompose the game into reasonably independent sub-games.

References

1. M. Arioka. Search in Shogi program KFEnd. In *Advances in Computer Shogi 4*, pages 18–40. Kyoritsu, 2003.
2. T. Cazenave. A Generalized Threats Search Algorithm. In *3rd Computers and Games Conference (CG 2002)*, (eds. J. Schaeffer, M. Müller, and Y. Björnsson), LNCS 2883, pages 75–87, Springer-Verlag, Berlin, 2003.

3. T. Cazenave. Iterative Widening. In *IJCAI-01 Proceedings*, volume 1, pages 523–528, 2001.
4. J.H. Conway. *On Numbers and Games*. Academic Press, 1976.
5. R. Grimbergen and T. Muraoka. What Shogi Programs Still Cannot Do - a New Test Set For Shogi. In *The 9th Game Programming Workshop in Japan*, pages 40–47, 2004.
6. H. Iida and F. Abe. Brinkmate Search. In *Game Programming Workshop in Japan '96*, pages 160–169, Kanagawa, Japan, 1996.
7. H. Iida, M. Sakuta, and J. Rollason. Computer Shogi. *Artificial Intelligence*, 134(1-2):121–144, 2002.
8. Y. Kawano. Using Similar Positions to Search Game Trees. In *Games of No Chance*, pages 193–202. Cambridge University Press, 1996.
9. A. Kishimoto and M. Müller. DF-PN in Go: An Application to the One-Eye Problem. In *10th Advances in Computer Games (ACG10), Many Games, Many Challenges* (eds. H.J. van den Herik, H. Iida, and E.A. Heinz), pages 125–141, Kluwer Academic Publishers, Boston, 2004.
10. M. Müller. Computer Go. *Artificial Intelligence*, 134(1-2):145–179, 2002.
11. A. Nagai and H. Imai. Application of DF-PN Algorithm to a Program to Solve Tsume-Shogi Problems. In *IPSJ Journal*, 43:1769–1777, 2002.
12. Y. Tanase. *Algorithm of IS Shogi*, pages 1–14. Kyouristu, 2000. (in Japanese).
13. K. Thompson. Retrograde Analysis of Certain Endgames. *ICCA Journal*, 9(3):131–139, 1986.
14. Th. Thomsen. Lambda-Search in Game Trees — With Application to Go. In *2nd Computer and Games Conference (CG 2001)* (eds. T. Marsland and I. Frank), LNCS 2063, pages 19–38, Springer-Verlag, Berlin, 2001.
15. H. Yamashita. YSS - Data Structure and Algorithm. In *Advances in Computer Shogi 2*, pages 112–142. Kyoritsu, 1998. (in Japanese).

Chunking in Shogi: New Findings

Takeshi Ito[1], Hitoshi Matsubara[2], and Reijer Grimbergen[3]

[1] Department of Computer Science,
University of Electro-Communications, Tokyo, Japan
ito@cs.uec.ac.jp
[2] Department of Media Architecture,
Future University Hakodate, Hokkaido, Japan
matsubar@fun.ac.jp
[3] Department of Informatics,
Yamagata University, Yonezawa, Japan
grim@yz.yamagata-u.ac.jp

Abstract. In the past, there have been numerous studies into the cognitive processes involved in human problem solving. From the start, games and game theory have played an important role in the study of human problem solving behavior. In Chess, several cognitive experiments have been performed and those experiments have shown that expert chess players can memorize positions very quickly and accurately. Chase and Simon introduced the concept of chunking to explain why expert chess players perform so well in memory tasks. Chunking is the process of dividing a chess position into smaller parts that have meaning. We performed similar experiments in Shogi with a set of next-move problem positions, collecting verbal protocol data and eye-movement data. Even though experiments in Chess indicated that expert chess players searched as wide and deep as non-expert players, our experiments show that expert Shogi players search more moves, search deeper and search faster than non-expert players. Our experiments also show that expert Shogi players cannot only memorize the patterns of the positions but also recognize move sequences before and after the position. The results suggest that other than the perceptual spatial chunks introduced in chess research, there are also chunks of meaningful move sequences. We call such chunks "temporal chunk". Our research indicates that Shogi players become stronger by acquiring these temporal chunks.

1 Introduction

Many studies of expertise derived from games have been conducted in the field of cognitive science. The study of Chess by De Groot [1] is a particularly famous example. He conducted experiments on the memory of expert chess players. His experiments revealed that an expert chess player can memorize the arrangement of pieces in a chess position accurately in a short time (time limit of 5 seconds). It also became clear that a top-class chess expert (Grandmaster) does not always look at more moves or searches deeper than a lower-class expert player [2].

Simon and Chase built on this work and explained the recognition ability of an expert by using the concept of "chunk". The important idea is that chess knowledge is

H.J. van den Herik et al. (Eds.): ACG11, LNCS 4250, pp. 140–154, 2006.
© Springer-Verlag Berlin Heidelberg 2006

the ability to recognize a typical spatial arrangement of pieces on the chess board as a pattern [6]. They called this a chunk, and reported that the size and volume of the stored chunks becomes larger as the player becomes stronger. Their assumption was that a chunk is an arrangement of pieces, stored as a perceptual and spatial unit. We will call these "spatial chunks" to distinguish this type of chunking from a different type that we will discuss in this paper.

To confirm the presence of chunks in Shogi, we have repeated the memory experiments for Chess in Shogi [4,5]. When the results of the experiments were examined in detail, a difference was observed between the memory ability of intermediate Shogi players (amateur Dan players) and those of advanced Shogi players (professional Shogi players). In the opening phase of the game, the intermediate Shogi players showed memory ability that was little different from those of the advanced Shogi players. However, this memory ability went down significantly in the middle game and the endgame. In contrast, the memory ability of the advanced Shogi players did not decline as the game progressed. If parts of the position are memorized as a spatial chunk like a still image, it is understandable that the opening is easier to memorize, because similar positions are encountered often. However, it is difficult to explain that middle-game and endgame positions can be remembered in this way by using only spatial chunks.

Our observation of advanced Shogi players was that they (1) recognized a (partial) position, (2) recognized how this position occurred and (3) how play would likely develop next. It seemed that they do not recognize positions as a separate entity, but the current position is part of the flow of the game.

To study this phenomenon further, we have used so called "next-move problems", where subjects need to find a good way to continue play in a certain position. These types of positions have not been used in past comparative studies, but we feel that they can be an important tool for examining both the thought processes and the memory processes of players of different playing strengths. As a tool for examination, we collected eye-movement data with an eye camera and simultaneously had the subjects think aloud. We used the same set of problems to collect data from players with different playing strength.

2   Experiments with Next-Move Problems

Below we will provide a full description of our experiments. In Section 2.1 the objective of our experiments will be given, followed by a description of the experiments in Section 2.2. The experiments resulted in three types of data: numerical data (discussed in Section 2.3), verbal data (discussed in Section 2.4), and eye-camera data (discussed in Section 2.5). In Section 2.6 a summary of the results is given.

2.1   Objective

As pointed out by Ito [3], the decision process of a move in Shogi is composed of "recognition of the position", "formation of candidate moves", "look-ahead", "evaluation", and "decision". This process is called a player script, and it was shown that a Shogi player obtains this player script as a beginner and that improvement in playing strength is a gradual refinement of this general script.

The experiment described below is aimed at the problem solving process of deciding a move in a certain Shogi position. In particular, we wish to examine the differences between players of different playing strength (from beginning Kyu players to professionals) in the "recognition of the position" and "formation of candidate moves" phases.

2.2 Method

In the experiment, the thought processes of the subjects were examined by using both verbal data from think-aloud protocols and the data from an eye camera[1] when solving next-move problems. For our next-move problems, we did not use problem positions that can be found in Shogi magazines. These positions usually require the discovery of a hidden tactical move, which only rarely happens in actual game play. Instead, we asked a strong player to create a number of positions satisfying the following two conditions: (1) a position that could appear in an actual game, and (2) a position that has multiple reasonable candidate moves. The problem positions that we used in our experiments can be found in the Appendix.

The problem positions were given to 10 subjects: two beginners (amateur 5-kyu and lower), three intermediate players (around 1-dan), two advanced players (amateur 4-dan or higher) and three top professional players (8-dan or higher). They were each asked to think aloud about the next move in the current position, which was recorded. Also, the movement of the eyes of each subject was recorded with an eye camera. We did not impose a time limit for solving the problem and gave the subjects instructions to take as much time for deciding on their move as they thought necessary. The problems were given in the same order as presented in the Appendix. Because the experiment as a whole could take several hours, we allowed a break after finishing a problem position. However, except for one intermediate player and one advanced player, all subjects did the full set of problem positions without taking a break.

In the analysis, numerical data such as thinking time and the width and depth of the look-ahead are given first. Then, the results of analyzing the thought processes of players of different strength using the think-aloud protocols are given. Finally, the thought processes of players of different strength during the position-recognition phase are analyzed using the combination of eye-camera data and think-aloud protocols.

2.3 Numerical Data

In Table 1 the answers of all subjects to the problem positions are given. N1 and N2 are the two beginners, M1 to M3 are the intermediate players, E1 and E2 the advanced players, and P1 to P3 are the three professional players. As can be seen from the table, in some problems (such as problem 1 or 4) even the professional players did not agree about the best move, so the problem positions are not easy.

In Table 2 the time for answering each problem position is given for all subjects. This is the time from the moment the problem was shown until the subject chose a move. As can be seen from the table, there are substantial differences between the subjects.

---

[1] The Eyemark Recorder EMR-8 of NAC Image Technology Inc.

Table 1. Answers of each subject

	Problem 1	Problem 2	Problem 3	Problem 4	Problem 5	Problem 6	Problem 7	Problem 8	Problem 9	Problem 10
P1	rook 2e	pawn 4e	knight 3g	silver 4f	pawn 2b	rook 2d	pawn 9d	king 7h	bishop 8f	silver 4h
P2	pawn 7g	pawn 4e	knight 3g	gold 6h	pawn 2b	rook 2d	pawn 9d	gold 6f	bishop 8f	silver 4h
P3	pawn 9e	pawn 5e	pawn 4e	pawn 4f	pawn 2b	bishop 8d	pawn 9d	gold 6g	bishop 8f	silver 4h
E1	gold 4g	pawn 4e	pawn 9f	gold 6h	pawn 4e	rook 2d	pawn 9d	gold 6g	bishop 8f	silver 4h
E2	pawn 9e	pawn 4e	bishop 8f	silver 4f	pawn 4e	rook 2d	pawn 9d	gold 3h	bishop 8f	silver 4h
M1	rook 2f	pawn 4e	pawn 4e	pawn 3e	rook 4a	rook 3f	pawn 9g	gold 4h	+rook 3c	silver 4h
M2	pawn 4e	pawn 4c	pawn 4e	silver 3g	pawn 2b	rook 3f	silver 4f	gold 4h	pawn 9e	silver 4h
M3	pawn 9e	pawn 4e	pawn 9f	pawn 4f	rook 2h	gold 3h	pawn 9g	gold 3h	+bishop 5d	silver 4h
N1	pawn 4e	pawn 4e	pawn 4e	gold 6h	pawn 2b	rook 3f	pawn 9d	gold 7h	pawn 9e	silver 4h
N2	pawn 4e	pawn 4e	knight 3g	silver 3g	pawn 4e	pawn 2d	pawn 9i	pawn 1f	pawn 2d	gold 2b

Table 2. Answer time of each subject in seconds

	Problem1	Problem2	Problem3	Problem4	Problem5	Problem6	Problem7	Problem8	Problem9	Problem10
P1	345	336	269	134	229	410	170	296	146	196
P2	63	102	188	41	78	95	135	91	184	272
P3	39	30	36	50	60	214	20	35	77	97
E1	191	143	189	77	112	256	68	142	139	265
E2	135	210	275	128	220	113	113	220	167	237
M1	254	133	377	104	192	147	210	149	207	544
M2	367	379	281	56	322	207	365	186	115	155
M3	672	143	502	72	332	825	100	255	56	330
N1	574	121	127	56	124	115	151	112	154	141
N2	157	157	124	92	98	69	122	131	89	168

In Fig. 1 the average time for deciding on the move in each position is given for the four different levels of playing strength. When looking at this data, we see that position 4 has a low average decision time for all players despite the number of different answers that are given. The reason for this is that this position resembles a standard opening position, which is known to all players, thus shortening the decision

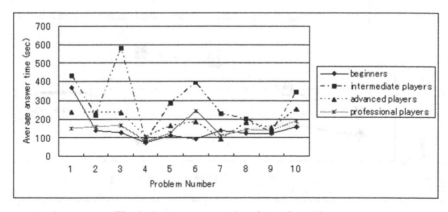

Fig. 1. Average answer time for each problem

time. Position 7 is a variation of a more advanced type of standard position; advanced and professional players agree upon the best move here. Furthermore, for those who are more familiar with this type of position, the decision time is very short.

Despite these individual and position differences, it seemed that the intermediate players were thinking quite long before deciding upon their move. To confirm this, we averaged the decision time for the four levels of playing strength over all problems. The results are given in Fig. 2. We see that intermediate players indeed seem to take longer than the players of other playing strengths.

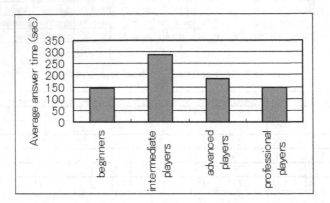

Fig. 2. Average answer time by playing level

Furthermore, we analyzed the think-aloud protocols to determine the look-ahead in the thought process, in particular the number of candidate moves that were considered and how deeply these moves were searched. Figure 3 shows the width and depth of the look-ahead of each subject, where the width is the number of candidate moves considered in the problem position and the depth the length of the longest variation that was considered. Moreover, Fig. 3 shows that intermediate players search slightly wider than the other players, and that the depth of the look-ahead increases with playing strength.

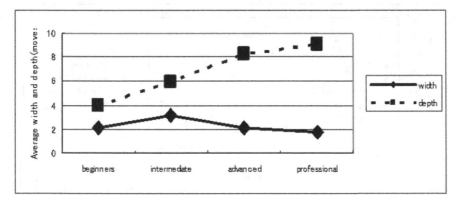

Fig. 3. Width and depth of look-ahead for different playing levels

We also compared the volume of the look-ahead between different levels of playing strength. To measure this volume, we used the number and the speed of the look-ahead. The "number" of the look-ahead is the total number of moves that was mentioned in the think-aloud protocol. The "speed" is acquired by dividing this number by the decision time in minutes. The results are given in Fig. 4. This figure shows that both number of moves and speed of the look-ahead increase with playing strength.

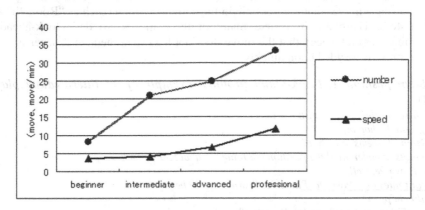

Fig. 4. Move number and speed in the look-ahead for different playing levels

2.4 Verbal Data

Below we provide some examples of think-aloud protocols and the thought patterns of subjects with different levels of playing strength. In the protocols, the comment in < > is an explanation that is not part of the verbal data and "..." indicates omitted data.

Problem position 1, think-aloud protocol example of a beginner (N2)
After rook+4d.
Hmm, what kind of position is this?
Well, there is a Rook, <u>the opponent's Rook is on 4d,</u>
<u>The opponent has a Bishop,</u> ...,
<For around 55 seconds there are attempts to understand the position>
<u>Rook 4h?</u> Rook 4h, ...,
The Pawn will not be taken if the Rook goes to 4h, but if so, here, Rook 2d will attack. ...,
<u>Pawn 4e,</u> Eh, where to escape. ...
<Some candidate moves to reply to the opponent's previous move are given.>
Rook 7d, oh, it is good, too. ..., hmm, well, um?
Pawn 4e might be better.
Well, I will play Pawn 4e.
<A move which is not obviously bad is selected by short look-ahead and elimination.>

The first thing that stands out in this read-aloud protocol of the beginner is that it takes quite a long time to understand the position. In total, it takes the subject around 55 seconds to get a clear idea about the positional features. On average, beginners took between 40 and 70 seconds for this stage. When combining this with the data in

Figs. 1 and 2, understanding the position takes around 50 seconds on average, while the average time to decide upon a move is about 150 seconds. Beginners therefore spend one third of their decision time to understand the position.

Also clear from the protocol is that understanding the position is characterized by the individual pieces in the position (such as "the opponent's Rook is on 4d" and "the opponent has a Bishop"). Understanding the position comes from examining each piece individually.

As far as look-ahead is concerned, the protocol seems to indicate that beginners focus on single moves that are possible in the current position (such as "Rook 4h" and "Pawn 4e"). There are of course many of these moves, so deep look-ahead is impossible. A quick check that the move does not lead to immediate disaster is made and the move is decided upon quickly.

Problem position 1, think-aloud protocol example of an intermediate player (M3)

Well, it seems the game is a little rough.
I let go the Bishop on 5f,
Yes, Hmm. It seems to be hard. ...
<Attempts to understand the position continue for about 30 seconds.>
From Pawn 9e, well,
If it continues like Pawn-9e, Pawn(9e, Pawn*9b, it'll be good.
Will this happen?
Well,...Will it go like "Pawn-9e, Pawn(9e"?
Is it really alright if the reply is Pawn-3f, Pawn(3f, Rook(4f,
... what will happen if I drop a Pawn on 4g? ...
<Several candidate moves are mentioned in short succession and look-ahead is performed.>
Let me see.
Rook 2e seems to be a problem, so I will first play Pawn 9e.
<Selecting a move is done by elimination just as the beginner did.>

In the case of intermediate players, the time taken for position recognition is considerably shorter than for beginners. In this case, position understanding took about 30 seconds. In general, position understanding takes between 20 and 40 seconds after which candidate moves are beginning to appear.

For intermediate players, candidate moves seem to pop up one after another. However, these candidates are based on the knowledge of local tactics, not based on an understanding of the complete position. In the read-aloud protocol example, "Pawn-9e, Pawn(9e, Pawn*9b" is a standard sequence of moves to play an edge attack. Also, the sequence "Pawn-3f, Pawn(3f, Rook(4f" is a well-known sacrifice to activate the Rook. However, there is a clear difference in the use of knowledge here when comparing this with the crude disaster check that beginners do before deciding upon their move.

Problem position 1, think-aloud protocol example of an advanced player (E1)

Hmm, It looks like an "Aigakari-formation".
Black doesn't lose a Pawn, or?
There is a Pawn here.
Oh, Black and White are reversed.
After Rook 4d, ...,
< There is established knowledge for understanding the position.>

Gold 4g is activating the most inactive piece. ...
Rook 2f seems not that solid. ...
Well, it is Gold 4g. ...
I would like to do opposing Pawn 7g somewhere.
<The candidate moves with evaluation are given to conduct look-ahead.>
Both Pawn 4e and Rook 2f need to be considered, I don't want to drop there.
Or play Gold 4g.
It can't be helped, so Gold 4g?
<The move is determined by evaluation of the piece activity based on look-ahead.>

The position recognition of advanced players is quite quick, generally taking about 10 seconds. There is no verbal data indicating that the position recognition is also done after this initial position-recognition phase. Also, the protocol has a reference to the strategic features of the whole position (the "Aigakari-formation" is one of the general opening strategies in Shogi). Also, the comment "Black and White are reversed" shows that previous knowledge about the full position is being used.

In the phase of forming candidate moves, the influence of a candidate move upon the whole position is given ("Gold 4g is activating the most inactive piece") and there is also verbalization of how the candidate move fits into the flow of the position ("Rook 2f seems not that solid"). These references show that the position is recognized considering subsequent developments.

Problem position 1, think-aloud protocol example of a professional player (P1)
I think I played this pattern before.
Uh, well ...
<The understanding of the position is immediate, and it is checked with existing knowledge instantly.>
Hmm, my first impression is to sacrifice a Pawn on 9e.
But, yes, this move is good.
Retreating Silver 6d against Pawn 9e can be used.
Pawn 3f, Pawn×3f, Rook×4f against Pawn 9e can also be used.
Well, if I push the Pawn to 4e, the Rook moves sideways to 3d.
Although I think this Rook is less active. ...
There is Pawn-3f, Pawn×3f, Bishop 4f,
So Pawn 4e is a little difficult to play. ...
<Subsequent moves are mentioned as well as the candidate moves.>

Problem position 1, think-aloud protocol example of a professional player (P2)
Well. I think I have seen this pattern before.
It is a position with experience.
<The understanding of the position is immediate, and it is checked with existing knowledge instantly.>
Uh, well, Black's front side is a little defensive,
so Pawn 4f seems to be taken,
But it is impossible to defend with Gold 4g or Rook 4h,
and moving the Pawn to 4e might be a little unpleasant.
<The move that might be bad is not read from the beginning.>
Well, Silver 8h forms a wall-silver, so how about the opposing pawn 7g, is it possible?
Uh, this, Do I have to read ahead only after Pawn 7g and Rook 4f?
If then Pawn 7g and Rook×4f, take Pawn 7f and Pawn 7g,

Pawn×7g+, Silver×7g and Rook×4f,
hum, such a development is possible.
<Candidate moves come up immediately, and look-ahead is conducted with a clear objective.>

When analyzing the verbal data connected to the position recognition phase of professional players, there are a high number of references to past experience. This not only concerns the current position, but also about how this position came about and how this position is likely to develop. In this position, comments such as "I think I have seen this pattern before" and "I think I played this pattern before" are examples of this use of past experience. Adding comments about how the position came about seems to indicate that positions are not recognized separately, but are embedded in a context of related knowledge.

As for look-ahead, the characteristic feature of the look-ahead protocols of professional players is that moves are evaluated as having good or bad prospects without any look-ahead. Look-ahead in most cases is only used to confirm that the initial judgment was correct. Adding evaluation to candidate moves during the move generation phase requires knowledge about the context of the position and knowledge about the prospects from the current position.

2.5 Eye-Camera Data

The results given in the previous section show that thought processes concerning position recognition and candidate-move generation are present in the think-aloud protocols and that this whole process takes at most 60 seconds after the initial presentation of the problem position. Here, we will look at how position recognition leads to candidate-move generation more closely by combining data from an eye camera with the verbal data given in the previous section.

In Fig. 5, the combination of eye-camera data and verbal data of a beginner is presented for problem position 1. The eye camera traces how long the eyes focus on a certain point. Whenever the eye stops moving for more than 0.2 seconds, this is recorded and represented with a ○. The size of the ○ is relative to the time of fixation.

The eye-camera data shows that beginners recognize the position by looking at a wide area of the board. The destination square of the previous move (Rook to 4d) is focused on for a long time. That this is important for the beginner is also clear from the verbal data. To understand the position, looking at many pieces on the board and understanding the move that lead to this position is important for the beginner.

Figure 6 shows eye-camera data and verbal data of intermediate player M3 for problem position 1. When compared to beginning players, the findings are that the area of examining the position becomes smaller, and the time for position recognition shortens. Although it is not clear from the data from this position, some references on tactics and castle formation can also be observed during position recognition, indicating that recognition of a chunk of pieces becomes possible. Also, moves based on tactical sequences appear during candidate-move formation. The sequence of "Pawn-9e, Pawn×9e, Pawn*9b" in this position is the tactical move sequence of an edge attack starting with the sacrifice of the edge pawn, and it can also be observed from the eye-movement data (within 40 seconds) that the three move tactical combination is remembered from the combination of the edge file and the diagonal controlled by the Bishop on 5f in this position.

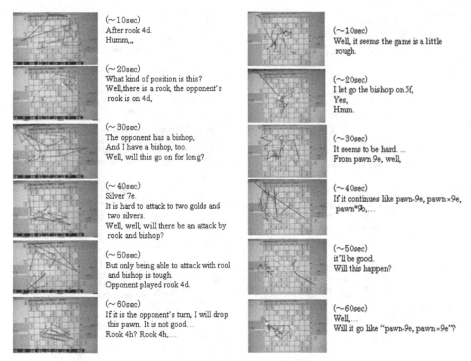

(~10sec) After rook 4d. Humm,,,	(~10sec) Well, it seems the game is a little rough.
(~20sec) What kind of position is this? Well, there is a rook, the opponent's rook is on 4d,	(~20sec) I let go the bishop on 5f, Yes, Hmm.
(~30sec) The opponent has a bishop, And I have a bishop, too. Well, will this go on for long?	(~30sec) It seems to be hard. ... From pawn 9e, well,
(~40sec) Silver 7e. It is hard to attack to two golds and two silvers. Well, well, will there be an attack by rook and bishop?	(~40sec) If it continues like pawn-9e, pawn×9e, pawn*9b,...
(~50sec) But only being able to attack with rook and bishop is tough. Opponent played rook 4d.	(~50sec) it'll be good. Will this happen?
(~60sec) If it is the opponent's turn, I will drop this pawn. It is not good... Rook 4h? Rook 4h,...	(~60sec) Well,... Will it go like "pawn-9e, pawn×9e"?

Fig. 5. Eye-tracking diagrams and verbal data of a beginning player (N2)

Fig. 6. Eye-tracking diagrams and verbal data of an intermediate player (M3)

Figure 7 shows the eye-camera data and verbal data of an advanced player (E3) for problem position 1. As explained, the position recognition of advanced players uses an understanding of how this position came about and about the strategic features of the position. Also, this is not a direct recognition of the piece arrangement as the comment "Black and White are reversed" shows. The eye-camera data shows that during the position-recognition phase the area where the eyes focus on becomes narrower and that the number of fixations becomes smaller than those observed for beginners.

When generating candidate moves, not only the direct meaning of the move for the current position, but also different meanings for future positions become important. In this problem position, the pawn push is not only considered because it can be taken on the next move, but also because it has a potential of pushing the Pawn up further in a later stage of the game. The positional drawbacks of this move (weakening the square that the Pawn left) are also taken into account.

An interesting observation when combining verbal data with eye-camera data is that advanced players do not always focus on the area that is important in their verbal comments. For example, at the 40 second mark the advanced player watches the opponent's King while the verbal data shows a comment about moving the Pawn to 4e and then to 4d. Although not present in the verbal data, this indicates that the player was thinking of advancing this Pawn in relation to an attack on the opponent King. This discrepancy between point of focus and verbalization is only found for advanced players.

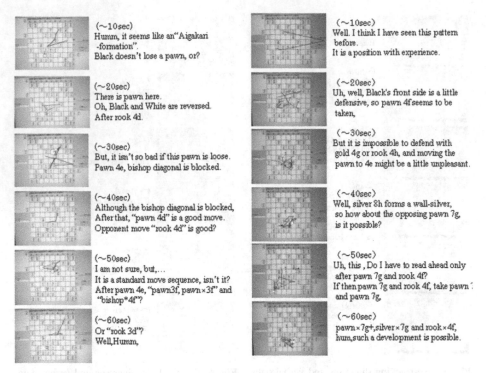

The following verbal data appears alongside the eye tracking diagrams:

(~10sec)
Humm, it seems like an"Aigakari
-formation".
Black doesn't lose a pawn, or?

(~10sec)
Well, I think I have seen this pattern
before.
It is a position with experience.

(~20sec)
There is pawn here.
Oh, Black and White are reversed.
After rook 4d.

(~20sec)
Uh, well, Black's front side is a little
defensive, so pawn 4f seems to be
taken,

(~30sec)
But, it isn't so bad if this pawn is loose.
Pawn 4e, bishop diagonal is blocked.

(~30sec)
But it is impossible to defend with
gold 4g or rook 4h, and moving the
pawn to 4e might be a little unpleasant.

(~40sec)
Although the bishop diagonal is blocked,
After that, "pawn 4d" is a good move.
Opponent move "rook 4d" is good?

(~40sec)
Well, silver 8h forms a wall-silver,
so how about the opposing pawn 7g,
is it possible?

(~50sec)
I am not sure, but,...
It is a standard move sequence, isn't it?
After pawn 4e, "pawn3f, pawn×3f" and
"bishop*4f"?

(~50sec)
Uh, this , Do I have to read ahead only
after pawn 7g and rook 4f?
If then pawn 7g and rook 4f, take pawn
and pawn 7g,

(~60sec)
Or "rook 3d"?
Well,Humm,

(~60sec)
pawn×7g+,silver×7g and rook×4f,
hum,such a development is possible.

Fig. 7. Eye tracking diagrams and verbal data of an advanced player (E1)

Fig. 8. Eye tracking diagrams and verbal data of a professional player (P2)

Figure 8 shows the eye-camera data and the verbal data of a professional player (P2) for problem position 1. A characteristic of professional players is that they recognize the position instantly and immediately add remarks about past experience and knowledge concerning the current position. There are comments like "I have seen this pattern before", but also more specific remarks about how the position came about such as "It is a very sharp position due to the early rook exchange" (problem position 5) or "This is a standard 'Yokofu-dori' opening" (problem position 6). Positions are not understood by focusing on every piece, but based on castle formation and strategy and on the flow of the game. Eye-camera data shows only a few points of fixation and only the central part of the board is scanned quickly.

The verbal data indicates that future developments are vital for candidate-move generation. Comments like "it is impossible to defend with Gold 4g or Rook 4h" and "moving the Pawn to 4e might be a little unpleasant" take into account the future developments, leading to quick discarding of the mentioned moves. In the end, the attention focuses on a move (Pawn 7g) that will solve a potential endgame problem.

2.6 Summary of the Results

Summarizing the results of the previous subsections, we obtain the following:

Result 1: The Numerical Data
- Intermediate players think longer before a move decision than beginners.
- Intermediate players look at more candidate moves than beginners, but look-ahead is shallow.
- Advanced players look more deeply and are better at narrowing the search.

Result 2: The Verbal Data
- A large part of the verbal data of beginners is concerned with position recognition.
- Beginners only look at single, unconnected moves that are possible in the current position.
- Intermediate players can do look-ahead based upon tactical and strategic combinations of moves.
- Advanced players take knowledge about how a position came about into account.
- Advanced players immediately evaluate the merits and drawbacks of the candidate moves.

Result 3: The Eye-Camera Data
- Beginners move their eyes over the whole board during position recognition.
- From intermediate to advanced player, the area of the board that is focused on gradually becomes smaller.
- Advanced players also sometimes focus on a part of the position unrelated to their comments.

3 Discussion

The results indicate that intermediate players have the longest thinking time (Fig. 2). This seems related to our observation that intermediate players search a wide range of candidate moves (Fig. 3). In Shogi, improvement in strength is accompanied by learning more tactical and strategic moves. Without the proper knowledge of when and how to apply these tactics, there are many positions in which these moves are possible. More possible moves and a better ability to do look-ahead is the reason for longer thinking times. However, when the skills improve, knowledge about how the position came about and how it is likely to develop is taken into account. This leads to a better understanding of the merits and disadvantages of moves, narrowing the number of possible candidate moves for which look-ahead is performed. Advanced players therefore do look-ahead more narrowly and more deeply.

Figure 4 shows that when the player's skill improves, the number and the depth of look-ahead increases. This is based upon verbal data, so it is possible that because translating thoughts in words is slow, the actual number of moves and look-ahead is considerably larger than the numbers given in the figures. However, the difference between the speed of thought and thought verbalization seems especially present in the case of advanced players. Therefore, the increase in total move number and speed seems a significant first result.

A second result is that there are clear differences in position recognition between players with different Shogi skills. This may be concluded from both the verbal data and the eye-camera data. It was observed that beginners focus on individual pieces or small groups of pieces during position recognition. Intermediate players are able to understand the position by adding "spatial chunks" about larger piece formations such as castles. Because of this ability to recognize piece configurations, they are also able to use combinations of moves rather than single moves. Professional players have a strong ability to connect to the flow of the game, i.e., how the current position came about and what the likely development is. This seems to be a different type of knowledge from "spatial chunks" and could be called "temporal chunk". The results indicated that temporal chunks are acquired by intermediate players in the form of tactical moves and standard move sequences. The sense of the flow of the game is then improved by learning how these move combinations influence a position. Therefore, intermediate players can only generate a limited number of candidate moves and advanced players are able to search more narrowly and deeply.

4   Conclusion and Future Work

The various results in our study seem to suggest that in Shogi the important issues are (1) spatial chunks and (2) time-related knowledge (temporal chunking). A different phenomenon are the comments by advanced players concerning the immediate assessment of candidate moves, which does not seem to be a result of look-ahead. It is hard to prove the existence of temporal chunks, but we believe that our results are a strong indication of the existence of time-related knowledge. Further experiments are needed to investigate the exact nature of this temporal knowledge.

We would also like to know (1) whether this way of thinking is specific to Shogi, (2) can be found in other games or (3) even can be considered a general feature of expertise. Similar experiments in other chess-like games or Go could answer this question.

Moreover, a different way of obtaining or analyzing eye-camera data might be needed. In the current setting, eye fixation is represented by a point, but humans are known to process data outside of where the eye focuses. As a future work, we would like to investigate how peripheral vision influences the information processing in Shogi and if there is a relation with Shogi skill.

There are also reasons to reconsider the set of problem positions. From Table 1 it may be concluded that for most positions the subjects chose a wide variety of moves (except for position 2 and 10). This met our requirement that the positions needed to have multiple reasonable candidate moves, but it might also be interesting to see by which thought processes players of different skill might end up with the same solution.

Furthermore, the number of problem positions and subjects might be considered too small to support our conclusions. Time constraints (especially for professional players) make it difficult to conduct larger experiments, but it might be possible to increase the data of all but the top players. The current set of positions is not well-balanced as far as difficulty is concerned and we also found some tactical problems in the positions that might influence the data. Increasing the problem set and having a better balance in the degree of difficulty is also a future work.

Finally, our experiments show that even professional players disagree on many occasions. Other than general Shogi skill, there are individual differences influencing the move choice. The amount of freedom for individual differences might be an important factor for how interesting a game is. This, together with the notions of width and depth in the thought process might be an indication of how profound and interesting a game is. In future work, we would like to investigate this connection further.

References

1. A.D. de Groot. *Thought and Theory in Chess.* Mouton Publishers, The Hague, The Netherlands, 1965.
2. A.D. de Groot and F. Gobet. *Perception and Memory in Chess.* Van Gorcum, Assen. 1996.
3. T. Ito. Human Cognitive Processes on Playing Shogi. *Game Programming Workshop in Japan '99,* pages 177–184, 1999.
4. T. Ito, H. Matsubara, and R. Grimbergen. The Use of Memory and Causal Chunking in the Game of Shogi. In *The Third International Conference on Cognitive Science*, pages 134–140, 2001.
5. T. Ito, H. Matsubara, and R. Grimbergen. Cognitive Science Approach to Shogi Playing Processes – Some Results on Memory Experiments –. *IPSJ-JNL*, 43(10): 2998–3011, 2002.
6. H.A. Simon and W.G. Chase. Skill in Chess. *American Scientist*, 61:393–403, 1973.

Appendix. Problems Used for the Next-Move Test Experiments

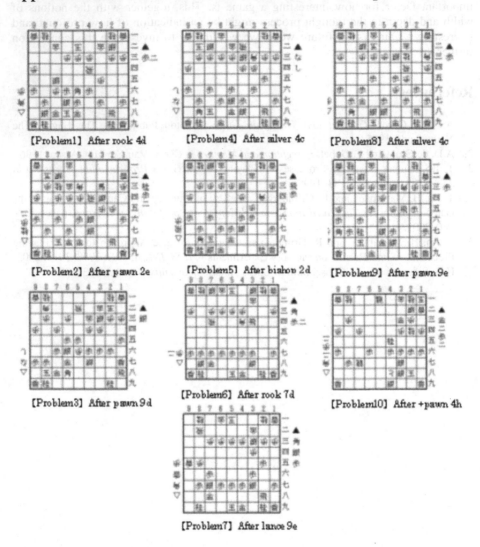

[Problem1] After rook 4d

[Problem4] After silver 4c

[Problem8] After silver 4c

[Problem2] After pawn 2e

[Problem5] After bishop 2d

[Problem9] After pawn 9e

[Problem3] After pawn 9d

[Problem6] After rook 7d

[Problem10] After +pawn 4h

[Problem7] After lance 9e

King Race

Alejandro González Romero

Departament de Llenguatges i Sistemes Informàtics,
Universitat Politècnica de Catalunya, Barcelona, Spain
aromero@lsi.upc.es, yarnalito@gmail.com

Abstract. In the last decade many games have been successfully approached by what is commonly known as brute force, i.e., searching as much as possible. This tendency of computer play has been criticized since it differs much from emulating human thoughts which once was one of the primary goals of computer game playing. An alternative way which is much closer to human game playing, is to discover rules automatically for a given game with the aid of Artificial-Intelligence techniques and human thought. Obtaining such rules can be done by building computer game programs. This research line may have a much broader scope (1) to understand better the ideas of a certain game, (2) to investigate how to build automatically learnt policies for a planning problem, and perhaps even (3) to understand a bit better human thinking. To illustrate these three points, we use a small game, that could be called King Race, and we aim at discovering rules to solve it. To discover these rules we use some human thought and a decision tree program. Then we prove mathematically that these automatically obtained rules indeed solve the game. The rules can aid in building computer-chess endgame programs that do not use brute force.

1 Introduction

In some chess endgames the outcome depends on whether a King will be able to occupy a certain square at the end of a king race. For example, in a rook-against-pawn endgame, with the white Rook at b8, the white King on the interval [a4,a6] (that is, on a4, a5, or a6), the black Pawn at a2 and the black King at e6 (see Diagram 1), with White to move, the game should continue as follows:

1. Re8 check (a lateral check at b6 would be a bad move).
1. ... Kd5 2. Re1 (or using a figurine notation: 1. ♖e8+ ♚d5 2. ♖e1).

Now, after a move of the black King to c4, the rook is unable to capture the Pawn safely so the outcome of the game depends on a king race: if the black King is able to occupy the square b3 Black draws; if not Black loses. If the white King is at a6 the black King will reach b3 and it is a draw; if the white King is at a5 or a4 the black King will not reach b3 and Black will lose.

H.J. van den Herik et al. (Eds.): ACG11, LNCS 4250, pp. 155–164, 2006.
© Springer-Verlag Berlin Heidelberg 2006

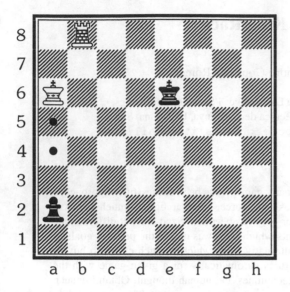

Diagram 1. Example of a chess ending where the game of king race is useful (the black dots indicate other positions of the white King)

More generally, consider an endgame with the same pieces, the white Rook WR on the interval [e1,h1], the white King not on the interval [a1,WR], the black Pawn at a2 and the black King at a distance less than 4 from a2 and at a distance more than 1 from row 1 (there are 2,324 such positions). Black to move. Black draws if the black King is able to reach b3; otherwise Black loses.

This suggests the general problem of deciding, if one has two opposing kings K1 and K2 on a rectangular board (not necessarily 8 × 8) and a square S has been determined, whether or not K1 will be able to occupy S.

2   King Race

Let us now define our little game which will be called *king race*. Assume we have a rectangular board (infinite boards are also allowed) where two opposing chess Kings K1 and K2 are placed and $d(K1,K2)>1$. An unoccupied square S on the board is also specified. The problem is:

> *Assuming K1 moves first, decide whether K1 will be able to occupy the square S.*

Call the horizontal lines rows and the vertical lines columns. The *horizontal* distance between two squares S1, S2, is the distance between the column L1 containing S1 and the column L2 containing S2, that is, the number of king moves needed to go from a square in L1 to a square in L2; we denote it by $d_h(S1,S2)$. For the vertical distance we analogously arrive at $d_v(S1,S2))$.

The distance $d(S1,S2)$ between S1 and S2, that is, the number of king moves needed to go from S1 to S2, is $\max\{d_h(S1,S2),d_v(S1,S2)\}$. An example is given in Diagram 2.

One suspects that, in the search for a criterion to decide whether or not a square is reachable, not only distances between squares are important but also horizontal and vertical distances between them. Below we list five attributes which seem important.

$A1 : d(K1,S) < d(K2,S)$
$A2 : d_h(K1,S) < d_h(K2,S)$
$A3 : d_v(K1,S) < d_v(K2,S)$
$A4 : d_h(K1,K2) \leq 1$
$A5 : d_v(K1,K2) \leq 1$

We now put a coordinate system and take S as the origin (0,0). In Table 1 we list 18 positions and, for each of them we indicate if S can be reached by K1 (the king which moves first) and if Ai is satisfied (+) or not

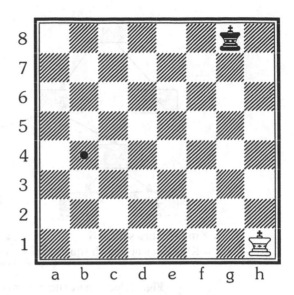

Diagram 2. Example of a position in the game of king race. The white King moves first and reaches square S (denoted by a black dot) in 6 moves, e.g.: 1. ♔g2 ♚f7 2. ♔f3 ♚e6 3. ♔e4 ♚d6 4. ♔d4 ♚c6 5. ♔c4 ♚b6 6 ♔b4

Table 1. Reachability of 18 Positions

Positions	Is (0,0) reachable?	A1	A2	A3	A4	A5
P1: K1=(3,–3),K2=(4,3)	Yes	+	+	–	+	–
P2: K1=(2,–3),K2=(4,3)	Yes	+	+	–	–	–
P3: K1=(3,–2),K2=(4,3)	Yes	+	+	+	+	–
P4: K1=(2,2),K2=(4,3)	Yes	+	+	+	–	–
P5: K1=(–3,3),K2=(4,3)	Yes	+	+	–	–	+
P6: K1=(–3,2),K2=(4,3)	Yes	+	+	+	–	+
P7: K1=(4,–3),K2=(4,3)	No	–	–	–	+	–
P8: K1=(4,–2),K2=(4,3)	Yes	–	–	+	+	–
P9: K1=(–3,4),K2=(4,3)	Yes	–	+	–	–	+
P10: K1=(–4,3),K2=(4,3)	No	–	–	–	–	+
P11: K1=(–4,2),K2=(4,3)	No	–	–	+	–	+
P12: K1=(–4,0),K2=(4,3)	No	–	–	+	–	–
P13:K1=(–4,–3),K2=(4,3)	No	–	–	–	–	–
P14: K1=(2,–4),K2=(4,3)	No	–	+	–	–	–
P15: K1=(3,–4),K2=(4,3)	No	–	+	–	+	–
P16: K1=(–3,3),K2=(3,4)	Yes	+	–	+	–	+
P17: K1=(–3,2),K2=(3,4)	Yes	+	–	+	–	–
P18: K1=(4,–3),K2=(3,4)	Yes	–	–	+	+	–

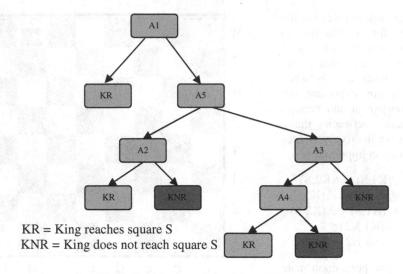

KR = King reaches square S
KNR = King does not reach square S

Fig. 1. Decision tree derived from Table 1

(–) (i = 1,2,3,4,5). (The examples could be modified so that they fit into an 8 x 8 board if so desired.) Using these examples and a decision tree program the tree in Fig. 1 is obtained. Here a terminal node labeled KR (or KNR) indicates that the square S is reached (or not reached) by the first King.

The examples and the tree suggest the following rule:

> A1 or (A2 and A5) or (A3 and A4) → S can be reached
> not (A1 or (A2 and A5) or (A3 and A4)) → S cannot be reached

It turns out that, assuming S is not on the edge of the board, this is a valid result, as we will prove in the next section.

3 Proving the Obtained Rules

Theorem 1. Assume S is not on the edge of the board. The square S can be reached from K1 with the second king at K2 if and only if

(C1.i) $d(K1,S) < d(K2,S)$, or
(C1.ii) $d_v(K1,S) < d_v(K2,S)$ and $d_h(K1,K2) \leq 1$, or
(C1.iii) $d_h(K1,S) < d_h(K2,S)$ and $d_v(K1,K2) \leq 1$.

Remarks. For the sufficiency part (the "if" part) the assumption that S is not on the edge of the board is not needed. At the end of the article we indicate the (slight) modifications needed if S is on the edge of the board. We will often omit the phrase "with the second King on K2" or even also "from K1" when the position of the second King, or of both Kings, is clear.

Lemma 1. If $n = d(K1,S) < d(K2,S)$ then S can be reached from K1 in n moves.

Proof by induction on n.

$n = 1$: Since $d(K2,S) > n=1$, K2 is not protecting S so K1 can occupy S in one move.

$n > 1$: Move K1 to a square K1' such that $d(K1',S) = n-1$. This is possible because K1' is not protected by K2, since if we had $d(K2,K1') \leq 1$, we would have $n < d(K2,S) \leq d(K2,K1') + d(K1',S) \leq 1+n-1 = n$ which is impossible.

Now K2 will move to a square K2' and $d(K2',S) > n-1$. Since $d(K1',S) = n-1 < d(K2',S)$, by induction we see that S can be reached from K1' in $n-1$ moves with the second King on K2'. Hence S can be reached from K1 in n moves with the second King at K2. $\qquad\square$

Lemma 2. If, with respect to some coordinate system, $S = (0,0)$, $K1 = (p,0)$ and $K2 = (m,n)$ with $p > 0$ and $m \geq p-1$, then S can be reached.

This lemma says that if K1 is to the right of S and on the same row, S will be reached, provided K2 is on a column contiguous to that of K1 or more to the right. Proof by induction on p.

$p = 1$: The squares at distance 1 from S and not protected by K1 have abscissa -1. However, $m > -1$ so $d(K2,S) > 1$. Hence K1 can occupy S.

$p > 1$: Move K1 to K1' $= (p',0)$ where $p' = p-1$. Then K2 will move to a square K2' $= (m',n')$ where $m' \geq m-1 \geq p'-1$.

By induction S can be reached from K1' with the second King on K2'. Hence S can be reached from K1 with the second King on K2. $\qquad\square$

We now prove the sufficiency part of Theorem 1. If (C1.i) holds then S can be reached, by Lemma 1. We now assume (C1.ii). (The argument is similar if we assume (C1.iii)). Choose the coordinate system so that $S = (0,0)$ and $K1 = (p,q)$ with $0 \leq q \leq p$. We have $q = d_v(K1,S) = |n|$ and $|p-m| = d_h(K1,K2) \leq 1$. Let K1' $= (p',0)$ where $p' = p-q$ and notice that $d(K1,K1') = q < |n| \leq d(K2,K1')$.

By Lemma 1, K1' can be reached from K1 in q moves with the second King at K2 $= (m,n)$. After q moves the second King will arrive at a square K2' whose abscissa m' is greater than or equal to $m-q$. Since $p-m \leq |p-m| \leq 1$, we have $m \geq p-1$ and therefore $m' \geq m-q \geq p-1-q = p'-1$. By Lemma 2, it follows that S can be reached from K1' with the second King on K2'. Hence S can be reached. This concludes the proof of the sufficiency part.

Now we start the necessity part. We collect some lemmas for the "only if" part. (The inequality $d((p,q),(m,n)) > 1$ indicates that the distance between Kings should be greater than 1.)

Lemma 3. Assume that $m \geq n > 1$, $d((p,q),(m,n)) > 1$, and that the following three conditions are satisfied:

$\qquad$ (C2.i) $\quad \max\{|p|,|q|\} \geq m$

$\qquad$ (C2.ii) $\quad |q| \geq n$ or $|p-m| > 1$

$\qquad$ (C2.iii) $\quad |p| \geq m$ or $|q-n| > 1$.

Then $d((p,q),(m-1,n-1)) > 2$.

Because of this lemma the second King will be able to move diagonally towards S, from (m,n) to $(m-1,n-1)$, if $n > 1$ and none of (C1.i), (C1.ii), and (C1.iii) is satisfied.

Proof. Assume $d((p,q),(m-1,n-1)) \leq 2$. Then, since $d((p,q),(m,n)) > 1$ and (C2.i) holds, (p,q) must be one of the following eight squares: $(m-3,n)$, $(m-3,n+1)$, $(m-2,n)$, $(m-2,n+1)$, $(m,n-3)$, $(m,n-2)$, $(m+1,n-3)$, or $(m+1,n-2)$. If it is one of the first four squares (C2.ii) is not satisfied (remember that $m>1$); if it is one of the last four squares then (C2.iii) is not satisfied. Hence, we must have $d((p,q),(m-1,n-1)) > 2$.   □

Now we deal with the case $n = 1$; in Lemma 4 for $m > 2$, and in Lemma 5 for $m = 2$.

Lemma 4. Assume $n = 1$, $m > 2$, $d((p,q),(m,n)) > 1$ and $\max\{|p|,|q|\} \geq m$. If $d((p,q),(m,n)) \leq 2$ then $(p,q) \in \{(m,-2), (m,-1), (m+1,-2), (m+1,-1)\}$.

Proof. One looks at the 25 squares at distance ≤ 2 from $(m-1,n-1)$ and sees that only if (p,q) is $(m,-2)$, $(m,-1)$, $(m+1,-2)$, or $(m+1,-1)$ the hypotheses of the lemma are satisfied.   □

In a similar way one verifies the following lemma.

Lemma 5. Assume $(m,n) = (2,1)$ and $d((p,q),(m,n)) > 1$. Assume also that (C2.i) and (C2.ii) of Lemma 3 hold. If $d((p,q),(m-1,n-1)) \leq 2$ then $(p,q) \in \{(2,-2), (2,-1), (3,-2), (3,-1), (1,-2), (0,-2), (-1,-2)\}$.

Lemma 6. If $\max\{|p|,|q|\} \geq m > 1$, $|q| < m$, $|p-m| > 1$, $|p'-p| \leq 1$, and $m' = m-1$ then $|p'-m'| > 1$.

Proof. Since $|q| < m$, the inequality $\max\{|p|,|q|\} \geq m$ implies $|p| \geq m$. Then $p \notin \{m-3,m-2\}$ because any of the two inequalities $|m-3| \geq m$, $|m-2| \geq m$ implies that $m \leq 1$.

Since $|p-m| > 1$ we have $p \notin [m-1,m+1]$ and, in fact, $p \notin [m-3,m+1]$ because $p \notin \{m-3,m-2\}$. Hence $|p-m'| = |p-(m-1)| > 2$ and therefore $|p'-m'| > 1$.   □

Lemma 7. If $\max\{|p|,|q|\} \geq m \geq n \geq 1$, $|p| < m$, $|q-n| > 1$, $|q'-q| \leq 1$, and $n' = n-1$ then $|q'-n'| > 1$ or $(m,n) \in \{(2,1),(1,1)\}$.

Proof. Since $|p| < m$, the inequality $\max\{|p|,|q|\} \geq m$ implies $|q| \geq m$. If $q = n-3$ we have $|n-3| \geq m \geq n \geq 1$ which implies (m,n) is $(2,1)$ or $(1,1)$. If $q = n-2$ we have $|n-2| \geq m \geq n \geq 1$ which implies (m,n) is $(1,1)$. If $q \notin \{n-3,n-2\}$ then $q \notin [n-3,n+1]$ because $q \notin \{n-1,n+1\}$ due to the inequality $|q-n| > 1$; hence $|q-n'| = |q-(n-1)| > 2$ and, therefore, $|q'-n'| > 1$.   □

We now prove the "only if" part of Theorem 1, that is, the following proposition.

Proposition. Assume S is not on the edge of the board and the following three conditions are satisfied:

(C3.i) $d(K1,S) \geq d(K2,S)$
(C3.ii) $d_v(K1,S) \geq d_v(K2,S)$ or $d_h(K1,K2) > 1$
(C3.iii) $d_h(K1,S) \geq d_h(K2,S)$ or $d_v(K1,K2) > 1$.

Then S cannot be reached. (C3.i) is the negative of (C1.i), (C3.ii) is the negative of (C1.ii), and (C3.iii) is the negative of (C1.iii).

Proof. Choose a coordinate system such that S = (0,0), K2 = (m,n) with $0 \le n \le m$ and K2 = (p,q). Thus $d_h(K2,S) = m$, $d_v(K2,S) = n$, $d(K2,S) = m$ and the conditions (C3.i), (C3.ii), and (C3.iii) of the proposition become:

(C4.i) $\max\{|p|,|q|\} \ge m$
(C4.ii) $|q| \ge n$ or $|p-m| > 1$
(C4.iii) $|p| \ge m$ or $|q-n| > 1$.

The proof will be by induction on $d(K2,S)$, that is, on m.

$m = 1$: Clearly S cannot be reached by the first King since S is protected by K2. Notice that once the second King is at distance 1 from S, it cannot be forced to move to a square at distance 2 from S.

$m > 1$: The rest of the proof will be divided into three cases: Case I: $n > 1$, Case II: $n = 1$, and Case III: $n = 0$. In all three cases the second King will be able to move to (m',n') with $m' = m-1$, decreasing its distance to S. Hence, by induction, it is sufficient, in order to complete the proof of the proposition, to show that, after a move of K1 to K1' = (p',q') (with $|p'-p| \le 1$ and $|q'-q| \le 1$) and a move of K2 to K2' = (m',n') = $(m-1,n')$, the conditions (C4.i), (C4.ii), and (C4.iii) are satisfied with p, q, m, and n replaced respectively by p', q', m', and n'. Accordingly, it is sufficient to prove the following three conditions:

(C5.i) $\max\{|p'|,|q'|\} \ge m'$
(C5.ii) $|q'| \ge n'$ or $|p'-m'| > 1$
(C5.iii) $|p'| \ge m'$ or $|q'-n'| > 1$.

knowing that (C4.i), (C4.ii) and (C4.iii) hold. Notice that when we change p into p' and q into q', $|p|$ decreases by 1, stays the same, or increases by 1. The same holds for for $|q|$ and $\max\{|p|,|q|\}$. In particular, we have $\max\{|p'|,|q'|\} \ge \max\{|p|,|q|\}-1 \ge m-1 = m'$ and so (C5.i) holds. Thus we only have to prove (C5.ii) and (C5.iii) in all the cases.

Case I. $n > 1$. After a move from K1 to K1' = (p',q'), with $|p'-p| \le 1$ and $|q'-q| \le 1$, K2 moves to K2' = (m',n') where $m' = m-1$ and $n' = n-1$. This is possible by Lemma 3. If $|q| \ge n$ then $|q'| \ge |q|-1 \ge n-1 = n'$ so (C5.ii) holds. If $|q| < n$ then we must have $|p-m| > 1$ and, by Lemma 6, $|p'-m'| > 1$ so, again (C5.ii) holds.

Similarly, if $|p| \ge m$ then $|p'| \ge |p|-1 \ge m-1 = m'$ so (C5.iii) holds; if $|p| < m$ then we must have $|q-n| > 1$ and, by Lemma 7, $|q'-n'| > 1$ so, again (C5.iii) holds. Consequently, if $n > 1$, both (C5.ii) and (C5.iii) are valid.

Case II. $n = 1$. If $m > 2$ (respectively $m = 2$) then, by Lemma 4 (respectively Lemma 5) either $d((p,q), (m-1,n-1)) > 2$ or $(p,q) \in \{(m,-2), (m,-1), (m+1,-2), (m+1,-1)\}$ (respectively $(p,q) \in \{((2,-2), (2,-1), (3,-2), (3,-1), (1,-2), (0,-2), (-1,-2)\}$).

If $d((p,q),(m-1,n-1)) > 2$ or if the first King moves from (p,q) to a square (p',q') different from $(m,-1),(m-1,-1)$, and $(m-2,-1)$ then the second King can move to K2' = (m',n') = $(m-1,0)$ and, by the same reasoning of Case I, (C5.ii) and (C5.iii) are valid.

If $d((p,q),(m-1,n-1)) \le 2$ and the first King moves to (p',q') = $(m,-1)$ or to (p',q') = $(m-1,-1)$ then move K2 to K2' = (m',n') = $(m-1,1)$ and one has $|q'| = 1 = n'$ and $|p' \geq m-1 = m'$ so (C5.ii) and (C5.iii) hold. Now, if $m>2$, K1 cannot move to $(m-2,-1)$ because p is m or $m+1$. If $m = 2$ and K1 moves to K1' = (p',q') = $(m-2,-1)$ then, again, move K2 to K2' = $(m'.n') = (m-1,1)$ and one has $|q'| = 1 = n'$ and $|q'-n'| = 2 > 1$ so (C5.ii) and (C5.iii) hold.

Case III. $n = 0$. If $m = 2$, then $\max\{|p|,|q|\} \geq m = 2$ so the first King will move to a square $K1' = (p',q')$ with $d(K1',S) \geq 1$. Then the second King can move to a square $K2'$ $= (1,n')$ with $|n'| \leq 1$ because S is not on the edge of the board (this is the only place in the proof where we use this hypothesis) and one can see that (C5.ii) and (C5.iii) hold. Assume finally, $m > 2$, then $\max\{|p|,|q|\} \geq m \geq 3$ so $|q| \geq m \geq 3$ or $|p| \geq m \geq 3$. If $|q| \geq m$ then $d(K1,(m-1,0)) = |q| \geq m > 2$; if $|p| \geq m$ then K1 will not be able to move to a square at distance 1 from $(m-1,0)$. Hence, no matter to which square $K1' = (p',q')$ the first King moves, the second King will be able to move to $K2' = (m',n') = (m-1,0)$. We have $|q'| \geq 0 = n'$ so (C5.ii) holds. Also, if $|p| \geq m$ we have $|p'| \geq |p|-1 \geq m-1 = m'$, and if $|q| \geq m$ we have $|q'| \geq |q|-1 \geq m-1 > 1$ so (C5.ii) holds. This completes the proof of the proposition and of Theorem 1.   □

Remark. The proof shows not only necessary and sufficient conditions ((C1.i), (C1.ii), and (C1.iii)) for S to be reachable but it also indicates how to reach S if they are satisfied and, for the second king, how to prevent K1 from reaching S when the three conditions (C1.i), (C1.ii), and (C1.iii) fail.

If (C1.i), (C1.ii), or (C1.iii) is satisfied then K1 should move diagonally towards S, reducing its horizontal and vertical distance from S until it arrives at the row or column of S, and then proceed towards S in that row or column. It will always be possible (under any of the conditions i, ii, iii) to make the corresponding moves. If (C3.i) (the negative of (C1.i)), (C3.ii) (the negative of (C1.ii)), and (C3.iii) (the negative of (C3.iii)) hold then K2 should also move diagonally towards S reducing its horizontal and vertical distance from S (this is possible, for example, if the horizontal and vertical distances from the second King to S are greater than 1). If such a move is not possible then reduce the horizontal (respectively vertical) distance from S keeping the same vertical (respectively horizontal) distance from S, if $d_h(K2,S) > d_v(K2,S)$ (respectively $d_v(K2,S) > d_h(K2,S)$). This will always be possible, except in the situation K2 = (2,0), K1 = (0,1), S = (0,0) (with respect to some coordinate system); in this situation move K2 to (1,−1) : this is possible if S is not on the edge of the board.

What happens if S is on the edge of the board? Assume S=(0,0), row −1 does not exist but column −1 exists (that is, S is on the edge of the board but not at a corner). Let S1 = (0,1). Then S is reachable if and only if:

> (C6.i) $d(K1,S1) < d(K2,S1)$, or
> (C6.ii) $d_v(K1,S) < d_v(K2,S)$ and $d_h(K1,K2) \leq 1$, or
> (C6.iii) $d_h(K1,S) < d_h(K2,S)$ and $d_v(K1,K2) \leq 1$.

Assume row −1 and column −1 do not exist and S = (0,0) (that is, S is at a corner). Let S1 = (0,1) and S2 = (1,0).Then S is reachable if and only if:

> (C7.i1) $d(K1,S1) < d(K2,S1)$, or
> (C7.i2) $d(K1,S2) < d(K2,S2)$, or
> (C7.ii) $d_v(K1,S) < d_v(K2,S)$ and $d_h(K1,K2) \leq 1$, or
> (C7.iii) $d_h(K1,S) < d_h(K2,S)$ and $d_v(K1,K2) \leq 1$.

4 Discussion and Conclusions

Racing situations occur in Chess and other games. A very simple one is the pawn-vs.-king race: the board has only two pieces, the white Pawn at P heading towards its promotion square S and the black King K trying to capture the Pawn; White promotes its Pawn safely, that is, without being captured if and only if $d(P,S) < d(K,S)$ ($d(P,S)-1 < d(K,S)$ if P is on the second rank). This is equivalent to the "square" rule which, of course, is discussed in many books on endings.

Less simple is the king-vs.-king race considered in the present article: the board has only two pieces, the white King at K1 and the black King at K2; an unoccupied square S is also specified. The problem is to characterize those triples (S,K1,K2) such that the white King, which moves first, can reach S. So far, I have not seen this problem discussed in the literature in a complete way.

One of my motivations for studying this problem is that many positions in the KRKa2 ending [3] reduce to king-vs.-king races (for example the 2,324 positions mentioned in the Section 1). This must be true for many more endgame positions.

Comparing the pawn-vs.-king race, in an instance (S,K1,K2) of the king-vs.-king race, the condition $d(K1,S) < d(K2,S)$ is sufficient for S to be reached by the white King (this is the content of Lemma 1) but not necessary, as Diagram 2 shows.

One might try to replace "king" distance by a different distance D such that S can be reached if and only if $D(K1,S) < D(K2,S)$. However, it is impossible to find such D because considering (S,K1,K2) = ((0,0),(3,0),(2,2)) one obtains $D((3,0),(0,0)) < D((2,2),(0,0))$ and considering (S,K1,K2) = ((0,0),(2,2),(3,0)) gives $D((2,2),(0,0)) < D((3,0),(0,0))$. Nevertheless there might be two distances D_1 and D_2 such that S can be reached if and only if $D_1(K1,S) < D_2(K2,S)$.

What is done in the present paper is to consider the horizontal and vertical distances $d_h(K1,S)$, $d_v(K1,S)$, $d_h(K2,S)$, $d_v(K2,S)$, $d_h(K1,K2)$, $d_v(K1,K2)$, and, because of Theorem 1 and the discussion at the end of the previous section for the cases where S is on the edge, these six numbers determine whether or not S can be reached (recall $d(X,Y) = \max\{d_h(X,Y),d_v(X,Y)\}$). Thus a concise characterization is obtained of all the triples (S,K1,K2) such that S can be reached. In addition, following the last remark in the previous section, an algorithm is described to play optimally (both sides).

In conclusion, we hope that, besides being relevant in the KRKa2 ending, Theorem 1 would be useful in the problem of characterizing (rather than building a database) the endings king-and-pawn vs. king and the endings king-and-pawn vs. king-and-pawn which are won by White. Here, of course, besides the racing ingredient appears the theme of opposition.

References

1. M.E. Bain. *Learning Logical Exceptions in Chess*. PhD Thesis, Department of Statistics and Modelling Science, University of Strathclyde, Scotland, 1994.
2. A. De Groot and F. Gobet. *Perception and Memory in Chess. Heuristics of the Professional Eye*. Van Gorcum, Assen, The Netherlands. (With R.W. Jongman), 1996.
3. A. González Romero and R. Alquézar. Learning Through the KRKa2 Chess Ending. 9th *Iberoamerican Congress on Pattern Recognition, CIARP 2004*, pages 208-215, 2004.

4. R. Khardon. Learning Action Strategies for Planning Domains. *Artificial Intelligence*, 113(1–2):125–148, 1999.
5. M. Martin and H. Geffner. Learning Generalized Policies in Planning Using Concept Languages. *Proc. 7th Int. Conf. on Knowledge Representation and Reasoning (KR 2000)*, Morgan Kaufmann, 2000.
6. A.H. Marroquín. *Chess Maya Font*. Http://www.enpassant.dk/chess/fonteng.htm, 1998.
7. R. Rivest. Learning Decision Lists. *Machine Learning*, 2(3):229–246, 1987.
8. A.D. Shapiro and T. Niblett. Automatic Induction of Classification Rules for a Chess Endgame. In *Advances in Computer Chess 3* (ed. M. R. B. Clarke). Pergamon, Oxford, pages 73–92, 1983.
9. A.D. Shapiro and D. Michie. A Self Commenting Facility for Inductively Synthesized Endgame Expertise. In *Advances in Computer Chess 4* (ed. D. F. Beal), Pergamon, Oxford, pages 147–165, 1986.
10. A.D. Shapiro. *Structured Induction in Expert Systems*. Turing Institute Press, Addison-Wesley. 1987.

The Graph-History Interaction Problem in Chinese Chess

Kuang-che Wu[1,*], Shun-Chin Hsu[2], and Tsan-sheng Hsu[3,*,**]

[1] Department of Computer Science and Information Engineering,
National Taiwan University, Taipei, Taiwan
kcwu@csie.org
[2] Department of Information Management,
Chang Jung Christian University, Tainan, Taiwan
schsu@mail.cju.edu.tw
[3] Institute of Information Science,
Academia Sinica, Taipei, Taiwan
tshsu@iis.sinica.edu.tw

Abstract. Chinese-chess rules for cyclic moves differ from Western-chess rules in two respects. First the outcome of a cyclic game can be a win, a loss, or a draw. Second, depending on the plies made inside a loop, there are up to 16 rules a player can violate when a loop occurs. However, the same rule has to be violated three times in a row, i.e., in three consecutive loops, in order to lose a game. Therefore, a player can violate different rules in three cycles and still achieve a draw. In contrast, Western-chess rules always define a game as a draw after three consecutive loops. This paper reports on an adequate implementation of the Chinese-chess rules used to decide the outcome of a game when it falls into loops. The rules are proposed by the Asia Chinese-Chess Association.

1   Introduction

It is important that strong Chinese-chess computer programs also have a good knowledge of the game's rules. Sometimes a player can win a game by forcing an opponent to violate the existing rules, instead of using chess tactics or strategies to capture the opponent King. However, it is not easy to apply the full set of Chinese-chess rules correctly without substantial performance degradation. There are two problems with regard to incorporating the rules into a game tree search algorithm. First, how to apply Chinese-chess rules accurately? Second, how to apply the complicated rules with a small (reasonable) performance degradation?

1.1   Chinese-Chess Rules

In addition to the basic rules of Chinese Chess, such as how to move a piece and how to capture an opponent's piece, and the main goal of the game, there are

---

* Supported in part by National Science Council (Taiwan) Grants 91-2213-E-001-027, 92-2213-E-001-005, and 93-2213-E-001-001.
** Corresponding author.

H.J. van den Herik et al. (Eds.): ACG11, LNCS 4250, pp. 165–179, 2006.

some rules for judging games that fall into cyclic positions. Since the additional rules are hard to deal with, many players avoid playing cyclic positions. This situation also applies to computer programs. They have often implemented only a very restricted subset of the rules. Obviously, knowledge of the full set of rules is essential for players who want to play perfectly.

The rule on repetition of positions in Western Chess and Chinese Chess is completely different. After three consecutive repeated sequences of a board position have occurred in Chinese Chess, the game can end in a draw, a loss, or a win, depending on whether the participants' moves repeatedly violate the rules in the three consecutive sequences. Many Western game researchers have discovered that cyclic positions may affect the normal search program and refer to it as a kind of *graph-history interaction* (GHI) problem [3]. We note that there are other kinds of GHI problems in Western Chess, for example the right of castling. The prevailing research question is: how to solve the GHI problem in cyclic games without sacrificing too much performance? [2,10,11].

In Chinese Chess, the only GHI problem is the cyclic game. Henceforth, unless otherwise stated, we refer to the GHI problem of cyclic games in Chinese Chess as the GHI problem. The problem is more difficult in Chinese Chess than in Western Chess for two reasons. The first reason is that there are complicated rules to decide the outcome of a game, which may be a win, a draw, or a loss depending on who has the advantage when the repeated pattern occurs. The intuition is that a player at a disadvantage will try to avoid losing the game by forcing an opponent to play repeated moves, so that the game results in a draw. The main strategy of such a player is to check the opponent, or threaten to capture the opponent's King in the next move. Because the King can only move inside a 3 by 3 square in Chinese Chess, it is easy to check this. Second, strategy is to threaten to capture a major unprotected piece belonging to the opponent on the next move. For brevity of notation, when a player threatens such a move, we say the player *chases* the opponent's piece. When a player threatens to capture the opponent's King, we say the player *checks* the opponent.

There are many versions of Chinese-chess rules, of which the most influential ones are the one used in the Mainland China [4] and the one published by the Asia Chinese-Chess Association [8]. The China version is complicated and tries to decide who has the advantage in many possible situations, which causes a number of inconsistencies [16]. The current version, published in 1999, is expected to be revised in the near future. Although the Asia version is also very complicated, it is much simpler than the China version and is considered more stable. The Asia version, which has not been changed substantially since 1989, is more straightforward because it is easier to tell if a game is a win or a loss.

In this paper, we use the implementation of the Asia rule set as described in [9]. According to this set, when a sequence of repeated board positions occurs, two flags are output, one for each player. Each flag indicates the rule violations made by the player in every ply he made in the repeated sequence. In general, there are two levels of rule violation. The more serious level of violation is to check the opponent's King at every ply. The other level, which contains 15 rules,

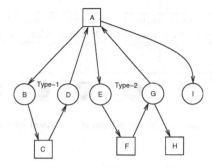

Fig. 1. Two cycles in which Red violates different rules

is to chase the opponent's non-King piece at every ply. If both players violate the same level of rules, or neither one of them violates any rules, the game ends in a draw. If only one player violates the more serious rule, he loses.

The second reason that Chinese-chess rules are more difficult is that the outcome of a cyclic game can only be decided when the same rule violation is made after (i) a sequence is repeated three times consecutively and (ii) both players want to play further; that is, plus two additional plies. Thus, we need to look at a sequence of plies consisting of at most three repeated sequences and an additional two plies. The Asia version allows a player to repeat the same position three times and change his mind without any penalty. Thus, if one player violates different rules in three consecutive sequences of a board position, it is not considered a violation of the rules. In contrast, in Western Chess both players need to engage in a cyclic sequence of moves three times in a row to call the game a draw; the outcome of a cyclic game is always a draw. Hence, it can be decided after only one occurrence of a repeated sequence. It is known that a repeated sequence consists of a minimum of 4 plies. In general, we can predict the outcome of a game by searching at least 14 plies for rule violations.

For example, in Fig. 1, it is assumed that: (1) the square nodes are positions played by Red, and the round nodes are positions played by Black; (2) A-B-C-D-A is a cycle in which Red threatens to capture one black piece in the plies A-B and C-D; (3) A-E-F-G-A is a cycle where Red threatens to capture another black piece in the plies A-E and F-G; and (4) Black does not threaten to capture any red piece in either cycles. In this scenario, both cycles violate a normal rule, but the rules are different. We note that a player needs to violate the same rule three times in order to lose. Hence, the sequence of board positions A-B-C-D-A-E-F-G-A-B-C-D-A-B-C is judged to be a draw. The game ends when Black plays the last ply B-C.

1.2   Motivating Examples

Knowing Chinese-chess rules is vital when playing against grand masters, as shown by the following examples.

Names	King	Guard, Assistant	Minister, Elephant	Knight, Horse	Rook, Car	Cannon, Gunner	Pawn
Red	帥	仕	相	馬	車	炮	兵
Black	將	士	象	馬	車	炮	卒

Fig. 2. Legend of Chinese-chess pieces used in this paper

Example 1. *In Fig. 3, in the repeated sequence, neither player violates the rules. Hence, the game ends in a draw if neither player wants to make other moves. In this game, a Chinese-chess master [6] comments that all other plies end in bad positions for both players, so both players want to play repeated sequences. The legend of pieces used in this paper is shown in Fig. 2.*

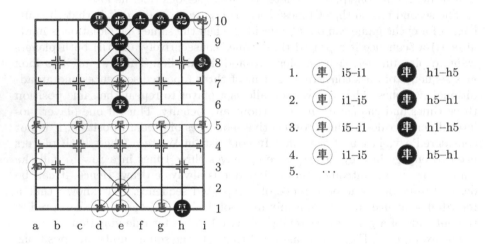

1.	車	i5–i1	車	h1–h5
2.	車	i1–i5	車	h5–h1
3.	車	i5–i1	車	h1–h5
4.	車	i1–i5	車	h5–h1
5.	...			

Fig. 3. Neither side violates the rules

Example 2. *In Fig. 4, in the repeated sequence, Red violates a normal rule by threatening an unprotected black Cannon in b9. Black, however, does not violate any rules, since it threatens the unprotected red Rook in f9 and the red Knight in g7 alternately. The game ends in a loss for Red, unless both sides play other moves. However, even if Red makes other plies, all other moves eventually lead to a loss of materials.*

Using non-basic rules to decide the outcome of a game happens in real games between computer-chess programs. Fig. 5 shows a game played between the bronze medal winner CONTEMPLATION (Red) and the Champion XIEXIE (Black) at the first World Computer Chinese-Chess Championship [17]. In the repeated sequence, Black threatens the red Knight on g3. However, since the red Knight is

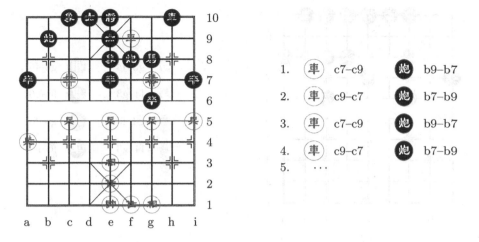

Fig. 4. Red violates a normal rule, but Black does not violate any rules

protected on g3 and unprotected on f5, Black does not violate any rules. Meanwhile, Red does not violate any rules, because it does not chase Black's unprotected piece. Thus, the game ends in a draw. A grand master has commented that all other moves for both sides would lead to bad positions [5]. Hence, both sides want to stay in the loop.

1.3 Previous Work

Although most human experts believe that mastery of the rules is essential to the playing skills of grand masters, there are very few publications on the subject.

The Chinese-chess computer program in [13] proposed some heuristics for dealing with repetition in general. Some partial treatments of Chinese rules are also proposed in the ELP program [18]. However, several problems remain unsolved, including the blocking of a Rook and a Cannon. We are not aware of any efficient full implementations of professional versions of Chinese-chess rules.

1.4 Structure of This Paper

We first give a detailed description of our Chinese-chess computer program CONTEMPLATION [15] and its current partial treatment of Chinese-chess rules in Sect. 2. We describe our approach in Sect. 3. We then describe our implementation, and the experimental results in Sect. 4. Finally, we provide concluding remarks in Sect. 5.

2   The Chinese-Chess Program CONTEMPLATION

In this section, we describe our Chinese-chess program CONTEMPLATION [15] on which we base our discussion. Many game-tree search algorithms are based

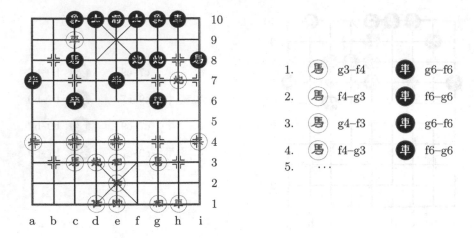

Fig. 5. CONTEMPLATION (Red) vs. XIEXIE (Black), at the first World Computer Chinese-Chess Championship, 2004

on the well-known $\alpha - \beta$ search algorithm. It performs a window search, like DFS, with two bounds, α and β, which are the lower and upper bounds of a minimax game tree respectively. The bounds are used to prune the search tree. Many researchers have enhanced the $\alpha - \beta$ search algorithm to further prune the search tree. (See [7,12] for some of these improvements.) However, using different window sizes and ranges on the root or on each node visited, move-ordering heuristics, iterative deepening, and caching calculated results into a transposition table are some of the improved search techniques not incorporated in modern Chinese-chess programs.

For reasons of rendability, we straightforward a simplified version of the search algorithm (see Algorithm 1)we currently use in our program [15]. In the next section, we present a revised version that deals with the GHI problem correctly.

Our search algorithm is performed recursively. The search score is relative to a player's turn — every player chooses the best move to maximize his[1] score. The terminal condition for searching occurs if one of the following is true.

1. The game's theoretical values, i.e., stalemate, checkmate, or game rule violations, have been reached.
2. There is a hit in the transposition table. Note that the transposition table is a look-up hash table in which the tree node is used as the hash key to retrieve searched results [19]. The results stored in the table are the search value; the trust level, i.e., the depth; and one flag indicating that the real node score is above, below, or equal to the stored value. Thus, we can reuse the search results stored in the table.
3. The depth limit is reached, in which case, we apply the evaluation function to this node.

---

[1] Whenever 'he or she' and 'his or her' are meant, we use 'he' and 'his'.

In the non-terminal condition, the search program enumerates all possible child nodes and performs the search recursively on each branch. It then picks the move that maximizes the score of the player using the node. After searching this sub-tree, the program stores the best score in the transposition table for later use.

Algorithm 1. Simplified search algorithm

VALIDRESULT(entry e, depth d, search bound (α, β))
1: {verify it is safe to use transposition table entry e}
2: **if** $d >$ depth(e) **then**
3:     **return false**
4: **else if** flag(e) = EXACT **then**
5:     **return true**
6: **else if** flag(e) = BELOW and value(e) $\leq \alpha$ **then**
7:     **return true**
8: **else if** flag(e) = ABOVE and value(e) $\geq \beta$ **then**
9:     **return true**

BASICSEARCH(position p, depth d, search bound (α, β))
1: **if** isEndInGameRuleApprox(p) **then**
2:     **return** GameRuleValue(p)
3: **else if** $d = 0$ **then**
4:     **return** EvaluationFunction(p)
5: **else if** validResult(probeTT(p), d, (α, β)) **then**
6:     **return** value(ProbeTT(p))
7: $r \leftarrow -\infty$
8: $ttflag \leftarrow$ BELOW
9: **for all** $m \in$ movegen(p) **do**
10:     $p' \leftarrow$ domove(p, m)
11:     $r' \leftarrow -$BASICSEARCH($p', d - 1, (-\beta, -\alpha)$)
12:     $r \leftarrow \max(r, r')$
13:     $ttflag \leftarrow$ EXACT
14:     **if** $r \geq \beta$ **then**
15:         $ttflag \leftarrow$ ABOVE
16:         **break**
17: storeTT($p, d, ttflag, r$)
18: **return** r

In our version of CONTEMPLATION, the game rules are dealt with heuristically by the function *isEndInGameRuleApprox* in Algorithm 1. When a cycle is encountered, we first examine whether we have checked the opponent continiously. We then examine whether the cycle matches a frequently encountered chasing pattern, e.g., our Rook threatens one of opponent's pieces. If we detect any of the above, we report the correct score that complies with the Chinese-chess rules. Otherwise, we return a very low value of -700, which, in our program, is slightly better than losing a Rook. Hence, our program prefers to stay in a cycle when all other plies lead to very bad scores, e.g., losing a Rook or worse, even if staying in the cycle is a disadvantage to us according to Chinese-chess rules. The heuristic value -700 may sometimes cause our program to act abnormally, in which case it may be better to repeat the cycle and end in a draw, or force the opponent to change his move.

3   Approach

Although some search results are stored in the transposition table, it is not always feasible to reuse those scores when the same position is encountered at a later stage. We only use the values in the transposition table if one of following two conditions holds.

(1) The paths are equal. Using the same concept as [11], we store the encoded path in the transposition table. If the path stored and the current path are the same, we can reuse the value in the transposition table.
(2) The value obtained by our program is not affected by the GHI problem.

First we define the notion *affected*. A subtree is affected by GHI if and only if there is a search path in the subtree that repeats the sequence of positions, but leads to a different search score. In Fig. 6 for instance, there are different paths p, p', and p'' that could lead to node n. Since there is a path p of subtree n in Figure 6a leading to a position in the history path p, the search score may differ depending on the repetition rules. We define this situation as *GHI affected* if the search result depends on the repetition score. Conversely, it is not affected if the search result is independent of the repetition score. There are no paths in Figure 6b leading to a position in the history p', so it is not affected. Though there is a path in Figure 6c leading to a position in the history p'', the score from the loop does not propagate; thus, it is not affected.

Since many subtrees are pruned using the α-β algorithm, some repetitions do not affect the search result. In such cases, we could still reuse the search results in another subtree.

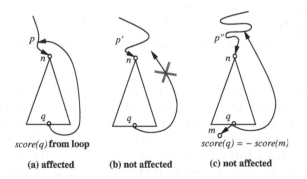

     (a) affected   (b) not affected   (c) not affected

Fig. 6. Illustrating how GHI affects the final evaluation scores

3.1   GHI Bounds

To determine whether a value returned by our search algorithm is affected by the GHI problem, i.e., cycles, we modify the algorithm as follows. Besides the original game-tree score, we add a *GHI-score bound* to the search score. This means that the low/high bound (b_l, b_h) of the score ranges indicates that the scores may be GHI affected.

In this way, the GHI-score bound is propagated along with the original search score. Because we always act from the maximizer aspect, we negate search scores v from children and exchange α and β. Similarly, the GHI-score bound is also exchanged and negated when propagated from children as follows.

$$(v, (b_l, b_h)) \leftarrow (-v', (-b_h', -b_l')) \,.$$

Depending on how the search score is obtained, the GHI-score bound is operated and maintained as follows. We distinguish five cases.

- **Case 1: the search score is a horizon score.** So, the value is returned from the terminal nodes by our evaluation function. We assume the horizon node does not have the GHI problem. But if it does, it will be considered in a next (deeper) search. For the moment, both bounds of the horizon score are the same as search score v.

$$\text{GHI bound} = (v, v) \,.$$

- **Case 2: the search score is obtained by Chinese-chess rules for cycles.** The score may be any value if the node is visited by different paths.

$$\text{GHI bound} = (-\infty, \infty) \,.$$

- **Case 3: the score is obtained from the transposition table.** If the stored path and the current path are equal, the bound can be fetched from the table directly and reused. Otherwise, the bound of the node is the same as the node's value, even if the GHI problem affected this node in advance. Since the GHI-bound heuristic does not help in the latter case, both the transposition score and bound may be incorrect.

$$\text{GHI bound} = \begin{cases} T[p].\text{bound} & \text{if paths equal,} \\ (-\infty, \infty) & \text{otherwise.} \end{cases}$$

- **Case 4: the score is obtained by merging the bounds of the child nodes.** In a negamax algorithm, we always choose the child with the highest score. The merged bound is thus the maximum value of the child nodes.

$$(b_l, b_h) \leftarrow (\max(b_l, b_l'), \max(b_h, b_h')) \,.$$

- **Case 5: the score is obtained by a beta-cut.** The bound is merged with $(-\infty, \infty)$, because we do not know the bounds of the child nodes that are yet to be explored.

$$(b_l, b_h) \leftarrow (\max(b_l, -\infty), \max(b_h, \infty)) \,.$$

Algorithm 2. Simplified search algorithm for dealing with GHI

VALIDRESULT(entry e, depth d, search bound (α, β), hash $path$)
1: **if** $d >$ depth(e) or (path(e) $\neq hash$ and GHIrelated(e)) **then**
2:    **return false**
3: **else if** flag(e) = EXACT **then**
4:    **return true**
5: **else if** flag(e) = BELOW and value(e) $\leq \alpha$ **then**
6:    **return true**
7: **else if** flag(e) = ABOVE and value(e) $\geq \beta$ **then**
8:    **return true**

GHISEARCH(position p, depth d, search bound (α, β))
1: **if** isEndInRule(p) **then**
2:    **return** $(RuleValue(\text{p}), (-\infty, \infty))$ {case 2}
3: **else if** $d = 0$ **then**
4:    $v \leftarrow$ EvaluationFunction(p)
5:    **return** $(v, (v, v))$ {case 1}
6: **else**
7:    $e \leftarrow$ probeTT(p) {case 3}
8:    **if** VALIDRESULT($e, d, (\alpha, \beta)$, encodepath()) **then**
9:      **if not** GHIrelated(e) **then**
10:       **return** (value(e), (value(e), value(e)))
11:      **else**
12:       **return** (value(e), (lbound(e), rbound(e))
13:    $(r, b_l, b_h) \leftarrow (-\infty, -\infty, -\infty)$
14:    $ttflag \leftarrow$ BELOW
15:    **for all** $m \in$ movegen(p) **do**
16:      $p' \leftarrow$ domove(p, m)
17:      $(r', (b'_l, b'_h)) \leftarrow$ negate(GHISearch($p', d - 1, (-\beta, -\alpha)$))
18:      $r \leftarrow \max(r, r')$;
19:      $b_l \leftarrow \max(b_l, b'_l)$; $b_h \leftarrow \max(b_h, b'_h)$ {case 4}
20:      $ttflag \leftarrow$ EXACT
21:      **if** $r \geq \beta$ **then**
22:       $ttflag \leftarrow$ ABOVE
23:       $b_l \leftarrow \max(b_l, -\infty)$; $b_h \leftarrow \max(b_h, \infty)$ {case 5}
24:       **break**
25:    storeTT($p, d, ttflag, r$, encodepath(), b_l, b_h)
26: **return** (r, b_l, b_h)

Algorithm 2 is the revised algorithm.

It is only safe to use the value from the transposition table if the path recorded and the current path are equal, or the value satisfies one of the following two cases: (1) If the search result a is equal to the upper bound, there is a child with a value equal to the search result a; thus, we can always use this value. (2) If the value is either above beta or below alpha, then the bound is outside the (α, β) range. In such a case, the value is not GHI-related and can be reused.

3.2   Discussion

Using this heuristic reduces the probability that the search result will be affected by GHI, but it does not eliminate the problem totally. Because repetition may be judged as win/draw/loss in Chinese Chess, we classify the GHI problem as cycle-first and cycle-last, instead of draw-first and draw-last as is done in [3]. Assume there are two lines leading to the current position; one forms a cycle, the other does not.

- **Case 1: cycle-first.** The line that forms a cycle is searched first. It is terminated according to the rules, and the search result is stored in the transposition table. The other line that does not repeat a previous sequence of positions will be searched later. In this case, our heuristic will detect that the transposition node may be GHI affected. Thus, the stored value will not be used, and the GHI problem is successfully handled.
- **Case 2: cycle-last.** In contrast to the previous case, the line with repetition will be searched later. The transposition node is not affected by the GHI problem, and we do not know whether this line will form a cycle when it encounters the transposition node. We will only know when we complete the search of this subtree. In this case, we will reuse the value from the transposition table, even though it may be incorrect.

4   Experiments

First, we describe the major components in our experiments. The test platform is our Chinese-chess program, CONTEMPLATION, which is a negascout search program with state-of-the-art Chinese-chess expert knowledge embedded. IG-NOREGHI is the original program with forward pruning disabled for comparison. WITHOUTTT is an implementation that does not use the transposition table. SAMEPATH is an implementation that only uses the transposition table when the current search path is identical to the path stored in the table. GHIBOUND is the algorithm that implements the GHI bounds described in this paper. WITH-OUTTT, SAMEPATH, and GHIBOUND return the same ply if the depth of the search is fixed. IGNOREGHI may pick some different plies, since it does not compute correctly on cycles during searching.

The experiments were run on AMD64 2.4G with a transposition table comprised of 2^{24} entries. In the first test suite, ASIARULE, 107 positions were chosen from examples illustrating the Asia Chinese-chess rules [8]. In order to make the right move in these examples, one needs to deal with the GHI problem correctly. In the second test suite, NETGAME, 150 positions were randomly chosen from 50 real games played by CONTEMPLATION on the Internet with 3 positions per game. The positions in the opening phase or with a material value difference larger than one Rook were ignored. For both tests, we searched 7 to 10 plies deep, and then compared the node count and time used. In the tables, entries with "-" took too much time to finish.

4.1   Performance on Positions in ASIARULE

Table 1 shows the experimental results of the positions in the test suite ASIARULE. All four programs selected the same move in each of the 107 test positions, except for 18 positions in which IGNOREGHI and GHIBOUND selected a different move. We note that each position was tested 4 times using four different search depths ranging from 7 to 10 plies. WITHOUTTT obviously searched many more nodes and spent up to 300% more time than IGNOREGHI on a 9-ply search, which is unacceptable. SAMEPATH spent up to 188% more time than IGNOREGHI, which is also unacceptable. GHIBOUND searched slightly more nodes than IGNOREGHI, given a fixed search depth. However, the average time spent on a node was larger for GHIBOUND than for IGNOREGHI. At the same search depth, GHIBOUND spent 7% more time than IGNOREGHI for a 10-ply search.

Table 1. Comparison of the node count (in millions) and time spent (in seconds) searching in the test suite ASIARULE for various fixed search depths. "61 ; 14" means 61 million nodes were searched in 14 seconds for all the positions in ASIARULE.

Method used	Depth			
	7	8	9	10
IGNOREGHI	61 ; 14	256 ; 65	961 ; 227	3,476 ; 904
WITHOUTTT	276 ; 67	1,424 ; 371	7,636 ; 1,934	–
SAMEPATH	125 ; 26	628 ; 146	3,187 ; 655	–
GHIBOUND	62 ; 15	259 ; 69	1,004 ; 249	3,576 ; 968

4.2   Performance on Positions in NETGAME

In Table 2, we report the performance results of testing NETGAME, i.e., 150 chosen positions in a real-world game previously played by our program. All four programs selected the same move in each position, except for 19 positions in which IGNOREGHI and GHIBOUND selected a different move. We note that each position was tested 4 times using 4 different search depths ranging from 7 to 10 plies. As in the ASIARULE test, WITHOUTTT and SAMEPATH searched more nodes and spent much more time than IGNOREGHI; this is unacceptable. GHIBOUND searched almost the same number of nodes in both methods given a

Table 2. Comparison of the node count (in millions) and time spent (in seconds) during the search of the test suite NETGAME at various fixed search depths. "254 ; 59" means 254 million nodes were searched in 59 seconds for all the positions in NETGAME.

Method used	Depth			
	7	8	9	10
IGNOREGHI	254 ; 59	1,097 ; 280	4,514 ; 1,095	20,368 ; 5,299
WITHOUTTT	1,038 ; 261	5,817 ; 1,525	32,076 ; 8,072	–
SAMEPATH	430 ; 93	2,255 ; 524	11,830 ; 2,491	–
GHIBOUND	255 ; 61	1,099 ; 289	4,542 ; 1,142	20,410 ; 5,486

fixed search depth. At the same search depth, GHIBOUND spent only 3.5% more time than IGNOREGHI. In comparison, GHIBOUND spent 7% more time than IGNOREGHI on ASIARULE. Since ASIARULE contains more positions in which loops may be encountered, GHIBOUND spent more time on loops.

5 Concluding Remarks

In this paper, we presented an algorithm that deals with most of the GHI problems encountered in Chinese Chess with acceptable performance degradation. On average, we used 3.5% more search time than our original version, but we improved the accuracy substantially. We believe that the extra time was well worth using it. We could only solve some positions correctly after incorporating GHI. One such an example is shown in Fig. 7, where Red moves next. In this game, if the red Cannon can move out of Black's palace, then it is well-known that Red can win the game. Hence, Black uses his King to block the red Cannon, which does not violate the rules. Our original program, IGNOREGHI, did not want to play cyclic positions and therefore lost the game. In our future work, we will use the revised program to determine if it improves the playing skills of our program significantly by testing it against top programs and by self-playing.

We note that our revised program can determine correctly whether a certain type of rule is violated by a player in a loop. However, the GHI bound used in our implementation is a heuristic that deals with the GHI problem. The GHI bound may be too restrictive in that it disallows the use of entries in the transposition table that could actually be used. The GHI bound may also cause

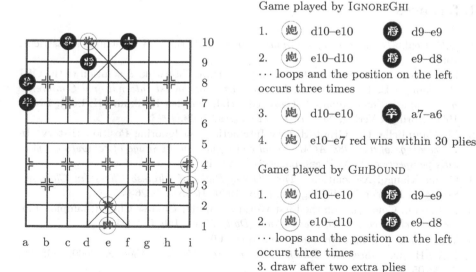

Game played by IGNOREGHI

1. (炮) d10–e10   (将) d9–e9

2. (炮) e10–d10   (将) e9–d8
··· loops and the position on the left occurs three times

3. (炮) d10–e10   (卒) a7–a6

4. (炮) e10–e7 red wins within 30 plies

Game played by GHIBOUND

1. (炮) d10–e10   (将) d9–e9

2. (炮) e10–d10   (将) e9–d8
··· loops and the position on the left occurs three times
3. draw after two extra plies

Fig. 7. Example game (our program plays Black)

other problems in very rare cases. However, GHIBOUND makes better selections than IGNOREGHI.

We discussed some of the problems in Subsect. 3.2; further details are given in [14]. Due to space limitations, we cannot cover all of them here. From the experiments given in this article, we may state that we have solved many of the cases that could not be solved previously. In further investigations we hope to assess whether these rare cases affect the strength of our program, and, if they do, whether we can find efficient implementations to handle these cases adequately.

One other possible avenue of future work would be to compare our results with other approaches. Our method introduces a so-called GHI bound into minimax search to determine whether the current best value could be affected by the GHI problem. The idea of using some kind of bound to guide searching is not new, see [1].

Acknowledgements

We wish to thank Haw-ren Fang, who introduced the problem in Chinese-chess rules to the third author; members of the IIS Computer Chinese-Chess Research Club; Chinese-chess masters Han-Sheng Huang and Wu-Chang Guo, for inspiring discussions; and Ming-Cheng Cheng, the author of the Chinese-chess computer program SHIGA, for his help. Most nontrivial games in this paper were designed and provided by Wu-Chang Guo. We are also grateful to Chia-Hsing Yu for providing a Chinese-chess prototype program in the C language.

References

1. D.F. Beal. *The Nature of Minimax Search*. PhD thesis, Universiteit Maastricht, 1999.
2. D.M. Breuker, H.J. van dan Herik, and J.W.H.M. Uiterwijk. A solution to the GHI Problem for Best-First Search. In *Proceedings of the 1st International Conference on Computers and Games (CG98)* (eds. H.J. van den Herik and H. Iida), LNCS 1558, Springer-Verlag, New York, NY, pages 25–49. 1998.
3. M. Campbell. The Graph-History Interaction: on Ignoring Position History. In *Proceedings of the 1985 ACM annual conference on the range of computing : mid-80's perspective*, ACM Press, pages 278–280, 1985.
4. China Xiangqi Association. *The Playing Rules of Chinese Chess (in Chinese)*. Shanghai Lexicon Publishing Company, 1999. IBSN: 7-5326-0556-6.
5. F.-L. Chu. Comments on the First World Computer Chinese Chess Championship. *Chinese chess column in Merit Times Daily News*, July 5, 2004.
6. W.-C. Guo. *private communication*. April 2005.
7. E.A. Heinz. *Scalable Search in Computer Chess*. Vieweg, 2000. ISBN: 3-528-05732-7.
8. Hong Kong Chinese Chess Association. Asia Xiang Qi rules.http://www.clubxiangqi.com/rules/asiarule.htm, 1989. English Translation by Eric Wu.

9. IIS Computer Chinese Chess Research Club. Private communication — an algorithmic definition of Asia Chinese chess rules (in Chinese). Institute of Information Science, Academia Sinica, Taiwan; 5 pages; unpublished manuscript, March 2004.

10. A. Kishimoto and M. Müller. A Solution to the GHI Problem for Depth-First Proof-Number Search. In *Proceedings of 2003 Joint Conference on Information Sciences*, pages 489–492, 2003.

11. A. Kishimoto and M. Müller. A General Solution to the Graph History Interaction Problem. In *Proceedings of Nineteenth National Conference on Artificial Intelligence*, pages 644–649, 2004.

12. A. Reinefeld. An Improvement of the Scout Tree Search Algorithm. *ICCA Journal*, 6(4):4–14, 1983.

13. K.-M. Tsao, H. Li, and S.-C. Hsu. Design and Implementation of a Chinese Chess Program. In *Heuristic Programming in Artificial Intelligence. The Second Computer Olympiad* (eds. D.N.L. Levy and D.F. Beal), pages 108–118. Ellis Horwood Ltd., Chichester, UK, 1991. ISBN: 0-13-382615-5.

14. K.-c. Wu. *Graph History Interaction Problem in Computer Chinese Chess* (in Chinese). Master thesis, Graduate Institute of CSIE, National Taiwan University, Taipei, Taiwan, 2005.

15. K.-c. Wu, T.-s. Hsu, and S.-C. Hsu. Contemplation Wins Chinese-chess Tournament. *ICGA Journal*, 27(3):172–173, 2004.

16. S.-Y. Xu. *Xiangqi Qili Yu Daipan Jumian De Caijue (Rulings of Chinese Chess Games that are not Clearly Stated in the Current Rules)*. People's Athelete Pulishing Co., 2000. (In Chinese); ISBN: 7-5009-1925-5.

17. S.-J. Yen, J.-C. Chen, and S.-C. Hsu. The 2004 World Computer Chinese-Chess Championship. *ICGA Journal*, 27(3):186–188, 2004.

18. S.-J. Yen, J.-C. Chen, T.-N. Yang, and S.-C. Hsu. Computer Chinese Chess. *ICGA Journal*, 27(1):3–18, 2004.

19. A.L. Zobrist. A New Hashing Method with Applications for Game Playing. Technical Report 88, Department of Computer Science, University of Wisconsin, Madison, USA, 1970. Also in *ICCA journal*, 13(2):69–73, 1990.

A New Family of k-in-a-Row Games

I-Chen Wu and Dei-Yen Huang

Department of Computer Science and Information Engineering,
National Chiao Tung University, Hsinchu, Taiwan
{icwu, teyen}@csie.nctu.edu.tw

Abstract. This paper contains three contributions. First, it introduces a new family of k-in-a-row games, *Connect*(m,n,k,p,q). In Connect(m,n,k, p,q), two players alternately place p stones on an $m \times n$ board in each turn, except for the start when the first player places q stones at her[1] first move. The player who first obtains k consecutive stones of her own first wins. The traditional game five-in-a-row, also called Go-Moku, in the free style is Connect(15,15,5,1,1). For brevity, Connect(k,p,q) denotes the game Connect(∞,∞,k,p,q), played on infinite boards.

Second, this paper analyzes the characteristics of these games, especially for the fairness. In the analysis of fairness, we first exclude the ones which are apparently unfair or solved. Then, for the rest of games, we argue that $p = 2q$ is a necessary condition for fairness in the sense that one player always has q more stones than the other after making a move. Among these games, Connect(6,2,1) is most interesting to this paper and is named *Connect6*.

Third, this paper proposes a threat-based strategy to play Connect(k, p,q) games and implements a computer program for Connect6, based on the strategy. In addition, this paper also illustrates a new null-move search approach by solving Connect(6, 2, 3) where the first player wins. The result also hints that for Connect6 the second player usually should not place the initial two stones far away from the first stone played by the first player.

1 Introduction

Traditionally, the game *k-in-a-row* is defined as follows. Two players, say *Black* and *White*, alternately place one stone, black and white respectively, on one empty *intersection*, henceforth called a *square*, of an $m \times n$ board; and Black plays first. The one who first obtains k consecutive stones (horizontally, vertically or diagonally) of her own wins. Such games are also called mnk-games in [11]. A well-known and popular game is *five-in-a-row*, also called *Go-Moku*. Go-Moku in the free style [1,11] is a (15,15,5)-game. For combinatorial analysis, researchers [7,15] investigated the games allowing each player to place p stones for one move.

This paper introduces a new family of k-in-a-row games, *Connect*(m, n, k, p, q). In Connect(m, n, k, p, q), two players alternately place p stones on an $m \times n$ board

---

[1] In this contribution we use 'she' and 'her' whenever 'she or he' and 'her or his' are meant.

H.J. van den Herik et al. (Eds.): ACG11, LNCS 4250, pp. 180–194, 2006.

at each move except for the start when the first player places q stones as her first move. Again, the player who first obtains k consecutive stones of her own wins. Games in the family are called *Connect games*. Obviously, Go-Moku in the free style is Connect(15,15,5,1,1). For brevity, *Connect*(k, p, q) denotes the game Connect$(\infty, \infty, k, p, q)$, played on infinite boards.

For Connect games, the major difference from traditional k-in-a-row games is to have an extra parameter q, a key that significantly affects the fairness. The higher q, the higher chances Black has to win. In [11], Herik, Uiterwijk, and Van Rijswijck gave the following definition of fairness. "A game is considered a *fair* game if it is a draw and both players have a roughly equal probability on making a mistake." From this, we argue that $p = 2q$ is a necessary condition for fairness, in the sense that one player always has q more stones than the other after making a move. Among these games, Connect(6,2,1) is most interesting to this paper and is named *Connect6*. More about fairness are discussed in Section 2.

This paper is organized as follows. Section 2 discusses the issue of fairness. Section 3 describes other characteristics of these games. Section 4 proposes a threat-based strategy to play Connect(k, p, q) games and implements a computer program for Connect6, based on the strategy. Section 5 illustrates a new null-move method by solving Connect(6,2,3) where Black wins. This result also indicates that White usually should not place the initial two stones far away from Black's first stone. Section 6 concludes our work.

2   Fairness

This section is organized as follows. Subsection 2.1 reviews the fairness problem of Go-Moku. Subsection 2.2 reviews the Connect games solved so far. It is based on combinatorial analysis, and also solves some more Connect games. Subsection 2.3 points out some unfair Connect games based on empirical experiments. Subsection 2.4 discusses the fairness of Connect6.

2.1   Fairness of Go-Moku

Fairness has been a major issue for Go-Moku, even though it is a popular game. In the free style of rules (without any restriction on Black), it has been well known that the game favors Black. To reduce the unfairness, the Japanese Professional Renju Association [13] added some rules to restrict the play of Black for professional players. For example, Black is prohibited to play double three and double four (see the definitions in [1,3]). The game with these restrictions is called *Renju*. In fact, Renju still favors B, from the experiences of professionals. Theoretically, it was proved that Black wins in the free style [1,3], and that Black wins even under these restrictions [21].

The Renju International Federation (RIF) changed the rules [16] to make the game more fair by imposing some opening rules for the first five moves. Furthermore, RIF requested for a proposal [17] for better opening rules again in

2003. These facts indicate that it is still hard to define a fair rule for the game. We note that adding more rules also makes it harder to learn the game.

The fairness problem for Go-Moku or Renju also causes an important side effect, viz. that of reducing the board size. It was argued in [18] that a larger board increases Black's advantage which resulted in the standard board size of 15 × 15. However, as described in Section 3, a smaller board reduces the complexity of the game. Consequently, it then becomes easier to solve the game.

2.2 Solved Games

In addition to [1,3,21] mentioned above, many researchers were engaged in studying the fairness of k-in-a-row. White ties [23] when $k \geq 8$. Many solved mnk-games are listed in [11,20].

For simplicity of combinatorial analysis, many researchers [6,7,14,15] followed an asymmetric version of the rules, called *Maker-Breaker*, where White is not allowed to win. This is because either Black wins or White ties in Connect(k, p,p) as proved in [7,14]. In contrast to Maker-Breaker, the version of normal rules is called *Maker-Maker*. Let δ denote $k-p$. In the Maker-Breaker version, it was proved in [15] that White ties under a condition, roughly like $\delta = \Omega(\log_2 p)$ (cf. Theorem 1 of [15]), and Black wins under a condition, roughly like $\delta = O(\log_2 p/ \log_2 \log_2 p)$ (cf. Theorem 2 of [15]). The above result implies that in the Maker-Maker version White still ties under the condition of $\delta = \Omega(\log_2 p)$, but it does not imply that in the Maker-Maker version Black still wins in the case of $\delta = O(\log_2 p/ \log_2 \log_2 p)$. We can easily extend Pluhar's result [15] to the following Corollary.

Corollary 1. *For Connect(k,p,q), let k and p satisfy the condition defined in Theorem 1 of [15]. For all q, where $1 \leq q \leq p$, White ties.*

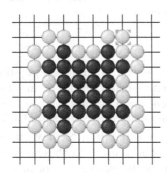

Fig. 1. Required defensive white stones when $q = 24$ and $\delta = 4$

Now, we want to investigate those Connect games that either Black or White wins. First, Black wins when $p < \lfloor q/\delta^2 \rfloor (4\delta + 4)$. For example, if Black places δ^2 stones on $\delta \times \delta$ squares as a group, White needs $4\delta + 4$ stones to defend the group. Thus, the above result is obtained when Black lets $\lfloor q/\delta^2 \rfloor$ groups be far away from one another.

If q is not a multiple of δ^2, we can possibly obtain tighter results. For example, for 4 × 4 squares, Black can add 8 additional stones as shown in Fig. 1 such that White needs one more white stone to defend for each additional black stone. Thus, if

(q mod δ^2) $\leq 8\lfloor q/\delta^2 \rfloor$, then Black wins when $p < \lfloor q/\delta^2 \rfloor(4\delta + 4) + (q$ mod δ^2). Otherwise, Black wins when $p < \lfloor q/\delta^2 \rfloor(4\delta + 4) + 8\lfloor q/\delta^2 \rfloor$. Thus, we obtain the following Corollary.

Corollary 2. *Let $\delta = k - p$. For Connect(k, p, q) game, Black wins when $p < \lfloor q/\delta^2 \rfloor(4\delta + 4) + \min(q$ mod $\delta^2, 8\lfloor q/\delta^2 \rfloor)$.*

For some cases, we can obtain some even tighter results, based on the above method. For example, for Connect(19,17,7), it does not satisfy the condition of Corollary 2, but Black still wins. In addition, it is also possible that White wins. For example, Connect(12,10,3). More results can be obtained based on the above principle and are not elaborated in this paper.

Since Connect games will become even more unfair when $q > p$, we usually assume $q \leq p$. Then, when $p < \lfloor q/\delta^2 \rfloor(4\delta + 4)$, we obtain $\delta < 2 + \sqrt{2}$ or $\delta < 4.82$. Thus, Corollary 2 becomes useless when $\delta \geq 5$ and $q \leq p$. For example, when $\delta = 5$ and $q = 25$, but White only needs $p=24$ to defend. An open problem is whether it can be proved that either Black or White wins when $\delta \geq 5$ and $q \leq p$.

2.3 Empirically Unfair Connect Games

This subsection makes some empirical experiments to investigate the fairness of some Connect games in the following two ways. First, we try to prove informally that either Black or White wins. Second, if we cannot prove, we try to see whether one player keeps obtaining initiatives leading to a win. If so, we call that the game *favors* that player.

Table 1. The empirical results for Connect games with $k \leq 9$ and $\delta = 3$

$q(\leq p)$	$k = 4$ $p = 1$	$k = 5$ $p = 2$	$k = 6$ $p = 3$	$k = 7$ $p = 4$	$k = 8$ $p = 5$	$k = 9$ $p = 6$
1	B	B	W	W	W	W
2		B	W	W	W	W
3			B	FB	FB	FW
4				B	B	FW
5					B	B
6						B

Our empirical experiments for $k \leq 9$ show that most games with $\delta \leq 3$ are unfair, as shown in Table 1 for $\delta = 3$. In this table, B (W) indicates that it is informally proved that Black (White) wins; and FB (FW) indicates that the game favors Black (White).

The game histories of these experiments are recorded in [12]. Among these game histories, the one for Connect(9,6,4) illustrated in Fig. 2 shows an interesting result. Even though Black has four stones initially, White still wins by

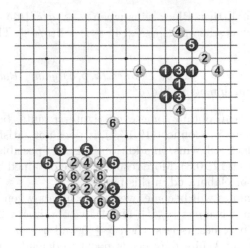

Fig. 2. Favoring White for Connect(9,6,4)

placing five of its six stones far away from Black and then keeping obtaining the initiative subsequently. The phenomenon of playing away from the major battle field is called *breakaway* in this paper. In Fig. 2, White makes an *initial breakaway*. If initial breakaways do not incur a penalty, the game may become unfair. For example, in Fig. 2, if Black plays in the lower left part subsequently, the game is like Connect(9,6,5) with White playing first.

2.4 Fairness of Connect6

In practice, based on our empirical experiments on Connect6 so far, we have not been able to identify which player the game favors. Following are two more arguments about its fairness.

(1) As described in Section 1, we argue that $p = 2q$ is a necessary condition for fairness, in the sense that one player always has q more stones than the other after making a move. Connect6 satisfies the condition.

(2) We argue that the following is a necessary condition for fairness: the initial breakaway does not apparently favor White. In Section 5, we prove that Black wins at Connect(6,2,3). This hints that the initial breakaway does not favor White for Connect6. We note that if Black does not win at Connect(6,2,3), White can place the initial two stones far away from the first black stone. Thus, if Black goes to defend White's two stones, the game is analogous to Connect(6,2,2) with White playing first, which favors White.

Surely, we expect to see more evidences, either fair or unfair, or more practical experiences in the near future.

3 Game Characterics

In this section, we investigate the characteristics of Connect6 and some other Connect games by following the definitions given in [11]. We list the characteristics of Connect6 below.

(1) The rules of Connect6 are simple to learn. Renju includes some prohibited moves and International Renju even includes some opening rules.
(2) Connect6 is potentially fair based on the arguments in Subsection 2.4.
(3) Connect6 is symmetric, if the first move by Black is not considered.
(4) Both state-space and game-tree complexities for Connect6 are very high as described below.

Connect6 has an infinite board, so both state-space and game-tree complexities are infinite too. In order to make it countable, we use a Go board for Connect6, instead, that is, Connect(19,19,6,2,1). Both state-space and game-tree complexities for it are still much higher than those in Go-Moku and Renju, in the sense that two stones per move make the branch factor increase by a factor of half of the board size. Based on the standard used in [11], the state-space complexity of Connect(19,19,6,2,1) is 10^{172}, the same as that of Go. If a larger board is used, the complexity is much higher.

Now, let us investigate the game-tree complexity. For Connect(19,19,6,2,1), assume that the averaged game length is still 30, the same as the estimation for Go-Moku [1]. The number of squares is about 300, and the number of choices for one move is about $(300 \times 300/2)$. Thus, the game-tree complexity is about $(300 \times 300/2)^{30} \sim 10^{140}$, much higher than that for Go-Moku. Also, if a larger board is used, this complexity is much higher.

The game-tree complexity grows much larger, as the value p increases. For example, consider Connect(8,4,2) assuming that it is a fair game. Based on the above calculation with a 19×19 Go board in use, the game-tree complexity grows up to 10^{260}. In fact, in our empirical experiments, a higher value p usually requires a larger board size, that makes the state-space complexity even higher.

4 Threat-Based Strategy

For Connect games, the threat-based strategy is a common strategy used to play. Subsection 4.1 describes the threats for Connect6, while Subsection 4.2 generalizes the threats for all Connect games. Subsection 4.3 briefly describes our Connect6 program.

4.1 Threats for Connect6

Definition 1. *For Connect6, assume that one player, say White, cannot connect six. Black is said to have t threats, if and only if White needs to place t stones to prevent Black from winning in Black's next move.*

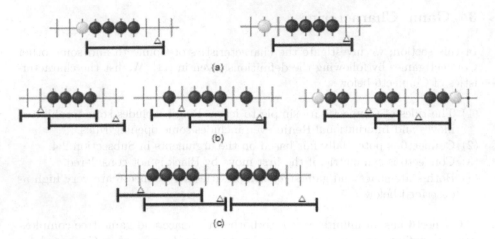

Fig. 3. Threat patterns for Connect6: (a) one threat, (b) two threats, and (c) three threats

For Connect6, we define threats in Definition 1. In Figs. 3(a), 3(b), and 3(c), Black has one, two, and three threats, respectively. In the case of three threats, Black wins since White needs three stones to defend but only has two stones for a move. Thus, the winning strategy of a player is to have at least three threats.

Now, let us investigate how to count the number of threats in one single line for simplicity. For example, at the right side of Fig. 3(a), Black has one threat (not two), because White only needs to place one stone at the square above "△". The algorithm to count the number of threats in one line is as follows.

1. For a line, slide a window of size six from the left to right.
2. Repeat the following step for each sliding window.
3. If the sliding window contains neither white stones nor marked squares and at least four black stones, add one more threat and mark all the empty squares in the window. Note that in fact we only need to mark the rightmost empty square. The window satisfying the condition is called a *threat window*.

In Fig. 3, the squares above "△" indicate the marked squares, if we only mark the rightmost empty one. Lemma 1 (below) shows that the algorithm is correct.

Lemma 1. *For Connect6, the above algorithm counts the number of threats correctly.*

Proof. First, any two threat windows found by the above algorithm do not cover the same empty squares, since one empty square will be marked at most once. Thus, for each threat window, White needs to place at least one stone to prevent Black from connecting six. That is, if the above algorithm finds t threat windows, then there are at least t threats.

Second, we want to prove that if the above algorithm finds t threat windows for B, then it suffices to defend by placing one stone at the rightmost empty square for each threat window, as illustrated in Fig. 3. Assume by contradiction that Black still wins after the t stones, that is, there still exists at least one sliding window that contains at least four black stones and no white stones. Then, according to the above algorithm, one of these sliding windows must be a threat window. Thus, the rightmost empty square of this window must be marked and be one of the t squares, contradictory to the assumption. □

Lemma 2. *In Connect6, consider one single line only. Placing one stone on that line increases threats by at most two.*

Proof. Let Black place a stone S on a line. It suffices to prove that the stone is covered by at most two threat windows from the above algorithm. Let T_1 be the first threat window covering S. If the threat window T_2 next to T_1 exists and covers S, the empty squares covered by T_2 must be to the right of S and the empty squares covered by T_1 must be to the left of S, since any two threat windows do not cover the same empty squares (as describe above). Thus, apparently, the next threat window T_3, if it exists, must not cover S, since the empty squares of T_2 is not to the left of S for the same reason. So, S is covered by at most two threat windows. □

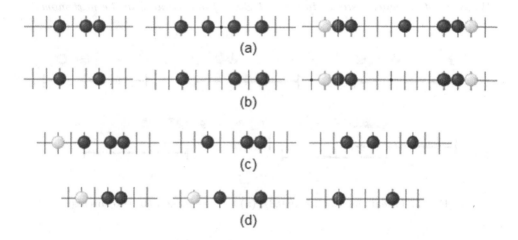

Fig. 4. Live-l and dead-l threats for Connect6: (a) live-3, (b) live-2, (c) dead-3, and (d) dead-2 threats

Lemma 2 shows that placing one stone on a line increases threats by at most two. From this lemma, we can evaluate the value of one line by counting how many stones one player must place subsequently in order to cause one threat or two threats. For example, in Go-Moku or Renju, a three is called *live-three* if it has two open ends and can create two threats by adding one stone, and *dead-three*, if it has only one open end and can create one threat only by adding one stone. Following the similar concept, we define *dead-l* and *live-l* threats in Definition 2 (below). For example, Fig. 4 illustrates the cases of live-3, live-2, dead-3, dead-2 threats.

Definition 2. *In Connect6, a line includes a dead-l threat for one player, say B, if Black only needs to add $(4-l)$ additional stones to create one threat. Similarly, a line includes a live-l threat for B, if Black only needs to add $(4-l)$ additional stones to create two threats.*

In Connect6, live-3, live-2, dead-3, and dead-2 threats are also important, since a move including two stones can make them become real threats. In particular, when players attack with real threats, it is better to associate these threats with more live and dead threats. This is a rather profitable strategy, also used in our Connect6 program in Subsection 4.3.

4.2   Threats for Connect Games

This subsection generalizes the work of Subsection 4.1 to Connect(k, p, q).

Definition 3. *In a line pattern of Connect(k, p, q), assume that one player, say White, cannot connect up to k. Black is said to have t threats, if and only if White needs to place t stones to prevent Black from winning in the next move.*

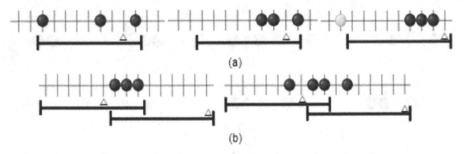

(a)

(b)

Fig. 5. Threats for Connect(9,6,3): (a) one threat and (b) two threats

Definition 3 defines the number of threats for general Connect games. For example, for Connect(9,6,3), Black has one threat in Fig. 5(a) and two in Fig. 5(b). In general, the winning strategy of a player is to have at least $(p+1)$ threats, since the opponent only has p stones to defend. For example, for Connect(9,6,3), one player needs to have 7 threats to win the game.

The algorithm in the previous subsection can be slightly modified to count the number of threats for Connect games, as follows.

1. For a line pattern, slide a window of size k from the left to right.
2. Repeat the following step for each sliding window.

3. If the sliding window contains neither white stones nor marked squares and at least $\delta(=k-p)$ black stones, add one more threat and mark all the empty squares (or the rightmost one only) in the window.

The above algorithm counts the number of threats correctly for Connect(k, p,q), as in Lemma 3 (below), of which the proof is similar to that of Lemma 1 and therefore omitted. Similarly, placing one stone on a line increases threats by at most two, as in Lemma 4 (below), whose proof is also omitted. Similarly, for Connect(k,p,q), dead-l and live-l threats are defined in Definition 4 (below).

Lemma 3. *For Connect games, the above algorithm counts the number of threats correctly.*

Lemma 4. *In Connect(k,p,q), consider one single line only. Putting one stone on that line increases threats by at most two.*

Definition 4. *In Connect(k, p, q), a line includes a dead-l threat for one player, say B, if Black only needs to add $(\delta - l)$ additional stones to have one threat. Similarly, a line includes a live-l threat for B, if Black only needs to add $(\delta - l)$ additional stones to have two threats.*

Now, let us go back to review Go-Moku, Connect(5,1,1). In Go-Moku, since players can place one stone ($p = 1$) only for each move, players cannot defend live-4 threats. Furthermore, in the case of not winning, players must defend a live-3 threat. Otherwise, the opponents can place one stone to make it a live-4 to win the game. Since players must defend live-3 threats in this case, live-3 threats can be viewed as *delayed threats*.

4.3   Programs for Connect Games

In this subsection, we will first describe the algorithm to generate moves for Connect games, and then describe the search techniques used.

Generating Moves. For Go-Moku or Connect games with $p = 1$, players usually order all the empty squares based on some criteria, e.g., the threats mentioned in Subsections 4.1 or 4.2, and then choose the best one to place a stone. However, for Connect games with $p \geq 2$, players cannot simply choose the best p squares to play. A common example for Connect6 is that the two squares may form two live-3 threats from two live-2 threats respectively. But, in most cases, it is better to use two stones to have one live-2 threat becomes two threats. That is, players need to consider the value of placing two stones together. For this problem, we use the following algorithm to generate moves for Connect6.

1. Order all the empty squares into a list L in a descending order according to the evaluated values.
2. Choose the first (best) w ones from L, $(s_1, s_2, ..., s_w)$.
3. For each square s_i, repeatedly do the following two steps.

4. Place a stone at s_i and then order all the empty squares into a new list L_i according to the re-evaluated values.
5. Choose the first w_i ones from L_i, and then, for each square s in the w_i squares, put a pair of squares (s_i, s) into the candidate list L_C. Note that if (s, s_i) is already in L_C, skip it.
6. Order the squares in L_C according to the re-evaluated values and choose the first w' pairs.

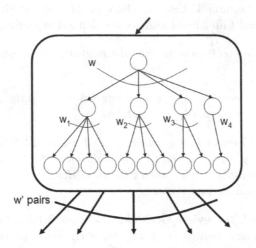

Fig. 6. Tree search for generating moves for Connect6

We suggest a monotonically-decreasing function for w_i, e.g., $w_i = w - i + 1$, as illustrated in Fig. 6. Note that in Fig. 6 the smaller circles are called *subnodes* to be distinguished from the bigger rounded rectangle, representing a move node. Now, we can see that the time complexity of a move node is quite high since one move node may include many subnodes. For example, if $w = 10$ and $w_i = w - i + 1$, the number of subnodes is 55; and if $w = 30$, it is about 500.

Amazons games [11] are the games that also require one player to do two operations for one move. So, programs for Amazons games may also require search trees with height two to generate moves for each move node. However, for Connect(k, p, q) with larger p, we need a search tree with higher height for each move node. This results in a much higher time complexity for larger p. Avetisyan and Lorentz [4] proposed a null-move technique for the first operation to generate moves. For Connect games, it is still an open issue to reduce the number of subnodes inside one node.

Search. Like Go-Moku, threat-based search is also important in Connect games. The search techniques, developed in the past, such as threat-space search and proof-number search [1], are also useful for Connect games. For Connect6, we use a two-level search technique, one level for normal alpha-beta search and the other for threat-space search as shown in Fig. 7. In our program, the depth of the alpha-beta search tree is about 3.

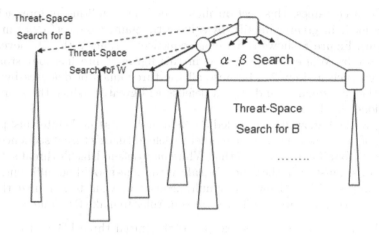

Fig. 7. A two-level tree search for Connect6

5   Null-Move Heuristic

The null-move heuristic [1,5,8,9,10,19] is usually used with threat-space search and proof-number search to determine implicit threats, such as in Go-Moku and Go. The basic idea is to let one player, say White, make a null-move and then apply the threat-space search to finding winning sequences of threat moves for Black. If the sequences are found, we need to determine the relevant zone for White to defend. The relevant zone is called *R-Zone* in [19]. However, for Connect games with $p \geq 2$, it is necessary to modify the above null-move heuristic.

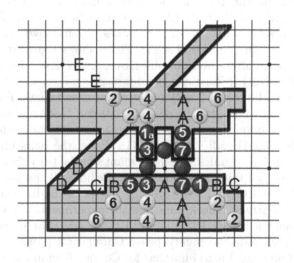

Fig. 8. A winning sequence of Black's threat moves after a null-move by White

For Connect games, this section illustrates a new null-move approach ([22] will describe it in greater detail), by solving the game Connect(6,2,3) in which Black wins. Figure 8 shows a winning sequence of Black's threat moves after a null-move by White. We note that in this figure White has four stones for each move, such as four 2's, because all the three kinds of defenses by White include the four squares, and such a defense will greatly reduce the search tree as mentioned in [1].

In Fig. 8, the shadowed zone, called $R1$-$Zone$, indicates that White must place at least one of two stones in the zone in order to defend Black's threat sequence possibly. Namely, for all moves (s_1, s_2) that White may defend Black's threat sequence, either s_1 or s_2 must be in the zone. For safety, the zone should be sufficiently large to cover all possibilities. However, simultaneously, we want to minimize the zone to reduce the cost of search. Following are our rules to make R1-Zone.

1. All black and white stones except for the initial three black stones are in R1-Zone, since it is possible for White to defend by placing a stone on any of these squares.
2. All defensive squares for the final threats are in the R1-Zone. For example, all As and Bs in Fig. 8. We note that for the single threat between the two Bs in Fig. 8, R1-Zone includes both Bs, but not both Cs for the following reasons. Since both Bs can be used to block the threat, both Bs are included. But, since the Cs cannot block the threat without the middle A, the Cs do not have to be in the zone (we note that the middle A is already in the zone).
3. For each pair of two empty squares, if two white stones placed on them can build a threat, the two squares are in the zone. For example, the Ds in Fig. 8. However, those Es are not in the zone, because for the upper left two 2's we actually place one stone only and thus two additional white stones at the Es cannot build a threat.

After determining the R1-Zone, for each square in the R1-Zone we run a semi-null-move process, illustrated by the square $1a$ in the R1-Zone in Fig. 8. First, White places one stone at $1a$, but, makes a "null-move", called *semi-null-move* in this paper, for the second stone. Fig. 9 shows the winning sequence of Black's threat moves after the semi-null-move, and a new relevant zone, called R2($1a$)-Zone. The R2($1a$)-Zone can be obtained based on the rules for the R1-Zone, except for Rules 2 and 3 with slight changes as follows. For Rule 2, for a single threat, we only consider the squares that can block one threat. For Rule 3, for one empty square, if one white stone placed on it can build a threat, the square is in the zone. For all s in the R2($1a$)-Zone, the pairs $(1a, s)$ are added into a defense list $L_{6,2,3}$, unless the redundant one $(s, 1a)$ exists already. After all semi-null-move processes are done, $L_{6,2,3}$ includes all the moves that White may defend Black's attack. Then, if all the moves in the list cannot be played to defend Black's attack, Black wins. In our experiments for Connect(6,2,3), there are 61 squares in the R1-Zone, and there are 1514 pairs in $L_{6,2,3}$. By going through each pair, we finally prove Corollary 3 (below). An important implication of Corollary 3 is to hint that for Connect6 an initial breakaway does not favor White as described in Subsection 2.4.

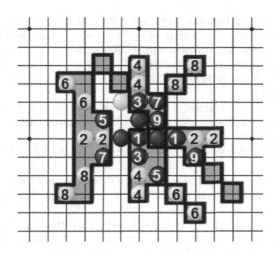

Fig. 9. A winning sequence of Black's threat moves after a semi-null-move

Corollary 3. *For Connect(6,2,3), Black wins.*

6   Conclusion

The contribution of this paper is summarized as follows.

1. This paper introduces a new family of k-in-a-row games, Connect(m, n, k, p, q). Among these games, Connect(6,2,1) or Connect6 is potentially fair, based on some arguments in Subsection 2.4. Thus, Connect6 has the potential to become popular.
2. This paper proposes a threat-based strategy to play Connect(k, p, q) games and implements a computer program for Connect6, based on the strategy.
3. This paper illustrates a new null-move search approach to solve Connect (6,2,3) with Black winning. This result hints that for Connect6 an initial breakaway does not favor White.

In addition, this paper also leaves several open problems, such as fairness, null-move heuristic, and reducing the time complexity for Connect games. We expect to see fruitful research related to these games in the future.

References

1. L.V. Allis. Searching for Solutions in Games and Artificial Intelligence. *Ph.D. Thesis*, University of Limburg, Maastricht, 1994.
2. L.V. Allis, M. van der Meulen, and H.J. van den Herik. Proof-Number Search. *Artificial Intelligence*, 66(1):91–124, 1994.
3. L.V. Allis, H.J van den Herik, and M. P. H. Huntjens. Go-Moku Solved by New Search Techniques. *Computational Intelligence*, 12:7–23, 1996.

4. H. Avetisyan and R. Lorentz. Selective Search in an Amazons Program. In *3rd Computers and Games Conference (CG 2002)* (eds. J. Schaeffer, M. Müller and Y. Björnsson), LNCS 2883, Springer-verlag, Berlin, pages 123-141, 2003.
5. D.F. Beal. Experiments with the Null Move. In *Advances in Computer Chess 5* (ed. D.F. Beal), Elsevier Science Publishers B.V., Amsterdam, The Netherlands, pages 65–79, 1989.
6. J. Beck. On Positional Games. *J. of Combinatorial Theory*, A–30:117–133, 1981.
7. L. Csirmaz. On a Combinatorial Game with an Application to Go-Moku. *Discrete Math*, 29:19–23, 1980.
8. T. Cazenave. Iterative Widening. *Proceedings of IJCAI-01*, Vol. 1, pages 523–528, 2001.
9. T. Cazenave. Abstract Proof Search. in *2nd Computer and Games Conference (CG 2001)* (eds. T. A. Marsland and I. Frank), LNCS 2063, Springer-verlag, Berlin, pages 39–54, 2001.
10. T. Cazenave. A Generalized Threats Search Algorithm. In *3rd Computers and Games Conference (CG 2002)* (eds. J. Schaeffer, M. Müller and Y. Björnsson), LNCS 2883, Springer-verlag, Berlin, pages 75–87, 2003.
11. H.J. van den Herik, J.W.H.M. Uiterwijk, and J. van Rijswijck. Games Solved: Now and in the Future. *Artificial Intelligence*, 134:277-311, 2002.
12. Internet Application Technology Lab. *Connect6 Homepage*. National Chiao Tung University, http://connect6.csie.edu.tw, 2005.
13. Japanese Professional Renju Association. *History of Renju Rules*. http://www.renjusha.net/database/oldrule.htm, 2003.
14. A. Pluhar. Generalizations of the Game *k*-In-A-Row. *Rutcor Res. Rep.*, pages 15–94, 1994.
15. A. Pluhar. The Accelerated *k*-In-A-Row Game. *Theoretical Computer Science*, 271(1–2):865–875, 2002.
16. Renju International Federation. *The International Rules of Renju*. http://www.renju.nu/rifrules.htm, 1998.
17. Renju International Federation. *MOM for the RIF General Assembly*. http://www.renju.nu/wc2003/MOM_RIF_030805.htm, 2003.
18. G. Sakata and W. Ikawa. *Five-In-A-Row*. Renju. Ishi Press, Tokyo, Japan, 1981.
19. T. Thomsen. Lambda-Search in Game Trees - with Application to Go. *ICGA Journal*,23(4):203–217, 2000.
20. J.W.H.M. Uiterwijk and H.J. van den Herik. The Advantage of the Initiative, *Information Sciences*, 122(1):43–58, 2000.
21. J. Wagner and I. Virag. Solving Renju, *ICGA Journal*, 24(1):30–34, 2001.
22. I-C. Wu and H.-C. Chang. Threat-based proof search for Connect6. *Technical report*, Department of Computer Science and Information Engineering, National Chiao Tung University, Hsinchu, Taiwan, 2006.
http://java.csie.nctu.edu.tw/~icwu/technical-reports/tr1-2006.pdf
23. T.G.L. Zetters. Problem S.10 Proposed by R.K. Guy and J.L. Selfridge. *Amer. Math. Monthly* 86, solution 87:575–576, 1980.

Exact-Bound Analyzes and Optimal Strategies for Mastermind with a Lie*

Li-Te Huang[1], Shan-Tai Chen[2], and Shun-Shii Lin[1,**]

[1] Graduate Institute of Computer Science and Information Engineering,
National Taiwan Normal University, Taipei, Taiwan, R.O.C.
linss@csie.ntnu.edu.tw
[2] Department of Computer Science, Chung Cheng Institute of Technology,
National Defense University, Tao-Yuan, Taiwan, R.O.C.

Abstract. This paper presents novel and systematic algorithms to solve a variant of the Mastermind game, which is called "Mastermind with a Lie". Firstly, we use the *k-way-branching*(KWB) algorithm to get an upper bound of the number of guesses for the problem. With the help of clustering technique, the KWB algorithm is able to obtain near-optimal results effectively and efficiently. Secondly, we propose a *fast backtracking*(PPBFB) algorithm based on the *pigeonhole principle* to get the lower bounds of the number of guesses. That is a computer-aided approach, which is able to estimate the depth of the game tree and to backtrack when the depth is larger than a predefined value. Moreover, we also develop two novel methods, named "volume-renewing" and "preprocessing". They can improve the precision in the estimation of the lower bound and speed up the game tree search. As a result of applying the KWB algorithm and the PPBFB algorithm, we are able to show that the upper bound is 7 and that is also the lower bound. Thus, the problem is solved completely and the exact bound of the number of guesses for the problem is 7.

1 Introduction

The deductive game called Mastermind, produced by Invicta Plastics Ltd., is quite popular all over the world since its appearance in 1972. It is a game for two players: a code-maker and a code-breaker. The code-maker chooses a secret code, e.g., (s_1, s_2, s_3, s_4), consisting of 4 digits with 6 possible symbols. Repeated symbols are allowed and thus, the number of possible codes is $6^4 = 1296$. The code-breaker does not know the choice the code-maker made, so he[1] has to guess some code, e.g., (g_1, g_2, g_3, g_4), and gets a hint $[B, W]$ from the code-maker repeatedly until the secret code is obtained exactly, where B means the number of "direct hits", i.e., the number of positions i such that $s_i = g_i$ and

---

* This research was funded by a grant NSC 93-2213-E-003-001 from National Science Council, R.O.C.
** Corresponding author.
[1] In this paper we use 'he' and 'his' whenever 'he or she' and 'his or her' is meant.

H.J. van den Herik et al. (Eds.): ACG11, LNCS 4250, pp. 195–209, 2006.
© Springer-Verlag Berlin Heidelberg 2006

W is the number of "indirect hits", i.e., the number of positions i such that $s_i \neq g_i$ but $s_i = g_j$ for some position $j \neq i$. For instance, if the secret code is $(1, 1, 2, 4)$ and the guess made by the code-breaker is $(2, 1, 4, 5)$, the hint given by the code-maker is then $[1, 2]$. The goal of the code-breaker is, based on the responses, to minimize the number of guesses needed. Note that there are 14 possible responses in Mastermind.

In 1976, Knuth [4] introduced a strategy for minimizing the number of guesses for Mastermind. His strategy requires at most five guesses in the worst case and 4.478 in the expected case. At the follow-up study, Irving [3] and Neuwirth [7] improved the bounds to 4.369 and 4.364, respectively, in the expected case. Finally, Koyama and Lai [5] demonstrated an optimal strategy for Mastermind, where the expected number of guesses is 4.34.

As a sequel to Mastermind, Renyi [8] and Ulam [9] proposed a two-person game with lies, which is called "Renyi-Ulam game". The nature of the game is similar to a fault-tolerant communication problem. In other words, the problem is how the receiver gets a message passed by the sender correctly through a noisy channel. A great deal of researchers therefore have paid much attention to this problem over the past decades. In this paper, we introduce a variant of the Mastermind game, which is called "Mastermind with a Lie". The concept of fault tolerance is added to the variant compared to the original version. In other words, it is the same as the original one but there is an additional rule: the code-maker is allowed to lie at most once. For example, it is a lie if the code-maker answers $[1, 0]$ instead of $[1, 2]$ in the above-mentioned example. Furthermore, the termination criteria of the Renyi-Ulam game substitutes for that of Mastermind in order to fit in with the area of fault tolerance. That is to say that the game will terminate when only one possible codeword remains.

Merelo et al. [6] showed that some combinatorial optimization problems – such as circuit testing, differential cryptanalysis, on-line models with equivalent queries and additive search problems – are related to solutions of the Mastermind game. Any results obtained in this problem may be applied to those optimization problems.

This paper is organized as follows. In Sect. 2, we describe some properties and definitions of the deductive games with lies. Section 3 introduces the k-way-branching algorithm for the "Mastermind with a Lie". In Sect. 4, we demonstrate the fast backtracking algorithm based on the pigeonhole principle to determine the minimal number of guesses required for "Mastermind with a Lie". Section 5 exhibits our conclusions.

2   Definitions of Deductive Games with a Lie

In this section, we show some definitions of deductive games with a lie by a simple number-guessing game, denoted $1 \times n$ games with a lie.

In the $1 \times n$ games with a lie, the code-maker chooses a secret code s, $s \in \{0, 1, 2, \ldots, n - 1\}$. After each guess g made by the code-breaker, the code-maker gives him a response r, $r \in \{<, =, >\}$, i.e., they stand for $g < s$, $g = s$, and $g > s$.

The code-maker is allowed to lie at most once in this game. The goal of the game is to obtain the secret code by using as few guesses as possible. The game can be viewed as a search problem intuitively. In other words, we can represent the process as game-tree search. For instance, a game tree for the 1×3 game with a lie consisting of terminal nodes and non-terminal nodes is shown in Fig. 1.

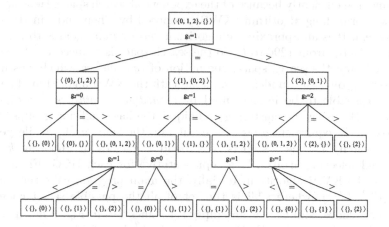

Fig. 1. A game tree for the 1×3 game with a lie

Definition 1. *While the code-breaker makes a guess, the **ordered pair of candidates** $\langle C^{(0)}, C^{(1)} \rangle$ consists of two sets, of which the elements are the remaining possible candidates at present. $C^{(0)}$ is the set of candidates which satisfy all previous responses and $C^{(1)}$ represents the set of candidates which satisfy all but one previous responses. For example, the root in Fig. 1 is $\langle \{0, 1, 2\}, \{\} \rangle$, which indicates that the elements in $C^{(0)}$ are 0, 1, and 2 while $C^{(1)}$ is an empty set.*

Definition 2. *An ordered-pair of candidates, $\langle C^{(0)}, C^{(1)} \rangle$, may be referred to a **state** $(|C^{(0)}|, |C^{(1)}|)$ which is a couple of natural numbers. The first number is the size of the set $C^{(0)}$, and the second number is the size of the set $C^{(1)}$. For instance, the state of the root in Fig. 1 is $(3, 0)$.*

Definition 3. *The **guess** g_i made by the code-breaker means that $(i - 1)$ guesses have been made previously. In Fig. 1, "$g_3 = 1$" means that it is the third guess and the guess number is 1.*

Definition 4. *A **final state** is a state s, where $s = (1, 0)$ or $(0, 1)$. In other words, only one possible codeword remains in the final state and the game is over.*

Definition 5. *The number of guesses required in the worst case for "deductive games with a lie" is **H**, where **H** is the height of the game tree, i.e., the length of its longest path from the root to a leaf. For example, **H** is equal to 3 in Fig. 1.*

3 k-Way-Branching Algorithm

For our problem, the code-breaker has 1296 possible guesses in each turn while the code-maker has 14 possible responses (it is called "classes" in the game tree). The height of the game tree is H and the search space for "Mastermind with a Lie" is $(14 * 1296)^H$ consequently. Traditional approaches are not able to search this game tree efficiently because of the exponential search space. Here, we adopt the k-way-branching algorithm (KWB) proposed by Chen and Lin [1] to solve the problem. It is an approximate algorithm and its main goal is to reduce the branching factor from 1296 to k. So the search space is reduced to $(14 * k)^H$.

For reducing the search space, the notion of *equivalence*, which was used by Neuwirth [7], offers us an idea to combine with the KWB algorithm. The KWB algorithm is able to be implemented by a *modified depth-first search* on the game tree. By using the clustering technique, the hash collision groups (HCG), the most likely equivalent guesses are put into the same group while performing the search. The next step is to expand k guesses by choosing k "best" HCGs and selecting arbitrarily a representative for each HCG. Figure 2 is a sketch of the KWB algorithm. Initially, the main program calls the function, $\text{KWB}\left(C^{(0)}, C^{(1)}\right)$, where $C^{(0)}$ is the set of all the possible secret codes, i.e., 1296 secret codes, and $C^{(1)}$ is an empty set. Lines 2–6 in Fig. 2 mean that the KWB algorithm generates HCGs by using the hashing function and chooses k best HCGs. Lines 8–18 show that a representative guess for each HCG is selected arbitrarily and is applied to expand the game tree recursively. Finally, the best one among the k results is returned. Even a small k, the KWB algorithm can obtain near-optimal results and save much time provided that the hashing function is designed appropriately. Therefore, we make use of the KWB algorithm with fine characters to investigate our problem, "deductive games with a lie".

With regard to Mastermind, the hashing function can classify the most likely equivalent guesses among 1296 possible secret codes into the identical groups and a representative guess is chosen from each of the k "best" groups. Some significant properties are described in [1,2].

We have to define a proper hashing function before using HCG. Thus, we should take the characters of our problem into account. Assume that for an ordered pair of candidates, the guess, g, made by the code-breaker partitions all possible secret codes in $C^{(0)}$ and $C^{(1)}$ into 14 classes, i.e., 28 sets. $\left\langle C^{(0)}, C^{(1)} \right\rangle_g$ stands for that the code-breaker makes the guess, g, when he encounters the ordered pair of candidates. $\left\langle C_g^{(0)} \right\rangle$ and $\left\langle C_g^{(1)} \right\rangle$ represent the ordered tuples of the size of the 14 sets produced by the partitions of $C^{(0)}$ and $C^{(1)}$ with g respectively. So we define the hashing function as

$$Hash\left(\left\langle C^{(0)}, C^{(1)} \right\rangle_g\right) = Str\left(\left\langle Sort\left\langle C_g^{(0)} \right\rangle, Sort\left\langle C_g^{(1)} \right\rangle \right\rangle\right) = Str\left(\left\langle C_g' \right\rangle\right),$$

where $Sort\left\langle C_g^{(0)} \right\rangle$ and $Sort\left\langle C_g^{(1)} \right\rangle$ are the new sorted tuples after sorting 14 elements in $\left\langle C_g^{(0)} \right\rangle$ and $Sort\left\langle C_g^{(0)} \right\rangle$, respectively, in non-increasing order. $\left\langle C_g' \right\rangle$

is the new tuple with 28 entries, which is formed by merging $Sort \left\langle C_g^{(0)} \right\rangle$ and $Sort \left\langle C_g^{(1)} \right\rangle$. The ability of the function, Str, is to convert the 28 integers in $\left\langle C_g' \right\rangle$ into 28 strings and concatenate them into a single string in order. This means, if $\left\langle C_g' \right\rangle = \langle n_{g,1}, n_{g,2}, \ldots, n_{g,28} \rangle$, $Str\left(\langle C_g' \rangle\right) = Str\left(n_{g,1}\right) Str\left(n_{g,2}\right) \ldots Str\left(n_{g,28}\right)$. This hashing function is a kind of string hashing function and its output is a string which means the index(id). The meanings of the hashing function is that, for any two guesses, g and g', they are most likely equivalent and are classified into the same HCG if $Hash\left(\langle C^{(0)}, C^{(1)} \rangle_g\right) = Hash\left(\langle C^{(0)}, C^{(1)} \rangle_{g'}\right)$.

KWB $(C^{(0)}, C^{(1)})$ {

01   **If** (final state is reached) **then Return** 0 ;   // final state indicates the game is over

02   **For** (each guess $g \in$ M) {   // M is the set of possible next guesses

03     $id = Hash(\langle C^{(0)}, C^{(1)} \rangle_g)$;   // classify possible next guesses to HCGs by a hashing

04   HCG$_{id} \leftarrow$ HCG$_{id} \cup \{g\}$;   function according to current ordered pair of candidates

05   }

06   B = {HCG$_{id}$ | HCG$_{id}$ is the top k groups when using the function f to evaluate} ;

          // B: the set of k selected HCGs

07   $s_{best} = \infty$;   // s_{id} is the minimal number of guesses found by KWB

08   **For** (each HCG$_{id} \in$ B) {

09     $g_{id} = Choose$(HCG$_{id}$) ;   // g_{id} is an arbitrarily selected representative for HCG$_{id}$

10   g_{id} partitions $\langle C^{(0)}, C^{(1)} \rangle$ into 14 classes named $\langle C_j^{(0)}, C_j^{(1)} \rangle$, $j = 1,\ldots,14$;

11     $s_{id} = 0$;

12   **For** (each $\langle C_j^{(0)}, C_j^{(1)} \rangle$) {

13       $s_{id,j} =$ **KWB**$(C_j^{(0)}, C_j^{(1)})$;

14   **If** $(s_{id,j} > s_{id})$ **then** $s_{id} = s_{id,j}$; // recursively k-way search to find the best solution

15   }

16   **If** $(s_{id} < s_{best})$ **then** $s_{best} = s_{id}$;

17   }

18   **Return** $s_{best} + 1$;   // return $s_{id} + 1$ to the parent node

   }

Fig. 2. The sketch of the KWB algorithm

Now the problem is how to choose k "best" HCGs and select a representative for each group to explore the game tree. Our method is based on the heuristic procedure introduced by Chen and Lin [1]. We made some modifications for this problem. We have to take the sizes of the 28 sets into account at the same time while choosing a guess. In other words, we should consider the distribution of the sizes, i.e., we have to choose the guess which can partition $C^{(0)}$ and $C^{(1)}$ as evenly as possible in order to minimize the height of the game tree. But there are different significances between the two distributions of $C^{(0)}$ and $C^{(1)}$ partitioned by g. Therefore, we make use of the idea, weighting, to determine which guess is better. We let $\langle C'_g \rangle = Sort \left\langle d_0 * \left\langle C_g^{(0)} \right\rangle, d_1 * \left\langle C_g^{(1)} \right\rangle \right\rangle = \langle n_{g,1}, n_{g,2}, \ldots, n_{g,28} \rangle$, where $\langle C'_g \rangle$ is a sorted tuple which is formed by sorting the 28 integers in $d_0 * \left\langle C_g^{(0)} \right\rangle$ and $d_1 * \left\langle C_g^{(1)} \right\rangle$, i.e., $n_{g,1} \geq n_{g,2} \geq \ldots \geq n_{g,28}$. In practice, we set the parameters, $d_0 = 3$ and $d_1 = 1$. Thus, we define an evaluation function as follows.

$$f(g, g') = \begin{cases} 1, & \text{if } n_{g,j} = n_{g',j} \text{ and } n_{g,i} < n_{g',i}, 1 \leq j < i \text{ for some } i \leq 28 \\ 0, & \text{if } n_{g,j} = n_{g',j}, \forall j \\ -1, & \text{if } n_{g,j} = n_{g',j} \text{ and } n_{g,i} > n_{g',i}, 1 \leq j < i \text{ for some } i \leq 28 \end{cases}$$

Clearly, g partitions $\langle C^{(0)}, C^{(1)} \rangle$ more evenly than g' if $f(g, g') = 1$. We believe that g is better than g' for this reason. In contrast, if "$f(g, g') = 0$", then it indicates that g and g' are in the same HCG. We take the function f to choose k "best" HCGs from all the HCG groups.

Figure 3 shows the game tree by applying the KWB algorithm to this problem. Among them, $\left\langle C_{i,j}^{(0)}, C_{i,j}^{(1)} \right\rangle$ is the j-th ordered pair of candidates, i.e., the j-th class, after the i-th guess. And $g_{i,j}$ is the j-th among the k best codes chosen by the KWB algorithm at the i-th guess. In the beginning, the state of the root in Fig. 3 is $(1296, 0)$ which means that there are total 1296 guesses satisfying all previous responses. Note that while the code-breaker makes the first guess, there are 5 non-equivalent guesses in 1296 possible codes according to the analyzes in [7]. They are "1111", "1112", "1122", "1123", and "1234". And "1123" is favorable for the first guess to get a promising upper bound of the number of guesses. We thus set the first guess, $g_{1,1}$, to "1123" in our algorithm. After that, there are 14 classes which have to be expanded since the code-maker has 14 possible responses. Then the code-breaker selects k best guesses to expand the game tree. The two steps take turn until the final state is met. At the final state, the program backtracks to its parent node and expands other branches continuously.

3.1   Experimental Results of the KWB Algorithm

When the program based on the KWB algorithm was well-written and tested, we ran it on an AMD Opteron 1.6GHz PC. The results are shown in Table 1. The larger the value of k is, i.e., the larger the search space is, the fewer the number of guesses required for the game is. But the time for running the program is relatively longer. Note that the strategy for "$H = 7$" is obtained by our algorithm when $k = 3$. This reveals that the KWB algorithm can obtain optimal (or near-optimal) results with a very small k. From Table 1, we are able to obtain a strategy for the game, "Mastermind with a Lie", by using the KWB algorithm and its number of guesses required in the worst case is 7. Hence, we have the following Lemma 1 evidently.

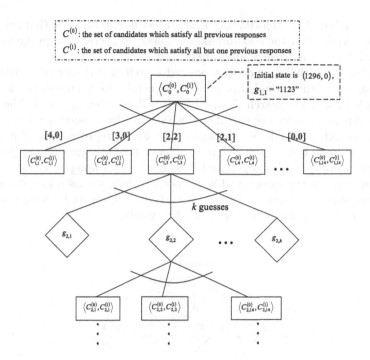

Fig. 3. The game tree expanded by the KWB algorithm

Table 1. The upper bound derived by the KWB algorithm

k	H: The number of guesses needed	Run time (minutes)
1	8	1.75
2	8	15.10
3	7	65.11
4	7	241.13
5	7	629.82

Lemma 1. *For the game, "**Mastermind with a Lie**", there exists a strategy such that the number of guesses required for the code-breaker to obtain the secret code is at most 7.*

We can regard Lemma 1 as an upper bound of this problem. In the following section we demonstrate that the pigeonhole-principle-based fast backtracking algorithm to prove that the lower bound of the game is also 7.

4   The Fast Backtracking Algorithm Based on Pigeonhole Principle

In this section, we will apply the pigeonhole principle to develop a fast backtracking algorithm for "Mastermind with a Lie". We call it the *pigeonhole-principle-*

based fast backtracking algorithm (PPBFB). We intend to prove that the number of guesses required for the code-breaker to guess the secret code in the worst case is at least 7, i.e., the lower bound of the game.

The idea of PPBFB is to do a worst-first search. It estimates the lower bound by making use of the extended pigeonhole principle and then backtracks as early as possible to save the search time. PPBFB is illustrated in Fig. 4. The concept of PPBFB is to consider the size of the two sets in the ordered pair of candidates when the search proceeds, so the squares in Fig. 4 represent the states. $g_{p,i}$ is the i-th candidate taken by the code-breaker at the p-th guess. $r_{p,max}$ means the class which results in the largest number of guesses among 14 classes after the p-th guess. q_{max} is the theoretical lower bound which means a lowest number of guesses required to approach the final state (i.e., $(1, 0)$ or $(0, 1)$) from the current state and h is the lower bound we intend to verify.

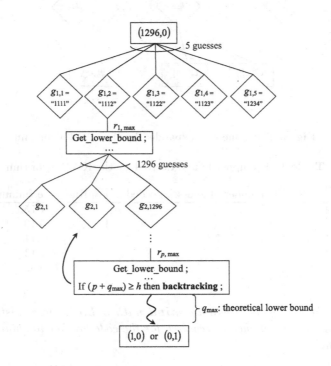

Fig. 4. The sketch of the PPBFB algorithm

We notice that the code-breaker has to explore all possible 1296 guesses at each guess except the first one. For the first guess, we only need to explore 5 representative guesses due to the equivalence property. For the code-maker, we only have to expand the worst case among the 14 classes. The so-called "worst case" denotes the class which will result in the most guesses needed by the code-breaker. We employ the *extended pigeonhole principle* presented by [1] to estimate the lower bounds of the guesses needed among 14 classes. The idea of the estimation of lower bounds is similar to that of the heuristic function in the A* algorithms. In other words, the actual number of guesses needed is more than or equal to the largest number, q_{max}, of lower bounds among 14 classes. Therefore, our verifying program is not necessary to search all the game tree. It can backtrack to the parent node to expand other branches if the condition holds: $p + q_{max} \geq h$, where we set $h = 7$.

The main idea of the estimation of lower bounds by using the extended pigeonhole principle is that the guess made by the code-breaker in each turn may divide the elements of the two sets in the ordered pair of candidates evenly. Hence, this strategy can minimize the height of sub-tree rooted in the current node. That is to say that there exists a "theoretical optimal" strategy for the code-breaker in the following guesses such that all the elements of the two sets in each node may be divided evenly. The actual number of guesses is thus more than or equal to the value of estimation.

The extended pigeonhole principle shows that we should have a set of the volumes, V. We know that different guesses result in distinct distributions of the remaining candidates in 14 classes. So we adopt the following approach to obtain a set of the volumes, V, in order to make the estimation of lower bounds for all nodes of the sub-tree rooted in the current node.

In Fig. 5, the set C represents one of the two sets, $C^{(0)}$ and $C^{(1)}$, in some ordered pair of candidates. $g_{p,i}$ is the i-th candidates at the p-th guess and $|C_{g_{p,i},j}|$ is the size of the j-th class in the distribution of the remaining candidates after $g_{p,i}$ partitions C. We make all 1296 possible guesses to get the distributions of remaining candidates. And then we select the maximal value of the sizes of the corresponding classes in 1296 distributions to acquire the set of volumes, $V = \{v_1, v_2, \ldots, v_{14}\} = \{|C_{g_{p,\max},1}|, |C_{g_{p,\max},2}|, \ldots, |C_{g_{p,\max},14}|\}$, where $v_j = |C_{g_{p,\max},j}| = \max_{1 \leq i \leq 1296} (|C_{g_{p,i},j}|)$.

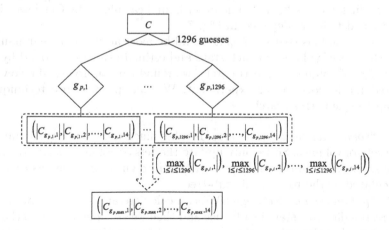

Fig. 5. The algorithm Get_volume(C) for obtaining the set of volumes, and its output is $V = \{v_1, \ldots, v_{14}\}$, $v_j = \max_{1 \leq i \leq 1296} (|C_{g_{p,i},j}|)$

The meaning of the set of volumes obtained by the approach is that every $\left|C_{g_{p,i,j}}\right|$ formed by each guesses is certainly less than or equal to v_j. Afterwards, the lower bound estimated with V is sure to be less than or equal to the actual guesses needed. This fits in with the requirements of PPBFB.

We discover some phenomena when using V to calculate the lower bound in practice. First, different C's will result in different V's. So the two sets of volumes corresponding to $C^{(0)}$ and $C^{(1)}$ in some ordered pair of candidates need to be calculated separately because they are disjoint sets. Second, the differences between the lower bounds calculated by V and the actual guesses needed become larger when searching the deeper part of the game tree if the V computed from some ordered pair of candidates are used by all the ordered pairs of candidates in its sub-tree. Thus, the estimated lower bound will be closer to the actual guesses required if the program updates V dynamically while visiting some ordered pair of candidates.

Figure 6 shows the sketch of the algorithm for calculating the lower bound in PPBFB. The parameters used in this algorithm are as following: s is the current state. V_0 and V_1 are the sets of volumes which are obtained from the two sets, $C^{(0)}$ and $C^{(1)} \cup C^{(0)}$, in the ordered pair of candidates corresponding to s. We ought to take $C^{(1)} \cup C^{(0)}$ into account in order to calculate V_1 since the elements in $C^{(0)}$ moves to $C^{(1)}$ continuously in the game. Lines 6–11 get the maximal number of remaining candidates in all classes of s_0 by employing V_0 and the extended pigeonhole principle. Lines 12–16 get the maximal number of remaining candidates in all classes of s_1 by employing V_1 and the extended pigeonhole principle. Note that if s_0 is not equal to 0, the function, *floor*, is applied to compute the variables avg_0 and avg_1 because the result we intend to calculate is a lower bound. Otherwise, the function, *ceiling*, is applied to avg_0 as a result of the consideration of the worst case if s_0 equals 0.

We illustrate Get_lower_bound with an example shown in Fig. 7. Assume that V_0 and V_1 are available and the state of some class is $\left(\left|C^{(0)}\right|, \left|C^{(1)}\right|\right) = (625, 671)$ in case the first guess is "1111". The lower bound calculated by Get_lower_bound is 5 and the details are depicted in Fig. 7.

An intuitive idea is that the result would be more accurate if we dynamically update the sets of volumes in each state and estimate the lower bound by using the new sets of volumes. Unfortunately, we found this approach will slow down the speed much more in our experiment. We thus propose two techniques to speed up the game tree search.

- *The "volume-renewing" technique*: It takes much time to update dynamically the sets of volumes. So we only update the sets of volumes before the second guess. The state in the sub-tree after the third guess uses the sets of volumes calculated by the root of this sub-tree.
- *The "preprocessing" technique*: We can calculate and store the lower bound corresponding to a state in advance and then we are able to look up the result directly when the same state is met again. In practice, when we update the sets of volumes at the second level of the game tree, we also produce all the possible states, calculate their lower bounds, and store the results into

```
     Get_lower_bound (state s = (s₀, s₁), volume V₀, volume V₁) {
01    lower_bound = 0 ;
02    If (s is a final state) Return lower_bound ;        // '0' is returned when s is a final state
03    While (s is not a final state) {
04     lower_bound++ ;
05     others = 0 ;            // others is the number of candidates which do not satisfy the current response
06     If (s₀ != 0) {
07      Sort V₀ to be  V₀′ = {v₁′,...,v₁₄′}, where  v₁′ ≤ ... ≤ v₁₄′ ;
08      Find the smallest k, such that  ⌊(s₀ - Σᵢ₌₁ᵏ vᵢ′)/(n-k)⌋ < v′ₖ₊₁ . And let  avg_0 = ⌊(s₀ - Σᵢ₌₁ᵏ vᵢ′)/(n-k)⌋ ;
09      others = s₀ - avg_0 ;
10      s₀ = avg_0 ;           // update s₀ for the next guess
11     }
12     If (s₁ != 0) {
13      Sort V₁ to be  V₁′ = {v₁′,...,v₁₄′}, where  v₁′ ≤ ... ≤ v₁₄′ ;
14      Find the smallest k, such that  ⌊(s₁ - Σᵢ₌₁ᵏ vᵢ′)/(n-k)⌋ < v′ₖ₊₁ . And let  ⎰ avg_1 = ⌊(s₁ - Σᵢ₌₁ᵏ vᵢ′)/(n-k)⌋ ,   if s₀ != 0 ;
                                                                                    ⎱ avg_1 = ⌈(s₁ - Σᵢ₌₁ᵏ vᵢ′)/(n-k)⌉ ,   otherwise
15      s₁ = avg_1 ;           // update s₁ for the next guess
16     }
17     s₁ = s₁ + others ;      // s₁ is the number of the set of candidates which satisfy all but one responses
18    }
19    Return lower_bound ;     // the function returns lower_bound finally
     }
```

$$
Get\_lower\_bound \ (state \ s = (s_0, s_1), \ volume \ V_0, \ volume \ V_1)
$$

Fig. 6. The algorithm Get_lower_bound for estimating the lower bound

hashing table, *lb_table*. For example, all the possible states, (x, y), $0 \leq x \leq 625$, $0 \leq y \leq 1296$, are produced if the current state is $(625, 671)$. The hashing function, *Hash_lb*, is therefore defined as follows.

$$
Hash\_lb \ (x, y) = Str \ (x) \ Str \ (y) \ ,
$$

where x and y are two integers and the output of this hashing function is a string. We can look up the lower bound by using the string as an index. Figure 8 gives an algorithm, Create_lb_table, for creating the hashing table, *lb_table*, where the parameters are the current state, s, and the current sets of volumes, V_0 and V_1. Lines 2–6 produce all the possible states in the sub-tree, estimate the lower bound, store the results to *lb_table* by using the hashing function *Hash_lb*, and return *lb_table* finally.

After some experiments, we found that the two techniques not only made the estimation accurate but also accelerated the speed of search. The speed of the new program using the above two techniques is more than ten times faster than that of the old one.

Figure 9 illustrates the details of the PPBFB which integrates all the concepts and techniques discussed in this section. At the beginning, the main program

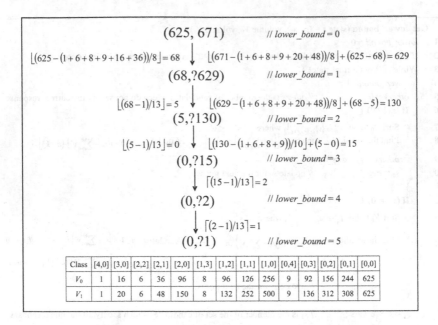

Fig. 7. An example for calculating the lower bound (the lower bound is 5 here)

```
    Create_lb_table (s = (s₀, s₁), V₀, V₁) {
01    lb_table = a hashing table with the hashing function Hash_lb ;
02    For (x = 0 to s₀) {
03      For (y = 0 to s₁) {
04        lb_table(x, y) = Get_lower_bound( s' = (x, y), V₀, V₁) ;
05      }
06    }
07    Return lb_table ;     // the function returns lb_table finally
    }
```

Fig. 8. The algorithm Create_lb_table for creating *lb_table*

makes the function call, PPBFB $(\langle C^{(0)}, C^{(1)} \rangle, 1, \varnothing)$, where $C^{(0)}$ is the set of all the 1296 possible guesses and $C^{(1)}$ is an empty set. Lines 9–27 calculate q_{max} and r_{max} after each guess by using the methods mentioned in the section. Lines 28–39 determine which one of the following cases is satisfied: the verification program fails, the program has to backtrack, or the program has to expand recursively. Line 40 returns "success!" if every $g_{1,i}$ successfully passes the verification program.

```
      PPBFB ( ⟨ C^(0), C^(1) ⟩ , p, lb_table) {
01    If (p = 1) {
02        M = {"1111", "1112", "1122", "1123", "1234"} ;    // |M| = 5 for the first guess
03    }
04    Else {
05        M = {"1111", "1112", "1113", ..., "6666"} ;        // |M| = 1296 for the second to h-th guess
06    }
07    For (each g_{p,i} ∈ M) {                               // M is the set of possible next guesses
08        g_{p,i} partitions ⟨ C^(0), C^(1) ⟩ into 14 classes named r_j = ⟨ C_j^(0), C_j^(1) ⟩ , j = 1,...,14 ;
09        If (p ≤ 2) {
10            q_max = 0 ;
11            For (each r_j) {
12                V_{0,j} = Get_volume(C_j^(0)) ;            // V_{0,j} is the set of volumes for C_j^(0)
13                V_{1,j} = Get_volume(C_j^(1)) ;            // V_{1,j} is the set of volumes for C_j^(1)
14                q_j = Get_lower_bound((|C_j^(0)|, |C_j^(1)|), V_{0,j}, V_{1,j}) ;
15                If (q_max < q_j) {
16                    r_max = r_j ; |C_max^(0)| = | C_j^(0)| ; |C_max^(1)| = | C_j^(1)| ;
17                    V_{0,max} = V_{0,j} ; V_{1,max} = V_{1,j} ; q_max = q_j ;
18                }
19            }
20            If (p = 2) {
21                lb_table = Create_lb_table((|C_max^(0)|, |C_max^(1)|), V_{0,max}, V_{1,max}) ;
                                // i.e., Get_lower_bound((|C_max^(0)|, |C_max^(1)|), V_{0,max}, V_{1,max}) = q_max
22            }
23        }
24        Else {
25            q_max = max_{1≤j≤14}(q_j), where q_j = lb_table(C_j^(0), C_j^(1)) ;

26            r_max = the class which results in q_max ;      // i.e., q_{r_max} = max_{1≤j≤14}(q_j)

27        }
28        If (p + q_max < h) {                                // h is the lower bound we would like to verify
29            If (q_max = 0) then terminate the program and output "failure!" ;    // final state is reached
30            Else {
31                If (p = 1) {                                // the first guess
32                    PPBFB(r_max, p + 1, ∅) ;                // recursively expand the lower levels
33                }
34                Else {
35                    PPBFB(r_max, p + 1, lb_table) ;         // recursively expand the lower levels
36                }
37            }
38        }
39    }
40    If (p = 1) then output "success!" ;
}
```

Fig. 9. The details of the PPBFB algorithm

Theorem 1. *Seven guesses are necessary and sufficient for the "Mastermind with a Lie" in the worst case*

Proof. The verification program based on the PPBFB algorithm was implemented and run on an AMD Opteron 1.6GHz PC to verify "Mastermind with a Lie" game. If we set the value of h, which is the lower bound we would like to verify, to 7, our program executed for about 2 days and outputted "success!" finally. In other words, the minimal number of guesses is at least 7 without respect to any strategies used by the code-breaker in the worst case. Thus we get that the upper bound is 7 and so is the lower bound. According to Lemma 1, the required optimal number of guesses for the problem is 7 in the worst case. This completes the proof. □

5   Concluding Remarks

In this paper, we demonstrate novel and systematic algorithms to solve a variant of the Mastermind game, which is called "Mastermind with a Lie". Firstly, we use the k-way-branching algorithm to get the upper bound of the number of guesses for the problem. With the help of clustering technique, the KWB algorithm is able to obtain near-optimal results effectively and efficiently.

Secondly, we propose a fast backtracking algorithm to get the lower bound of the number of guesses. That is a computer-aided algorithm, which can explore all the possible strategies elegantly by using the estimating-and-backtracking approach. Moreover, we also develop two novel techniques, named *volume-renewing* and *preprocessing*. They can improve the precision in the estimation of the lower bound and substantially speed up the game tree search.

As a result of applying the KWB algorithm and the PPBFB algorithm, we obtain that the upper bound is 7 and so is the lower bound. Thus, the problem is solved completely and the exact bound of the number of guesses for the problem is 7.

The games, "deductive games with multiple lies", are worth studied further in the future. Furthermore, we hope that the proposed algorithms and techniques can be applied to related problems and provide some directions to other researchers.

References

1. S.T. Chen and S.S. Lin. Novel Algorithms for Deductive Games. In *Proceedings of the 2004 International Computer Symposium on Artificial Intelligence*, TA-E1, 2004.
2. S.T. Chen, S.S. Lin, and L.T. Huang: Two-Phase Optimization Algorithm for Hard-Combinatorial Problems. In *Proceedings of the Fifteenth Australasian Workshop on Combinatorial Algorithms*, pages 248–259, 2004.
3. R.W. Irving. Towards an Optimum Mastermind Strategy. *Journal of Recreational Mathematics* 11(2):81–87, 1978–79.

4. D.E. Knuth. The Computer as Mastermind. *Journal of Recreational Mathematics*, 9(1):1–6, 1976.
5. K. Koyama and T.W. Lai. An Optimal Mastermind Strategy. *Journal of Recreational Mathematics*, 25:251–256, 1993.
6. J.J. Merelo, J. Carpio, P. Castillo, V.M. Rivas, and G. Romero (GeNeura Team). Finding a Needle in a Haystack Using Hints and Evolutionary Computation: The Case of Genetic Mastermind. *Genetic and Evolutionary Computation Conference*, Late breaking papers books, pages 184–192, 1999.
7. E. Neuwirth. Some Strategies for Mastermind. *Zeitschrift für Operations Research*, 26:257–278, 1982.
8. A. Renyi. On a Problem of Information Theory. *MTA Mat. Kut. Int. Kozl.* 6B:505–516, 1961.
9. S. M. Ulam. *Adventures of a Mathematician*. Charles Scribner's Sons, New York, NY, 1976. ISBN 0-684-14391-7.

Player Modeling, Search Algorithms and Strategies in Multi-player Games

Ulf Lorenz and Tobias Tscheuschner

Department of Computer Science,
Paderborn, Germany
{flulo, chessy}@upb.de

Abstract. For a long period of time, two person zero-sum games have been in the focus of researchers of various communities. The efforts were mainly driven by the fascination of special competitions such as DEEP BLUE vs. Kasparov, and of the beauty of parlor games such as Checkers, Backgammon, Othello, and Go.

Multi-player games, however, have been investigated considerably less, and although literature of game theory fills books about equilibrium strategies in such games, practical experiences are rare. Recently, Korf, Sturtevant and a few others started highly interesting research activities. We focused on investigating a four-person chess variant, in order to understand the peculiarities of multi-player games without chance components. In this contribution, we present player models and search algorithms that we tested in the four-player chess world. As a result, we may state that the more successful player models can benefit from more efficient algorithms and speed, because searching more deeply leads to better results. Moreover, we present a meta-strategy, which beats a paranoid α-β player, the best known player in multi-player games.

1 Introduction

We start with a short overview about two-person games (1.1), and about recent developments in the area of multi-player games (1.2).

1.1 Two-Person Games

In two-person zero-sum games, game-tree search is the core of most attempts to make computers play games. Typically, a game-playing program consists of three parts: a move generator, which computes all possible moves in a given position; an evaluation procedure which implements a human expert's knowledge about the value of a given position or an automatically generated heuristic evaluation function (in both cases, the values are quite heuristic, fuzzy, and limited), and a search algorithm which organizes a forecast.

For most of the interesting board games, we do not know the correct evaluations of all positions. Therefore, we are forced to base our decisions on heuristic or vague knowledge. At some level of branching, the complete game tree (as defined by the rules of the game) is therefore cut, the artificial leaves of the

H.J. van den Herik et al. (Eds.): ACG11, LNCS 4250, pp. 210–224, 2006.

resulting subtree are evaluated with the heuristic evaluations, and these values are propagated to the root [5,11,1] of the game tree as if they were real ones. For two-person zero-sum games, computing this heuristic minimax value is by far the most successful approach in computer-games history, and when Shannon [13] proposed a design for a chess program in 1949 — which is in its core still used by all modern game- playing programs — it seemed quite reasonable that deeper searches lead to better results. Indeed, the important observation over the last 40 years in the chess game and some other games is: *the game tree acts as an error filter.* Therefore, the faster and the more sophisticated the search algorithm, the better the search results! This is not self-evident, as some theoretical analyses show [2,10,4,6], but is the most crucial point in the success story of the forecast-based game- playing programs.

1.2   Multi-player Games

Most of the material concerning general strategic games comes from mathematical game theory. For an excellent introduction, we refer to [8]. In order to describe a strategic game, you need

1. the players (Who is involved?),
2. the rules (Who moves when? What do players know, when they move? What can they do?),
3. the outcomes (for each possible set of actions by the players, what is the outcome?), and
4. the payoffs (What are the players' utilities over the outcomes?).

In contrast to two-person zero-sum games, we cannot determine an optimal strategy in n-person zero-sum games independently from our opponent. In general, no so-called *dominating* strategy exists. Instead, we must be satisfied with some kind of equilibrium, mostly the so-called Nash- or User-equilibrium [9]: A set of n strategies $S_1, \ldots, S_n$ (one for each player) will be called in Nash-equilibrium, if none of the n players can improve his profit by changing his strategy S_i to S_i' under the assumption that all other $n-1$ players keep their strategies as they are. Such an equilibrium is called a *pure Nash equilibrium*. If we assume that the players do not select just one strategy, but that they select a random distribution over all their possible strategies, such a set of distributions will be called to be in equilibrium if none of the players can rise his expected profit by changing his probability distribution under the assumption that all the others keep their distributions. This kind of equilibrium is called a *mixed Nash-equilibrium*. For every finite game, at least one mixed equilibrium does exist.

The game that we inspect within this paper is an n-person game with complete information and alternating right to move. This means, at each point of time exactly one player has the opportunity to move, and all players can observe all other players' actions. Therefore, also a pure equilibrium does always exist [8]. For the special case that all outcomes at the leaves of the game tree are different from each other, the max^n-algorithm finds such an equilibrium [3]. Otherwise, it

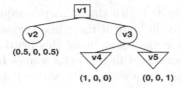

Fig. 1. Tree example with non-unique profit for player 1 at node v_1

cannot be predicted what is most rational for a player, respectively, the outcome is unstable. You can see this in Fig. 1.

Let us assume that we have three players, A,B, and C. A, to move at node v_1, can achieve 0.5, when he moves to the left to node v_2. He can also achieve either 1 or 0 when he goes to the right, but in the given example, B at node v_3 can decide whether it will be 0 or 1. It is not possible to assign unique numbers to positions in order to evaluate them. Instead, we need either sets of numbers in order to describe the value of a position, or we need a function that combines the numbers to one value.

Therefore, opponent models become an important factor. Sturtevant and Korf [16] examined the α-β algorithm in the paranoid model in that each player assumes that all the others play against him, as well as the max^n-algorithm in the model in that each player believes that all players try to maximize their own profit.

2   Organization of the Paper

We are going to describe the application of our interest in Section 2. In Section 3, we present the four player types which we used for our experiments: (1) the paranoid player, (2) the max^n player, (3) the careful max^n player, (4) the coalition player, and in particular (5) the coalition-mixer player. We were mainly interested in the following questions.

- Which opponent model is strong in multi-player games without dice? Can we acknowledge Sturtevant's and Korf's observations?
- Most importantly: can the error-filter effect of game trees (as we described above) be observed in n-person games? We are convinced that only player models which can benefit from Moore's law have a chance to survive in the long run. Are there differences between the known player models concerning this issue?
- How can we use our knowledge about error filtering forecasts and our observations about the previously mentioned players, in order to construct stronger players? Are there efficient algorithms for the strong player models?
- One important step in the progress of two-person chess was the introduction of quiescence searches. Are quiescence searches useful in four-person chess?

In Section 4 we discuss the results of our experiments and we also report about some attempts which failed. Section 5 ends the paper with some conclusions.

2.1 The Four-Person Chess Game

A four-person chess game is the game of interest in this paper. It has the following four interesting properties.

(1) The fact that four players fight against each other instead only two, changes the character of the game completely, and offers a wide research field. Is α-β search still useful? Is forecasting still useful? What is a good playing strategy?
(2) The board and the pieces of the game are similar to the traditional chess ones. Therefore, we may assume that the heuristic board evaluation can be kept similar to the chess evaluation: piece values, mobility, king safety... We see good chances that we can carry over our expertise from two-person chess.
(3) For players already familiar with chess, it is not difficult to understand the extra rules for four-player chess.
(4) Four player chess has no dice and contains all the problems of market models usually inspected in game theory. Therefore, we are optimistic that results that we achieve in this little game will have impact on the wide field of economics.

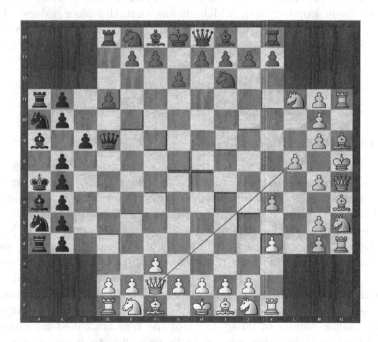

Fig. 2. The four-person chess board

All participants — usually called White, Black, Blue, and Yellow, respectively South, West, North and East — play against all others, fighting for one full point. A draw between four players brings a quarter point, between three players a third of a point etc. The chessboard is a Chessapeak Challenge(R)[1] design, but we tried to keep the rules as near to normal chess as possible.

As you see in Fig. 2, the board consists of an 8 × 8 chess board with 4 additional 3 × 8 partial boards connected to the four sides of the middle chess board. In the initial position, the Kings are always placed on the right of their Queens. All pieces move like in traditional Chess, only the Pawns have some extra opportunities. The squares on the main diagonals of the 8 × 8-mid-board are marked. There, Pawns can change their direction such that their distance from their original square to a promotion square remains 12 steps. Each Pawn can do that only once per game.

The game is artificially made finite with the help of the draw-by-repetition rule and the 50-moves rule, which are defined analogously to the classic chess rules. A move consists of 2, 3, or 4 partial moves, depending on how many players are on the board. If a player is mated or his king can be taken, his pieces are taken from the board.[2]

3   The Player Types

The evaluation procedure is for all player-types the same. It is measured by a relative game portion $S(i)$, which means that the piece values $M(i)$, the static and dynamic piece-square-values $SPT(i)$, and $DPT(i)$, plus the king safety $KS(i)$ of player i is divided through the sum of the values of all players:

$$S(i) = \frac{M(i) + SPT(i) + DPT(i) + KS(i)}{\sum_{j=1}^{n} M(j) + SPT(j) + DPT(j) + KS(j)}. \tag{1}$$

3.1   The Paranoid Player Type

Let $G = (V, E, H)$ be a game tree with the set of nodes V and edges E. The nodes shall correspond to game positions and the edges to moves from one position to the next. The α-β algorithm is a depth-first search algorithm, which runs into the search tree down to a predetermined level d. The leaves of the tree are evaluated by some (heuristic) evaluation function $H : V \rightarrow [0, 1]$, that assigns values (here for the sake of simplicity $[0, 1]$) to positions of the game. An appropriate game tree for the α-β algorithm consists of two disjoint subsets of nodes, so called min-nodes and max-nodes. At max-nodes, the max-player must move and he[3] builds values of inner nodes by computing the maximum over its successor values. The analog is valid for the min-player who minimizes over successor values. Values of inner nodes of the tree are called minimax values. The α-β standard algorithm can easily be extended to n-person games if the paranoid model is used.

---

[1] http://chessapeak.com/chess.html

[2] The detailed rules can be found at http://wwwcs.uni-paderborn.de/fachbereich/ AG/monien/PERSONAL/FLULO/4PChess.html

[3] In this paper we use 'he' and 'his' whenever 'he or she' and 'his or her' are meant.

value alphabeta (Position v, value α, value β, remaining depth d, player i)
  // Let player 1 be the max-player, let the others be min-players.
  compute the feasible successor positions $v_1, \ldots, v_b$ of v.
  // Let $H(v)$ be the evaluation function for leaves.
  if (d $==$ 0 **or** b $==$ 0) return $H(v)$;
  for j $:=$ 1 to b
    if (max-player has to move) {
      $\alpha := \text{maximum}(\alpha, \text{alphabeta}(v_j, \alpha, \beta, d-1, (i+1) \bmod n))$;
      if $\alpha \geq \beta$ return α;
      if $j == b$ return α;
    } **else** {
      $\beta := \text{minimum}(\text{alphabeta}(v_j, \alpha, \beta, d-1, (i+1) \bmod n), \beta)$;
      if $\alpha \geq \beta$ return β;
      if $j == b$ return β;
    }

At the best, the algorithm finds out the minimax value of the root position and needs to examine only $O(b^{t \cdot (n-1)/n})$ leaves [16], n being the number of players, b a uniform branching factor, and t the search depth.

Hash tables, history heuristic, killer moves, null moves, zero-window search, and iterative deepening are important additional techniques that may enhance an α-β algorithm [12], such that the best case can be nearly achieved in practice.

3.2   The max^n-Player Type [7]

Again, let (V, E, H) be a game tree, but V being partitioned into n disjoint subsets for n different players. Let H be the heuristic evaluation function. In contrast to what we saw before, let it return a vector of profits. $H : V \to [0,1]^n$, with $\sum_{i=1}^{b} H_i(v) = 1$. Let $F_i(v)$, the ith component of a current max^n-vector of a node v, describe the profit of player i in position v.

profit-vector max^n (position v, lower bound on predecessor's profit
        $m = F_{i-1}(\text{predecessor}(v))$, remaining depth d, player i)
  compute the feasible successor positions $v_1, \ldots, v_b$ of v.
  // Let $H(v)$ be the evaluation function for leaves.
  if (d $==$ 0 **or** b $==$ 0) return $H(v)$;
  profit-vector $a := (-1, \ldots, -1)$;
  for j $:=$ 1 to b
    $F(v_j) := \text{max}^n \ (v_j, a, d-1, (i+1) \bmod n))$;
    if $(F_i(v_j) > a_i)$ $a := F(v_j)$;
    if $a_i > 1 - m$ return a; // shallow pruning
    if $j == b$ return a;

The algorithm performs a so-called shallow pruning, i.e., pruning which is caused by the predecessor of a cutoff node. Without speculation, this is the only possible pruning [15]. The following example shows that so-called deep pruning is not possible.

In Fig. 3, the third component of the profit vector of node (e) has the value 0.5. This is larger than $1 - (F_1(a)) = 1 - 0.6 = 0.4$, and therefore, player 1 will

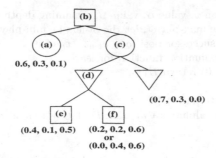

Fig. 3. Example for dangerous deep pruning

avoid that node (e) occurs on the board. Therefore, we might be willing to cutoff the node (f). This procedure is called 'deep pruning', because the grandfather node (b) of the node (d) causes the cutoff.

However, although the profit vector of the node (f) cannot reach the root (b) when the max^n-algorithm is used, (f) can change the root profit vector. First, let us assume that (f) has the vector $(0.2, 0.2, 0.6)$. In this case, player 3 will propagate the vector to node (d). As a consequence, player 2 will choose the vector $(0.7, 0.3, 0.0)$ for node (c). After all, player 1 assigns this vector $(0.7, 0.3, 0.0)$ to the root.

Now, let the value of node (f) be $(0.0, 0.4, 0.6)$. Player 3 will choose this for node (d) as well, but player 2 will propagate $(0.0, 0.4, 0.6)$ instead of $(0.7, 0.3, 0.0)$. Player 1 will see that node (a) is better for him than node (c), and the value vector of the root becomes $(0.6, 0.3, 0.1)$.

The paranoid player can also be interpreted as a special case of the max^n-player with an appropriate evaluation function. In this sense, our distinction between paranoid and max^n-player may look a bit artificial. We believe, however, that the strength of the 'special case' is severe enough that we feel encouraged to deal with it as a separate case.

3.3   The Careful max^n ($cmax^n$) Player Type

As we will see in the experimental section, the paranoid player does quite a good job. We tried to find out, what goes wrong with the max^n player, and we tried to find other updating rules for inner node values in order to beat the paranoid player.

The intuition behind the *careful max^n ($cmax^n$) player type* is the conjecture that the weakness of the max^n player comes from the fact that the players do not exactly know, what a real value vector of a position is, and that they do not know, how the other players estimate a position. Therefore, we assume that a player p, who has to move at node v, examines all successors, takes the observed profit vectors of the successors, and computes a weighted average over the successor profit vectors. We assumed it to be reasonable that a successor's weight becomes higher, the better it looks for player p. We assumed two advantages over the max^n updating rule. First, every position gets a unique value vector. Second, it is modeled that all players want to maximize their own profit, but that they are not sure about the opponents incentives. One possibility to realize this concept is as follows.

Let v be a node and $v_1, \ldots, v_b$ its successors, n the number of players. Let $p_1, \ldots, p_b = (p_1^1, \ldots p_1^n), \ldots, (p_b^1, \ldots, p_b^n)$ be the profit vectors of the nodes $v_1, \ldots, v_b$. Let us furthermore assume that player P has to move. Then we define weights $w_1^P, \ldots, w_b^P$ for the b successors with

$$w_i^P := \frac{(p_i^P)^2}{\sum_{j=1}^b (p_j^P)^2}. \tag{2}$$

The new profit vector for node v is then $(p_1, \ldots, p_b)$ with $p_i := \sum_{k=1}^b w_i^P \cdot p_i^k$. As the experimental section will show, this player type is a misconception.

3.4 Coalition Players

The idea of the *coalition player* is as follows. During the game, either all players are more or less equally strong, or one gets the best position, one player the worst, and two are somewhere in between. The three non-leading players have a certain interest to stop the strongest player, but the weakest cannot sacrifice anything. He has the least interest to attack the leading player. The strongest player knows that the others want to bring him back, and must therefore play against all three. He should especially play against the weakest player, in order to mate him, before his strength is reduced to an average level. The strongest player plays against all others. The weakest player plays against the strongest in order to defend himself. The other two players can start coordinated attacks against the strongest player, they can even sacrifice some game portion. We created a player, who assumes this coalition scheme and who tries to benefit from the coalition by examining fewer moves than the normal paranoid player. The player with the lowest game portion ignored moves of the coaliting players and the coaliting players ignored the moves of the player with the lowest game portion. This attempt is too restrictive and neither successful.

3.5 Coalition-Mixer Player Type

The basic idea of the *comixer player* and the resulting comixer algorithm is to improve the behavior of the paranoid α-β player. The paranoid player is a good defender of a position, but shows some lacks in the offense.

In order to add the idea of cooperative attacks to the paranoid player, we provide the comixer player with minimax values of various coalitions. All the help searches for minimax values were made with the help of the α-β algorithm, of course. Although the many small help searches can only be performed with a relative small search depth, this player type beats an efficient paranoid player type.

The Algorithm. For a rough description of the algorithm, we present the following pseudo code.

(value of position v, move (v, w)) comixer (position v, player M, depth parameter d)
   generate moves from v to the successors of v, named $v_1, \ldots, v_b$
   // Let b be the number of moves.
   // Let a paranoid value $p_i(v)$ be the minimax value that arises when player i
   // plays against all the others, preserving the move rights.
1 compute the paranoid values $p_i(v)$ for all players, using α-β searches with depth d
2 Let T be the player with best $p_T(v)$ with $M \neq T$.
   Select a set C of plausible coalitions. (Coalitions against T only.)
3 For all coalitions $c \in C$
4   For all moves (v, w_i) from v to successor $w_i, i \in \{1 \ldots b\}$
5   compute minimax value $e_c(v, w_i) :=$ alphabeta$(w_i, 0, 1, d)$
      // Needs modification of p.4 algorithm.
      // This leads to an total search depth of $d + 1$.
6 For all moves (v, w_i) from v to successor $w_i, i \in \{1 \ldots b\}$
7   compute the benefit value of the move (v, w_i): comix(v, w_i).
   //Comix is a piecewise linear function which combines the move values of the
   //various coalitions to one move value.
8 return the best move with its value.

Explanation of the Algorithm. In the case of three remaining players, there are three possible coalitions, in which two players can cooperate with each other. In the case of four players we have four coalitions in which one single player fights against the remaining players and three two-by-two coalitions. For our four-person Chess game this is sufficient, but you can easily generalize the approach for a n-player game by calculating the corresponding possible coalitions.

In order to reduce the number of coalitions to be examined, the comixer algorithm does not compute the values of all possible coalitions. Instead, we concentrate on coalitions against the player with the best position. Let M be the player, who has the right to move. The comixer selects that player $T \neq M$ as a target, who possesses the greatest portion of the game (see line 2 of the algorithm) except M itself. This decision is made by performing the paranoid α-β procedure with a low searching depth (see line 1 of the algorithm). After this, the comixer calculates all coalitions with M and T not simultaneously being a member (see line 3 of the algorithm). For example, in the four-player game it calculates the paranoid coalition against himself (all against M), the coalition against player T (all against T) and the two remaining two-by-two coalitions, in which T is not in the team of M. Note, that these coalitions can also be computed by the α-β algorithm, because the game is reduced to a two-player game.

For getting minimax values for the moves of player M at the present node v, the α-β algorithm is not started at the root, but at its successors (see line 4-5 in the algorithm).

The Mixing Functions. The goal is to make the paranoid player cooperate with other players when it seems necessary. For instance, in the situation that one player gets a relatively high game value, he may be able to dominate all other players and win the game. To prevent this, the cooperating coalitions get higher weights, the more player T's game value reaches the 50% border.

When we reach line 6 of the comixer algorithm, we know how good the moves, starting from v, are in the specific coalitions. In order to make the benefits of the moves comparable to each other, we divide the value of a move by the value of the best move in the specific coalition. Note, that player M minimizes in the paranoid coalition against T. Thus, the best move has a value, that is not greater than the value of the considered move. So here you have to divide the value of the best move by the value of the given move. Altogether you get the following definition.

Let $c \in C$ be the observed coalition and $h \in \{1, \ldots, b\}$ be the index, for which $e_c(v, w_h) \geq e_c(v, w_j) \; \forall j \in \{1, \ldots, b\}$. Then the relative strength of a move (v, w_i) in a coalition c is defined as

$$r_c(v, w_i) := min\{\frac{e_c(v, w_i)}{e_c(v, w_h)}, \frac{e_c(v, w_h)}{e_c(v, w_i)}\}. \tag{3}$$

Now we can mix the values of the different coalitions (this is where the name of the algorithm comes from). The total value $comix(v, w)$ of a given move (v, w) is calculated with the help of a set of k weight functions $f_1, \ldots, f_k$, which depend on the game portion of the player S with the highest game portion ($S = T$ or $S = M$), computed in line 1 of the comixer algorithm, whereas k is the number of the observed coalitions. A property of the f_i is, that

$$\sum_{c=1}^{k} f_c(x) = 1, \quad \forall x \in [\frac{1}{n}, 1]. \tag{4}$$

Then the comix function for a move with index i is defined as

$$comix(v, w_i) := \sum_{c=1}^{k} f_c(p_S(v)) \cdot r_c(v, w_i). \tag{5}$$

We split the description of the functions in two important parts.

(1) M is determined to be the player with the highest value in line 1 of the algorithm. In this case, the mixing functions are very easy. The paranoid coalition gets a continously very high weight of between 80% and 90% depending on the number of the remaining players. The more players there are, the less is the importance of the paranoid coalition, because more coalitions participate. The residual 10-20% are shared by the other coalitions.

(2) The more interesting case will occur, when M is determined to be not the player with the highest value. Here we use piecewise linear functions to combine the weights of the coalitions. If there are three remaining players (see Fig. 4) we only need to combine two coalitions, as mentioned above.

When player T has a value of 33% of the game, the paranoid coalition against M has 90% weight and the paranoid coalition against T has 10% weight. This ratio switches linearly until player T has 50% of the game. From this rate on up to 100%, the weights are constant.

If there are four players in the game, the ratios between the paranoid coalitions are similar (see Fig. 5).

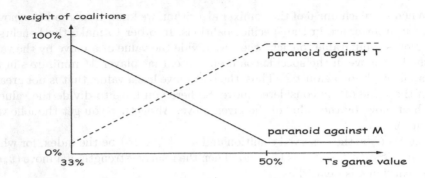

Fig. 4. Comix weight function for 3 players

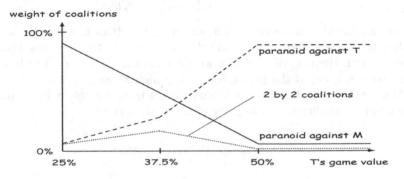

Fig. 5. Comix weight function for 4 players

Only the two two-by-two coalitions are added. Both begin with a value of 5%, when T's game value is 25% and reach their peak of 15% weight at 37.5% of T's game value. After that they fall down to 2% each at 50% of T's value. After this 50% line the functions again keep their values constant. Reasoned by the addition of the values of the two-by-two coalitions the values of the two paranoid functions (in particular the one against T) are decreased.

4   Experiments

Our generic four-person chess program consists of a move generator, a heuristic evaluation function which assigns a profit-vector to each position, and a search procedure. The evaluation function uses piece values for all pieces, similar as used in classic computer chess. A Knight, however, is not as much worth as a Bishop or three Pawns. Its value is only two Pawns. Moreover, we use mobility as a feature, and piece square tables in order to provide the players with a long-term idea where to place the pieces. The king cover consists of the existence/non-existence of the pawn shield, as well as of the distance between the pawn-shield and the King. The values are in relation to a player's portion of the game, i.e., the sum over all components of a profit vector is 1.

The search procedure depends on the player model used. In the case of the paranoid player, we use the α-β algorithm, enhanced with the killer heuristic, hash tables and iterative deepening. In all other cases, we use plain versions of the algorithms. Typical search data are the following. On a 2.4 GHz Pentium processor, the α-β algorithm searches about 1.1 million nodes per second without quiescence search and about 0.6 million nodes per second with quiescence search. Within 3 minutes computing time, it traverses search trees with depths between 7 and 8 without quiescence search, about 6 partial moves with quiescence search.

In order to play games, we set up 16 starting positions by hand. Six games were played with each of these 16 opening positions. Four players of two different types (i.e., two groups of two are identical players each) lead to six different orders, how the players can sit at the table. The two players of one group can sit at NE (North and East), SE, NW, SW, EW, or NS. The South-player starts the game. Thus, each contest consists of 96 games. In order to shorten the games, we adjudicate a game as soon as one player reaches more than 65% of the game portion. The semantic change of the players' actions can certainly be negotiated, and with this modification, a typical game takes between 200 and 300 quarter-moves. Mostly, at the end two players remain on board with one of them having more than 65% of the game portion.

4.1 Results

Table 1 presents some pairings with the paranoid player type and the max^n-player type involved, and the results. '+Qs' means quiescence search was used, '−Qs' means that it was not used. 'm/M' means minutes per move and tells us how many minutes time each side had for each move. We played 96 games per pairing, starting on a collection of 16 start positions. When a player type A scores more than 64 percent of the points against player type B, we may assume that this is not the effect of randomness.

Quiescence Search. Quiescence search is an important feature in traditional computer chess. The idea is that you should evaluate only so called quiet positions with the help of the static evaluation function. A position will be quiet if the player to move has no further taking moves. Quiescence search seems to be

Table 1. Results of pairings between the paranoid and the max^n-player type

Player 1	Player 2	Result
α-β +Qs , $3m/M$	α-β −Qs , $3m/M$	$46\frac{1}{6} : 46\frac{5}{6}$
max^n +Qs	max^n −Qs	$27\frac{5}{6} : 62\frac{1}{6}$
α-β +Qs , $3m/M$	max^n −Qs , $3m/M$	$86 : 10$
α-β +Qs, $3m/M$	max^n −Qs, $2m/M$	$84\frac{1}{2} : 11\frac{1}{2}$
α-β +Qs, $2m/M$	max^n −Qs, $3m/M$	$81\frac{5}{6} : 14\frac{1}{6}$
α-β +Qs, $3m/M$	α-β +Qs, $30s/M$	$64\frac{1}{2} : 31\frac{1}{2}$
max^n −Qs, $3m/M$	max^n −Qs, $30s/M$	$54 : 42$

less important in the 4-player chess game, because the α-β with quiescence search ($+$Qs) lost against the α-β without quiescence search ($-$Qs) with $46\frac{1}{6} : 46\frac{5}{6}$, and more distinct, the max^n $+$Qs lost against the max^n $-$Qs $27\frac{5}{6} : 62\frac{1}{6}$. For further experiments we use the quiescence search paranoid players, but not for the max^n players.

Paranoid α-β vs. max^n-Player. Although Sturtevant [14,15] already stated that the paranoid player was strong in several other games, the result of $86 : 10$ is astonishingly clear. The following example is shown in order to show the different behavior of the different player types.

In the position of Fig. 6, only three quarters of a move are played, but Yellow (East) is in danger already. If East continues with a careless move (e.g. if he moves his queen-bishop pawn one step forward), White (South) can play Bg3 (South's bishop on the file of the South's queen) and Blue (North) can already mate East. Only the max^n player moves South's bishop to g3, because the paranoid player assumes that North will not take East out of the game as from the paranoia's point of view North and East are in a coalition against South. If a paranoid player leads East's pieces, he will interestingly handle East's position correctly. If he leads North's pieces, he will also correctly eliminate East. In simple words, the paranoid player participates in cooperative attacks only in the terminator role, but he never initiates such attacks as we will see below.

We repeated the games with a predetermined search depth instead of restricting the time per move, in order to find out whether structural reasons or the

Fig. 6. Early elimination of a careless player

superior efficiency of the α-β algorithm are responsible for the max^n disaster. The result was 88 : 8 for the paranoid player, and therefore, we may conclude that the real reason for the weaknesses of the max^n player lies in the fact that he relies on the other players' actions to be performed as he does expect them. Let us again inspect the position of Fig. 2. Assume in Fig. 2 the max^n player has South's pieces and is to move, he may initiate a cooperative attack by playing Qxl8+ (South's Queen takes East's Pawn). Assume now that (1) West helps South and plays Qxm9+ (West's Queen takes East's Pawn) and that (2) North would not help East by playing Bxl8 (North's Bishop takes South's Queen), but helps South and West, East is mated and may leave the board.

If West is a max^n player he will indeed help South to mate East. Although West's Queen will vanish (because North will not (neither if it is a max^n nor if it is a paranoid player) help East with Bxl8, and South will thereafter take West's Queen), West has a certain advantage, because East is completely out of the game. If, however, West is a paranoid player, he will never participate in this attack, because he believes North prevents the elimination of East after Qxm9+ by moving Bxl8, whereafter he would simply lose his Queen, because the only move East has, is taking West's Queen with his Knight Nxm9. But if West does not participate in the attack, South will just lose his Queen, because East will take it in the next move. In other words: it is the existence of the paranoid player which makes the game of the max^n player wrong.

Differences in Thinking Times. In the way as we expected it, the α-β player type with 3 minutes per move wins against the α-β player type with 30 seconds per move *clearly* with $64\frac{1}{2} : 31\frac{1}{2}$. Again, the max^n player faces problems. The player who has more time per move available wins only by 54 : 42.

The Careful max^n ($cmax^n$) Players and the Coalition Player. Both player types could not gain any benefit from their models. The careful max^n player lost $81\frac{5}{6}$ to $14\frac{1}{6}$, and the coalition player $87\frac{5}{6}$ to $8\frac{1}{6}$ against the paranoid players.

The Comixer Algorithm. The comixer played 96 games against the paranoid α-β as well. The depth parameter of the comixer was fixed at 4, and the time that the paranoid player had available depended on the time which the comixer needed for its calculation. The α-β player then received the arithmetic middle of the last two times the comixer needed. With the help of this rule, the paranoid players reached searching depths between 7 and 9. The result of the match was $58\frac{1}{3}$ to $37\frac{2}{3}$ for the comixer. The main disadvantage with this approach is that we do not yet know how its performance scales with increasing machine power.

5   Conclusion

We made experiments with various player models in the area of the four-person chess game. We created a couple of candidate player models, and examined their relative strengths. Moreover, we investigated in how far computing power plays a role for the playing strength of the player models, and we investigated the effect of quiescence searches.

The paranoid player is quite strong and for a long time, we did not find any competitive other player model. After all, however, we were able to present the so-called comixer player type which performs better than the paranoid standard player. This proves that a simple worst-case analysis of the position is not the best choice, although being astonishingly effective. Moreover, some player models can better benefit from increasing computing power than others.

References

1. A. de Bruin, A. Plaat, J. Schaeffer, and W. Pijls. A Minimax Algorithm Better than SSS*. *Artificial Intelligence*, 87:255–293, 1999.
2. I. Althöfer. Root Evaluation Errors: How They Arise and Propagate. *ICCA Journal*, 11(3):55–63, 1988.
3. A.J. Jones. Game theory: Mathematical Models of Conflict. *West Sussex, England: Ellis Horwood*, 1980.
4. H. Kaindl and A. Scheucher. The Reason for the Benefits of Minmax Search. In *Proc. of the 11 th IJCAI*, pages 322–327, Detroit, MI, 1989.
5. D.E. Knuth and R.W. Moore. An Analysis of Alpha-Beta Pruning. *Artificial Intelligence*, 6(4):293–326, 1975.
6. U. Lorenz and B. Monien. The Secret of Selective Game Tree Search, When Using Random-Error Evaluations. *Proceedings of the 19th Annual Symposium on Theoretical Aspects of Computer Science (STACS), (H. Alt, A. Ferreira), Springer LNCS*, pages 203–214, 2002.
7. C.A. Luckhardt and K.B. Irani. An Algorithmic Solution of n-Person Games. *Proceedings AAAI-86, Philadelphia, PA*, pages 158–162, 1986.
8. A. Mas-Colell, M.D. Whinston, and J.R. Green. Microeconomic Theory. *Oxford University Press*, 1995.
9. J.F. Nash. Non-Cooperative Games. *Annals of Mathematics, 54*, pages 286–295, 1951.
10. D.S. Nau. Pathology on Game Trees Revisited, and an Alternative to Minimaxing. *Artificial Intelligence*, 21(1-2):221–244, 1983.
11. A. Reinefeld. An Improvement of the Scout Tree Search Algorithm. *ICCA Journal*, 6(4):4–14, 1983.
12. J. Schaeffer. The History Heuristic and Alpha-Beta Search Enhancements in Practice. *IEEE Transactions on Pattern Analysis and Machine Intelligence*, PAMI-11(1):1203–1212, November 1989.
13. C.E. Shannon. Programming a Computer for Playing Chess. *Philosophical Magazine*, 41:256–275, 1950.
14. N.R. Sturtevant. A Comparison of Algorithms for Multi-Player Games. *Proceedings of the 3rd Computers and Games conference (CG2002), (eds. J. Schaeffer, M. Müller and Y. Björnsson), LNCS 2883, Springer-Verlag, Berlin*, pages 108–122, 2003.
15. N.R. Sturtevant. *Multi-player games: Algorithms and Approaches*. PhD thesis, University of California, Los Angeles, 2003.
16. N.R. Sturtevant and R.E. Korf. On Pruning Techniques for Multi-Player Games. *Proceedings AAAI-00, Austin, Texas*, pages 201–207, 2000.

Solving Probabilistic Combinatorial Games

Ling Zhao and Martin Müller

Department of Computing Science,
University of Alberta, Edmonton, Alberta, Canada
{zhao, mmueller}@cs.ualberta.ca

Abstract. Probabilistic combinatorial games (PCGs) are a model for Go-like games recently introduced by Ken Chen. They differ from normal combinatorial games since terminal positions in each subgame are evaluated by a probability distribution. The distribution expresses the uncertainty in the local evaluation. This paper focuses on the analysis and solution methods for a special case, 1-level binary PCGs. Monte-Carlo analysis is used for move ordering in an exact solver that can compute the winning probability of a PCG efficiently. Monte-Carlo interior evaluation is used in a heuristic player. Experimental results show that both types of Monte-Carlo methods work very well in this problem.

1 Introduction

Heuristic position evaluation in game-playing programs should be a measure of the probability of winning. However, in point-scoring games such as Go or Amazons, evaluation functions usually approximate the score of the game instead of the winning probability. This can lead to serious blunders when programs make a risky move to gain even more territory, instead of playing conservatively to preserve a comfortable lead [8,7].

One approach to dealing with uncertainty of evaluation is to replace scalar evaluation by partially ordered values, such as probability distributions. For the case of minimax search such approaches were developed in [2,9]. Chen [7] extends this approach to the case of combinatorial sums of independent games [3], and defines the framework of *probabilistic combinatorial games (PCGs)*. Figure 1 shows an idealized example of how PCGs might be applied to Go. A board is split into three independent areas, and the result of local search in each area is represented by a PCG. Terminal nodes in the local search are evaluated by a probability distribution.

Chen [7] defined the PCG model and gave a high-level solution algorithm. This paper presents first computational results. The main contributions are as follows.

- Efficient exact and heuristic solvers for the special case of sums of 1-level binary PCGs, called *SPCGs*.
- An analysis of the search space of SPCGs.
- A Monte-Carlo move-ordering technique used in alpha-beta search for the exact SPCG solver.

H.J. van den Herik et al. (Eds.): ACG11, LNCS 4250, pp. 225–238, 2006.

- A Monte-Carlo based heuristic evaluation technique for SPCGs with performance close to that of the optimal player.
- An extensive experimental evaluation of the two solvers.

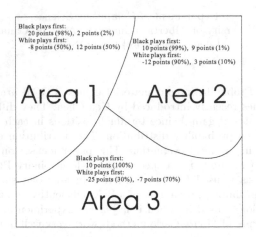

Fig. 1. PCG example: sum of three subgames

The remaining part of the paper is organized as follows. Section 2 gives definitions and notation for SPCGs, and introduces the main problems addressed in this paper. Section 3 analyzes game trees in SPCGs and describes methods to evaluate terminal and interior nodes. Section 4 develops the design of an exact solver and a strong heuristic player. Section 5 presents detailed experimental results and analysis. Future work is discussed in Section 6.

2   Definitions and Problem Identification

The following definition of a probabilistic combinatorial game (PCG) is adapted from [7].

1. A terminal position, represented by a probability distribution $d = [(p_1, v_1),$ $(p_2, v_2), \cdots, (p_n, v_n)]$, is a PCG. In d, the outcome is v_i with a probability of p_i, with $1 \leq i \leq n$, $0 \leq p_i \leq 1$ and $\sum_{i=1}^{n} p_i = 1$.
2. If A_1, A_2, $\cdots$, A_n and B_1, B_2, $\cdots$, B_n are all PCGs, then $\{A_1, A_2, \cdots, A_n \mid B_1, B_2, \cdots, B_n\}$ is a PCG. As in combinatorial games, the players are called Left and Right. A_1, A_2, $\cdots$, A_n are the Left options and B_1, B_2, $\cdots$, B_n are the Right options.
3. A sum of PCGs is a PCG. Summation is understood in the sense of combinatorial game theory [3]: a move in a sum game consists of a move to an option in exactly one subgame and leaves all other subgames unchanged.

PCGs are played as follows.

(1) A move can be played in any non-terminal subgame.
(2) If all subgames are terminal, the game itself is terminal, and its probability distribution is the sum of the distributions of all subgames.
(3) A game is won for Left if the value is greater than 0, and won for Right otherwise. From a terminal game, the probability of winning is computed by adding the probabilities of all values greater than 0.

The PCG model contains combinatorial games as the special case where all probability distributions in terminal positions have a single point distribution of the form $d = [(1, v)]$.

This paper focuses on a different special case of PCGs which models uncertainty but keeps the combinatorial games as simple as possible. A Simple PCG (SPCG) is a PCG which obeys the following constraints.

1. An SPCG consists of $n \geq 1$ subgames.
2. Each subgame has exactly one option for Left and one option for Right.
3. Each option immediately leads to a terminal position represented by a probability distribution.
4. Each distribution d in a terminal position has only two values with associated probabilities: $d = [(p_1, v_1), (p_2, v_2)]$.

An SPCG of n subgames is of the form $\sum_{i=1}^{n} \{d_i^L \mid d_i^R\}$, where both d_i^L and d_i^R are 2-valued probability distributions.

In SPCGs, a move can be simply indicated by the number of the subgame. In the experiments and analysis in this paper, without loss of generality, Left always moves first. This paper focuses on two basic problems concerning SPCGs.

1. Efficient exact solution: What is the winning probability of Left in a given SPCG?
2. Approximate solution: When solving an SPCG is too slow, how to play as well as possible under time constraints?

3   Game Tree Analysis of SPCGs

Game trees in SPCGs contain many regularities that can be exploited in an efficient solver. Assume a game G has n subgames. There are n choices of a subgames for the first move, $n-1$ choices for the next move, and the kth move has $n - k$ choices. If the root of a game tree is defined as a node at depth 0, then any node at depth k has exactly $n-k$ children, and there are $n!/(n-k)!$ nodes in total at depth k. The total number of nodes in the game tree is $\sum_{k=0}^{n} n!/(n - k)! = \sum_{k=0}^{n} n!/k!$, and there are $n!$ terminal nodes.

Since all subgames are independent, the order of play of the subgames chosen by the same players does not change the evaluation. Each position can be represented by three sets: the *Left set (Right set)* contains the indices of all subgames played by Left (Right), and the *open set* contains the indices of all subgames

Table 1. Statistics for SPCGs with 12 to 15 subgames

Subgames	All nodes	Terminal nodes	All distinct nodes	Distinct terminal nodes
12	1,302,061,345	479,001,600	143,365	924
13	16,926,797,486	6,227,020,800	414,584	1,716
14	236,975,164,805	87,178,291,200	1,201,917	3,432
15	3,554,627,472,076	1,307,674,368,000	3,492,117	6,435

that have not been played yet. For example, the sequence of subgames chosen $(1, 2, 3)$ is equivalent to the sequence of $(3, 2, 1)$. If $n = 5$, then the Left set after both these sequences would be $\{1, 3\}$, the Right set $\{2\}$, and the open set $\{4, 5\}$.

For a position after k moves (a node at depth k in the game tree), k subgames have been played by the two players, $\lceil k/2 \rceil$ by Left and $\lfloor k/2 \rfloor$ by Right. The number of distinct nodes at depth k is equal to the number of possible combinations of left and right sets, There are $\binom{n}{\lfloor n/2 \rfloor}$ distinct terminal nodes, and the total number of distinct nodes in the game tree is $\sum_{k=0}^{n} \binom{n}{k}\binom{k}{\lfloor k/2 \rfloor}$.

As an example, Table 1 lists the number of nodes for SPCGs with 12 to 15 subgames. The total number of distinct nodes in a game increases roughly by a factor of 3 per subgame. The number of distinct terminal nodes increases by a factor close to 2 (exactly 2 if n is odd, $2 - \frac{2}{n+2}$ if n is even). There is a very large number of transpositions in the game tree, and there are many more distinct interior nodes than terminal nodes.

3.1   Terminal Node Evaluation

In an SPCG with n subgames, the value of each terminal node T in its game tree is a sum of n probability distributions:

$$T = \sum_{i=1}^{n} [(p_{1,i}, v_{1,i}), (p_{2,i}, v_{2,i})]. \tag{1}$$

T itself can be expressed as a single, complex probability distribution over sums of values $v_{1,i}$ and $v_{2,i}$. The winning probability of T for Left is the sum of all probabilities of values greater than 0 in T. The following formula expresses the winning probability P_w of T. Let $q(i) \in \{1, 2\}$ $(1 \le i \le n)$ be such that the result value $v_{q(i),i}$ is chosen at subgame i. Then

$$P_w = \sum \{p_{q(1),1} p_{q(2),2} \cdots p_{q(n),n} | \sum_{i=1}^{n} v_{q(i),i} > 0\},$$

$$\forall 2^n \text{ choices of } q(1), q(2), \cdots, q(n). \tag{2}$$

There are at least three methods to evaluate P_w.

1. Direct evaluation of all 2^n combinations above.
2. A dynamic programming algorithm, representing the distribution as a list of bins and adding one distribution at a time. This is effective when the

values v are all integers within a small range. However, in the worst case (real numbers or large integers), the final distribution can contain up to 2^n distinct non-zero entries, and this algorithm does not improve on the direct one.

3. Using the fact that the probability density of a sum is the convolution of the probability density functions of summands. Again, in the worst case the result will contain up to 2^n non-zero components.

In the experiments reported in this paper, for the range of n tested, methods 1 and 2 perform similarly well. Method 3 was not implemented. Method 2 can be further improved by noting that the final result required is only the value P_w, not the exact distribution. All intermediate results that are far enough from 0 to ensure a win or loss can be bagged and removed from further processing. This improves the efficiency. In experiments with $n = 14$ subgames, optimized dynamic programming was about twice as fast as the direct approach for a range of values $v \in [-1000, 1000]$ but twice as slow for a larger range $v \in [-5000, 5000]$. Since the dynamic programming version is efficient only when the range is limited, method 1 was used for all further experiments and discussion in this paper.

It is costly or even infeasible to compute the exact winning probability of a terminal node when n is large. In such a situation, it is desirable to have a good approximation method with controllable statistical error. In cases when a quick estimate is sufficient, approximation can improve the performance of an SPCG solver or player.

Monte-Carlo sampling is used to approximate winning probabilities. For each distribution d, a value is generated. For example, if $d = [(p_1, v_1), (p_2, v_2)]$, then v_1 is generated with probability p_1 and v_2 with probability $p_2 = 1 - p_1$. In an SPCG terminal node T, n values are generated from the n distributions of subgame outcomes, and the sum of these n values is the result of the sample. Left wins if and only if the sum is greater than 0. The fraction of wins P_w^k from k independent samples is used to approximate the winning probability in T.

For each sample, the winning probability is P_w, and $1 - P_w$ is the probability of a loss. The mean of the distribution is P_w, and the standard deviation is $\sqrt{P_w(1 - P_w)}$. According to the Central Limit Theorem, the mean of random samples drawn from a distribution tends to have a normal distribution. When k is sufficiently large, P_w^k has a normal distribution with mean P_w and standard deviation $\sqrt{\frac{P_w(1-P_w)}{k}} \leq \frac{1}{2\sqrt{k}}$. This inequality can be used to select the minimum value of k for a given required accuracy. For example, according to the normal distribution, almost for sure (99.7%) the difference between P_w and P_w^k is no more than $\frac{3}{2\sqrt{k}}$. With 10,000 samples, the difference is 99.7% likely to be within 0.015.

Experiments shown in Table 2 list the difference between P_w and P_w^k in random terminal nodes, and compare it with the theoretical bounds. For games consisting of n subgames, 100 terminal nodes are randomly chosen as test cases. For each terminal node, probabilities were uniformly generated from between 0%

and 100% with granularities of 1% and 0.001%. Values were uniformly generated between -1000 to 1000 with a granularity of 1.

For each terminal node, the error estimate, which is computed by $P_w - P_w^k$, is recorded for k from 3 to 33333. The standard deviation of the estimated error from the 100 terminal nodes is compared with its theoretical bound $(\frac{1}{2\sqrt{k}})$ in Table 2 for n is 14, 17, and 20, respectively.

Table 2. Standard deviation with probability granularity of 1% and 0.001%

Granularity	# Samples								
1%	3	10	33	100	333	1,000	3,333	10,000	33,333
$1/2\sqrt{k}$	0.2887	0.1581	0.0870	0.0500	10.0274	0.0158	0.0087	0.0050	0.0027
14 subgames	0.1768	0.1118	0.0488	0.0320	0.0189	0.0118	0.0070	0.0050	0.0042
17 subgames	0.2294	0.1176	0.0717	0.0348	0.0203	0.0121	0.0074	0.0059	0.0052
20 subgames	0.2427	0.1020	0.0580	0.0314	0.0246	0.0121	0.0077	0.0059	0.0045

Granularity	# Samples								
0.001%	3	10	33	100	333	1,000	3,333	10,000	33,333
$1/2\sqrt{k}$	0.2887	0.1581	0.0870	0.0500	0.0274	0.0158	0.0087	0.0050	0.0027
14 subgames	0.1893	0.0977	0.0655	0.0322	0.0190	0.0107	0.0053	0.0034	0.0019
17 subgames	0.1923	0.1093	0.0619	0.0419	0.0206	0.0125	0.0064	0.0038	0.0019
20 subgames	0.2091	0.1154	0.0634	0.0329	0.0199	0.0130	0.0064	0.0034	0.0019

The experimental results match the prediction very well when there is a fine granularity for randomly generated probability, but the model is not appropriate for coarse granularity when large sample sizes are used.

The term *Monte-Carlo terminal evaluation* (MCTE) is used to refer to the Monte-Carlo method discussed in this subsection. MCTE estimates the winning probability of a terminal node.

3.2   Monte-Carlo Sampling for Heuristic Interior Node Evaluation

Left's winning probability at an interior node, which is represented by its Left, Right and open sets, can be computed by a complete minimax search, using the exact evaluation of Subsection 3.1 in all terminal nodes of the search. However, when full search is too slow, Monte-Carlo sampling is useful for improving heuristic evaluation as well. Such methods are popular in games with incomplete information such as Poker [4], and also in games with complete information such as Go [6]. Abramson's expected-outcome evaluation [1] evaluates a node in a search tree by averaging the values of terminal nodes reached from it through random play. This method is adapted here. Similar ideas are developed in [10,2].

From an interior node, the sequence of alternate moves by both players to reach a terminal node is simulated by randomly choosing each move of the sequence among all its legal choices with equal probability. Such a simulation is iterated k times, and the average winning probability of the k sampled terminal nodes is an approximation of the winning probability of the interior node. The winning probability at terminal nodes can be either accurately computed by using Formula (1) in Subsection 3.1, or approximated by methods such as MCTE.

In SPCGs, the order of moves chosen by the same player is irrelevant. For efficiency, a move sequence simulation can be replaced by uniformly randomly selecting k nodes out of all descendant terminal nodes below this interior node. Such sampling is only meaningful if k is (much) smaller than the total number of descendant terminal nodes of an interior node.

Experiments measure the difference between P_w and P'_w. For an interior node, its exact game value P_w and the approximate value P'_w estimated by the average winning probability of all its descendant terminal nodes are computed. The test set consists of starting positions from 100 randomly generated games with $n = 14$ subgames. The mean of $|P_w - P'_w|$ is 0.148 in the experiments. Since the difference is large, P_w can not be approximated well using P'_w. However, for a set of interior nodes at the same depth, their errors are highly correlated, and the relative order of their P'_w values is a very good approximation of the order of their P_w values. Subsection 5.4 gives detailed results. P'_w is an excellent move ordering heuristic.

In contrast to MCTE, the term *Monte-Carlo interior evaluation* (MCIE) is used to refer to the Monte-Carlo sampling discussed in this subsection. MCIE estimates the winning probability at an interior node. *Monte-Carlo move ordering* uses MCIE for move ordering at interior nodes in a search.

Monte-Carlo sampling is used in both the Monte-Carlo terminal evaluation and the interior evaluation, but they serve different purposes, and have different control parameters. These two types of evaluation are used in an exact SPCG solver and a heuristic SPCG player, which will be discussed in the next two sections.

4   Exact Solver and Heuristic Player for SPCGs

In order to compute the exact winning probability of a game, a complete solver was implemented based on alpha-beta search. Standard enhancements including transposition tables and move ordering using the history heuristic [11] are used.

The optimized solver spends more than 90% of its time on the accurate evaluation of terminal nodes. Since an SPCG game tree has far more interior than terminal nodes, most terminal nodes must be evaluated in order to solve the game. Unless a more efficient method can be found to evaluate terminal nodes, it is difficult to further improve the overall performance of the solver.

For cases when it is infeasible to solve an SPCG, or when it is desirable to play the game fast with reasonable strength, a heuristic player was designed as follows. A fixed depth alpha-beta search is performed. Non-terminal frontier nodes of the search are evaluated using MCIE. Within MCIE, terminal nodes are evaluated by MCTE for efficiency.

A heuristic Monte-Carlo player with search depth of 1 is the same as the expected-outcome player in [1]. It performed well in experiments. This player chooses its move as follows: it finds all legal moves from the starting position, generates a depth-1 interior node for each move, and compares these nodes using MCIE. The move that leads to the highest-valued depth-1 node is chosen. This simple technique performs very well in SPCGs. An advantage of this 1-ply Monte-Carlo sampling method is that it is friendly to time control. Each sampling process costs almost the same amount of time, and it can be performed continuously until time runs out. Experimental results for the solver and heuristic player are given in the next section.

5   Experimental Results and Analysis

This section summarizes experimental results for different configurations of the SPCG solver and the Monte-Carlo SPCG player, and measures how Monte-Carlo move ordering performs in games.

5.1   Experimental Setup

All experiments were run on Linux workstations with AMD 2400MHz CPUs. The compiler was gcc 3.4.2. The transposition table has 2^{20} entries, using about 34MB memory. Since the evaluation of terminal nodes is expensive, they can never be overwritten by interior nodes in the transposition table.

A set of 100 randomly generated games with $n = 14$ subgames are used for testing. Probabilities were uniformly generated between 0% and 100% with a granularity of 1% and values between -1000 to 1000 with a granularity of 1. Table 3 contains statistics for solving these instances. Each cell is of the form: mean value ± standard deviation. The first player has a big advantage in SPCG, so the average winning probability is much larger than 50%.

Table 3. Statistics of solving the 100 random games

Time:	8.2 ± 0.8 sec
Total nodes visited:	$(2.2 \pm 1.2) \times 10^5$
Terminal nodes visited:	$(2.5 \pm 0.2) \times 10^3$
Cache hits:	$(1.3 \pm 0.8) \times 10^4$
Winning probability:	$70.0\% \pm 25.8\%$

5.2   Performance of the Exact Solver

Monte-Carlo move ordering has two important parameters: the percentage of the total number of descendant terminal nodes to be sampled, n_t (see Subsection 3.2), and the number of value combinations to be sampled for approximate evaluation of each terminal node, n_c (see Subsection 3.1). The Monte-Carlo move ordering uses MCIE, which contains two levels of Monte-Carlo sampling: on the top level, a number of an interior node's descendant terminal nodes are randomly chosen. On the bottom level, each of these terminal nodes is evaluated by the average result of a number of value combinations sampled from the value distributions of the node. Finally, the evaluation of the interior node is the average evaluation of those terminal nodes sampled.

The solver has a depth limit for Monte-Carlo move ordering (d_m). Nodes at depth up to d_m use MCIE for move ordering, and others use the history heuristic. The total time for solving the 100 test games for different parameter combinations is given in Fig. 2.

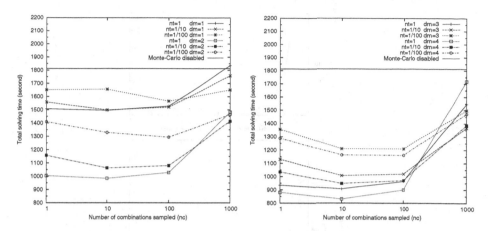

Fig. 2. Solving time with respect to parameters in Monte-Carlo move ordering

In the figure, the straight lines parallel to the x-axis denote the solving time when Monte-Carlo move ordering is disabled, and the solver uses only the history heuristic for move ordering. The large difference between these lines and the best solving time demonstrates that Monte-Carlo move ordering is superior to the history heuristic.

Even when only a small number of value combinations are sampled (n_c = 1, 10, 100), the terminal node evaluation seems to provide a reasonably good estimate of the approximate value of interior nodes. $n_c = 1000$ is too slow to be practical.

In the range of n_c from 1 to 100, for each depth limit d_m the solving time almost always increases when n_t decreases. This indicates that the more descendant terminal nodes sampled, the better the performance will be. Thus it is best to sample all those terminal nodes whenever possible.

The depth limit for Monte-Carlo move ordering is an intricate parameter that needs a trade-off between search efficiency and overhead computation. Figure 3 shows the solver's performance for d_m from 0 to 14 for fixed $n_t = 1$ and $n_c = 10$. With $d_m = 0$ only the history heuristic is used for move ordering, and with $d_m = 14$ only Monte-Carlo move ordering is used.

Figure 3 clearly shows that for $d_m \geq 4$, the solver's performance is relatively stable in terms of solving time and number of nodes visited. This fact suggests that the Monte-Carlo move ordering is more powerful for nodes close to the root of the game tree. It is also evident in the bottom graph of Fig. 3 that the number of terminal nodes gradually decreases as d_m increases, which proves that the Monte-Carlo interior evaluation is more accurate than the history heuristic for move ordering in this game.

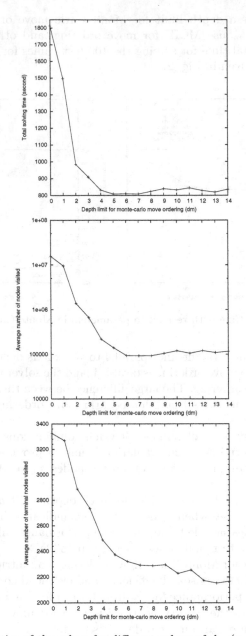

Fig. 3. Statistics of the solver for different values of d_m ($n_t=1$, $n_c=10$)

5.3   Performance of the Monte-Carlo Player

The relative strength of two players is tested by playing 100 random games twice, switching colors for the second game. A player's winning probability of a game is the average of its winning probabilities in the two rounds. The perfect player

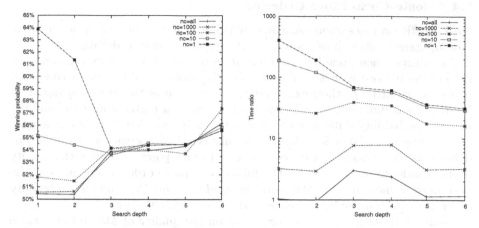

Fig. 4. The perfect player against the Monte-Carlo player: winning probability and time ratio

always achieves 50% winning probability against itself, and at least 50% against any player.

The Monte-Carlo player has two parameters: the search depth (at least 1), and n_c for the approximate evaluation of terminal nodes. It was tested with different configurations of these two parameters, and against the perfect player. Figure 4 illustrates the winning probability of the perfect player as well as the ratio of the time the perfect player spent to that of its Monte-Carlo counterpart. The parameter d_m is set to the search depth (meaning only MCIE is used), and $n_t = 1$. In the figure, $n_c = all$ means that all terminal nodes are evaluated exactly.

The left graph in Fig. 4 reveals an interesting property of the Monte-Carlo player: increased search depth does not necessarily lead to better winning probability, and sometimes even weakens the player. The Monte-Carlo player with accurate terminal node evaluation is best against the perfect player, but it uses almost the same amount of time, so it is not practical.

When the number of value combinations sampled, n_c, is from 100 to 1000 while the search depth is very shallow (1 or 2), the Monte-Carlo player performs very well. Especially when $n_c = 1000$ and the search depth is 1, the perfect player only achieves a winning probability of 50.56%. The Monte-Carlo player is just about 1% weaker.

The Monte-Carlo player with $n_c = 1000$ needs time comparable to the perfect player (about 25% to 30%), and is much slower than the player with $n_c = 100$. The latter player seems to offer a good compromise between time and accuracy: it plays games quickly, using 4% of the perfect player's time, and is only about 3% weaker.

It is not surprising that players with small n_c do not perform well. Insufficient sampling leads to large errors. The graph on the right side in Fig. 4 suggests that a deeper search can make up for insufficient sampling to some extent.

5.4   Monte-Carlo Move Ordering

Since MCIE provides good estimates, it is important to provide a quantitative measurement to show how good it is with regard to move ordering.

As an experiment, one interior node at each depth is randomly selected from each of the 100 test games. For each node, all pairs of legal moves are compared to test if the order of their exact values is the same as the order suggested by MCIE. If it is different, then Monte-Carlo evaluation makes a mistake, and the winning probability difference of the two moves is recorded. This value denotes the winning probability lost due to the mistake. The *average probability error*, defined as the average of this value over all move pairs, measures the quality of move ordering for this node. The influence of move ordering on choosing the best move is measured by the *worst probability error*, the winning probability difference between the best move and the move suggested by MCIE.

Figure 5 illustrates the influence of n_c on the quality of Monte-Carlo move ordering. Data points with probability error of 0 are not shown due to the logarithmic scale used. Again, n_c = all means that accurate terminal node evaluation is performed. The two graphs in Fig. 5 clearly show that Monte-Carlo evaluation provides nearly perfect move ordering with marginal error if there are a substantial number of value combinations sampled. Even when n_c is small, 10 or 100, the worst probability error is still less than 2%. The error becomes larger when the node is closer to terminal nodes.

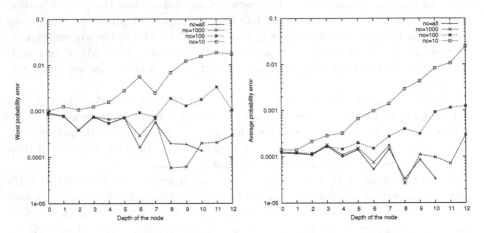

Fig. 5. Probability error due to incorrect move ordering

6   Conclusions and Future Work

This paper investigates SPCGs, sums of 1-level probabilistic combinatorial games, and discusses methods to solve them, as well as strong heuristic players. Properties of this game are analyzed, and experimental results clearly show that Monte-Carlo methods are useful for evaluation of both terminal and interior nodes.

Quite interestingly, comparing the average winning probability of all descendant terminal nodes of an interior node is a good indicator of the relative value as measured by winning probability. Experimental results show convincingly that this heuristic performs well in move ordering for this game, and that is why it could bring a big performance increase to the complete solver and also why the Monte-Carlo player based on it performs well against the perfect player. It is interesting to investigate when and why Abramson's simple expected-outcome evaluation provides a good heuristic in games, and the SPCG seems to be a good abstract model to study.

The SPCG solver and the Monte-Carlo player still have room for improvement. The bottleneck of the solver lies in the accurate evaluation of terminal nodes. An improvement on sampling strategy would be to incorporate progress pruning and selective search [4,6,5]. In the current Monte-Carlo player, each legal move is sampled at the same frequency. However, it is more efficient to use an adaptive strategy such that most of the effort is spent on those moves that have a high chance of being selected.

An SPCG solver could be incorporated in a Go program. The solver would be used on the high level to instruct the program to maximize its winning probability. As proposed in [7], such an approach might help to improve the playing strength of current Go programs. It would present a significant new application of Monte-Carlo methods.

Acknowledgements

The authors would like to thank Markus Enzenberger and David Silver for their valuable comments on this paper. This work has been supported by the Natural Sciences and Engineering Research Council of Canada (NSERC), the Alberta Informatics Circle of Research Excellence (iCORE), and the Alberta Ingenuity Fund.

References

1. B. Abramson. Expected-outcome: A General Model of Static Evaluation. *IEEE transactions on PAMI*, 12:182–193, 1990.
2. E. Baum and W. Smith. A Bayesian Approach to Relevance in Game-Playing. *Artificial Intelligence*, 97(1-2):195–242, 1997.
3. E. R. Berlekamp, J. H. Conway, and R. K. Guy. *Winning Ways for your Mathematical Plays*. Academic Press, 1982.
4. D. Billings, A. Davidson, J. Schaeffer, and D. Szafron. The Challenge of Poker. *Artificial Intelligence*, 134(1-2):201–240, 2002.
5. B. Bouzy. Associating Shallow and Selective Global Tree Search with Monte Carlo for 9x9 Go. In *4th Computer and Games Conference (CG 2004)* (eds. H. J. van den Herik, Y. Björnsson, and N. S. Nethanyahu), LNCS 3846, Springer-Verlag, Berlin, Pages 67–80, 2006.
6. B. Bouzy and B. Helmstetter. Monte Carlo Go Developments. In *Advances in Computer Games conference (ACG-10)* (eds. H.J. van den Herik, H. Iida, and E.A. Heinz), Kluwer Academic Publishers, Boston, pages 159–174, 2003.

7. K. Chen. Maximizing the Chance of Winning in Searching Go Game Trees. *Information Sciences*, 175(4):273–283, 2005.
8. M. Müller. *Computer Go as a Sum of Local Games: An Application of Combinatorial Game Theory*. PhD thesis, ETH Zürich, Diss. ETH Nr. 11.006, 1995.
9. M. Müller. Partial Order Bounding: A New Approach to Evaluation in Game Tree Search. *Artificial Intelligence*, 129(1-2):279–311, 2001.
10. A. Palay. *Search with Probabilities*. Morgan Kaufmann, 1985.
11. J. Schaeffer. The History Heuristic and the Performance of Alpha-Beta Enhancements. *IEEE Transactions on Pattern Analysis and Machine Intelligence*, 11(11):1203–1212, 1989.

On Colored Heap Games of Sumbers[*]

Kuo-Yuan Kao

Department of Information management,
National Penghu University, Taiwan
stone@npu.edu.tw

Abstract. A sumber is a sum of ups, downs and star. Sumbers can describe the positions of many partisan infinitesimal games. Earlier, we provided a simplification rule [6] that can determine whether a game G is a sumber or not, and if it is, determine the exact sumber of G from its left and right options, G^L and G^R. This article extends the previous result and presents three variations of colored heap games; each of them can be solved by sumbers.

1 Introduction

We are concerned with combinatorial games and follow the notations and conventions of *Winning Ways* [1]. We also assume the readers are familiar with *numbers* [3] and *nimbers* [2,4]. Numbers and nimbers are well-known game subgroups with the following two properties.

1. There exists a simple rule to determine the outcome of any game in the subgroup.
2. There exists a simplification rule that can simplify games in the subgroup.

In March 2005, we presented another subgroup, *sumbers* [6], having the above properties. This section briefly reviews the definitions and major results about sumbers.

Definition 1. *For any number d, define*

$$\uparrow(d) = \{\uparrow(d^L), *|*, \uparrow(d^R)\}, \tag{1}$$

where $* = \{0|0\}$ (pronounced star). $\uparrow$ followed by the empty set is the empty set.

Each $\uparrow(d), d > 0$, is called an *up*. The negation of an up is called a *down*. We use the notation $n{\cdot}\uparrow(d)$ to denote the sum of n copies of $\uparrow(d)$.

Definition 2. *A sum of ups, downs, and star is called a sumber. A sumber S can be written in the standard form:*

$$S = s_0 \cdot * + \sum_{k=1,n} s_k{\cdot}\uparrow(d_k), \tag{2}$$

where $s_0 = 0$ or 1, $s_k \neq 0, 0 < k \leq n$, and $0 < d_k < d_{k+1}, 0 < k < n$. $\sum_{k=1,n} s_k$ is called the net weight of S.

---

[*] This paper is sponsored by National Science Council, Taiwan. [93-2213-E-346-006].

H.J. van den Herik et al. (Eds.): ACG11, LNCS 4250, pp. 239–246, 2006.

The outcome of a sumber can be determined by theorem 1 [6].

Theorem 1. *Let S be a sumber in the above standard form. Then,*

- $S > 0$ *if and only if* $(\sum_{k=1,n} s_k > s_0)$ *or* $(\sum_{k=1,n} s_k = s_0$ *and* $s_1 < 0)$;
- $S < 0$ *if and only if* $-S > 0$;
- $S = 0$ *if and only if* $n = 0$ *and* $s_0 = 0$;
- $S \mid 0$, *otherwise.*   □

Definition 3. *Let S be a sumber in the above standard form. For each $m \in \{0, d_k : 1 \leq k \leq n\}$, define*

$$S^m = \sum_{k=1,n; \ d_k \geq m} s_k \cdot \uparrow(d_k) \quad - \sum_{k=1,n; \ d_k \geq m} s_k \cdot \uparrow(m). \quad (3)$$

We say S has a cut at m if $S^m \leq 0$.

Each sumber has at least one cut. We are only concerned with the *minimum cut*: the smallest number in $\{0, d_k : 1 \leq k \leq n\}$ which is a cut. When S is a sumber, and m is the minimum cut of S, we call S^m the *upper section* and

$$S_m = S - S^m \quad (4)$$

the *lower section* of S.

Sums of ups and downs (excluding $*$) are totally ordered. If G^L and G^R are sets of sumbers, then G has at most two non-dominated options, one contains $*$ and the other does not, in each of G^L and G^R. In other words, G can be simplified as

$$G = \{A, B \mid C, D\}, \quad (5)$$

where A, B are the non-dominated options in G^L and C, D are the non-dominated options in G^R.

Definition 4. *Let $G = \{A, B \mid C, D\}$, where A, B, C, D are sumbers and the net weight of C is less than or equal to the net weight of D. The critical section $X(G)$ of $G = \{A, B \mid C, D\}$ is defined as the set of numbers $x \geq m$ satisfying all the following inequalities:*

$$\uparrow(x) \mid > A - C_m + *,$$
$$\uparrow(x) \mid > B - C_m + *,$$
$$\uparrow(x) < \mid C - C_m + *,$$
$$\uparrow(x) < \mid D - C_m + *,$$

where m is the minimum cut of C.

Theorem 2 [6] can simplify games with sumber options.

Theorem 2. *Let $G = \{A, B \mid C, D\}$, where A, B, C, D are sumbers and the net weight of C is less than or equal to the net weight of D.*

- If $A < |0$, $B < |0$ and $C| > 0$, $D| > 0$ then $G = 0$.
- If $A \parallel *$, $B \parallel *$, and $C \parallel *$, $D \parallel *$ then $G = *$.
- Otherwise (without loss of generality, we may assume either $G > 0$ or $G > *$), G is a sumber if and only if $X(G)$ is not empty. Moreover, when G is a sumber,

$$G = C_m + * + \uparrow(p) \qquad (6)$$

where m is the minimum cut of C and p is the simplest number in $X(G)$. □

Sumbers can describe the positions of many partisan infinitesimal games. In each of the next three sections, we study one variation of a colored heap game.

2   Up-Down Game

The Up-Down game was first proposed by K. Y. Kao [5]. It is played on a number of heaps of counters. Each counter is colored either black or white. Left and Right move alternatively and their legal moves are different.

- When it is L's turn to move, he[1] can choose any one of the heaps and repeatedly removes the top counter until either he removed a white counter or the heap has become empty.
- When it is R's turn to move, he can choose any one of the heaps and repeatedly removes the top counter until either he removed a black counter or the heap has become empty.

The player who removes the last counter is the winner.

Let S be a heap, we use the notation $S : B$ (or $S : W$) to denote the heap by adding a black (or white) counter on top of S.

Proposition 1. *Each colored heap in Up-Down game is a sumber (or the negation of a sumber) of the form:*

$$S = s_0. * + \sum_{k=1,n} s_k \cdot \uparrow(d_k), \qquad (7)$$

where $s_0 = 0$ or 1, $s_k > 0$, $0 < k \leq n$ and $0 < d_k < d_{k+1}, 0 < k < n$.
Moreover, a heap with one single counter has the value $*$ (in this case, we can assume $n = 0$ and $d_0 = 0$.)
When S has at least two counters, and the color of the counter next to the bottom counter is black,

$$S : B = \{S^L | S\} = S + \uparrow(d_n) + *, \qquad (8)$$
$$S : W = \{S | S^R\} = S - \uparrow(d_n) + \uparrow(d_n + 1). \qquad (9)$$

---

[1] In this paper, we use 'he' and 'him' wherever 'he or she' and 'his or her' are meant.

When S has at least two counters, and the color of the counter next to the bottom counter is white,

$$S : B = \{S^L | S\} = S + \uparrow(d_n) - \uparrow(d_n + 1), \qquad (10)$$

$$S : W = \{S | S^R\} = S - \uparrow(d_n) + *. \qquad (11)$$

Proof. The proof is by induction. We assume S is a sumber in the standard form. Consider the case where S has at least two counters, and the color of the counter next to the bottom counter is black. According to the rule of the game, we have $S : B = \{S^L | S\}$. Since d_n is the minimum cut of S, the critical section of $S : B$ is

$$X(S : B) = \{x \geq d_n : S^L - S + * < | \uparrow(x)\}$$
$$= \{x \geq d_n : \uparrow(d_n - 1) - \uparrow(d_n) + * < | \uparrow(x)\} = \{x \geq d_n\}. \qquad (12)$$

The simplest number in $X(S : B)$ is d_n. Thus, according to theorem 2,

$$S : B = S + \uparrow(d_n) + *. \qquad (13)$$

According to the rule, $S : W = \{S | S^R\}$. The critical section of $S : W$ is

$$X(S : W) = \{x : S - S^R + * < | \uparrow(x)\}$$
$$= \{\uparrow(d_n) < | \uparrow(x)\} = \{x > d_n\}. \qquad (14)$$

The simplest number in $X(S : W)$ is $d_n + 1$. Thus,

$$S : W = S^R + \uparrow(d_n + 1) + * = S - \uparrow(d_n) + \uparrow(d_n + 1). \qquad (15)$$

The case where the color of the counter next to the bottom counter is white can be proven in a similar way. $\qquad \Box$

Proposition 1 can help us figuring out the exact sumber of any Up-Down heap. The color of the counter next to the bottom counter determines whether the sumber is positive or not. If it is black then the sumber is positive, otherwise (white) the sumber is negative. When the sumber of a heap is positive, the number of black counters (other than the bottom counter) equals the net weight of the sumber; the number of white counters plus 1 equals the highest order of the up terms in the sumber.

Example 1. Consider the heap $BBBWWB$ (from bottom up).
The color of the counter next to the bottom counter is black. By repeatedly applying proposition 1, we have

$$B : B = * : B = \uparrow(1),$$
$$BB : B = BB + \uparrow(1) + * = 2 \cdot \uparrow(1) + *,$$
$$BBB : W = BBB - \uparrow(1) + \uparrow(2) = \uparrow(1) + \uparrow(2) + *,$$
$$BBBW : W = BBBW - \uparrow(2) + \uparrow(3) = \uparrow(1) + \uparrow(3) + *,$$
$$BBBWW : B = BBBWW + \uparrow(3) + * = \uparrow(1) + 2 \cdot \uparrow(3). \qquad (16)$$

Thus,

$$BBBWWB = \uparrow(1) + 2 \cdot \uparrow(3). \qquad \Box$$

3   Up-Down Game II

The setup of Up-Down II is the same as the Up-Down game, but the legal moves are different.

- When it is L's turn to move, he can choose any one of the heaps and
    1. repeatedly removes the top counter until a white counter is removed, or
    2. removes all the counters from the heap.
- When it is R's turn to move, he can choose any one of the heaps and
    1. repeatedly removes the top counter until a black counter is removed, or
    2. removes all the counters from the heap.

The player who removes the last counter is the winner.

Each colored heap in Up-Down II corresponds to a Hackenbush number [1].

1. First, *translate the heap into a binary string*. From bottom up and ignoring the bottom counter, each black counter is translated into digit 0; each white counter is translated into digit 1. If the string starts with digit 1, add digit 1 to the end of the above string; if the string starts with digit 0, add digit 0 to the end of the above string.
2. Next, *translate the binary string into a Hackenbush number*. A string starting with digit 1 is translated into a positive Hackenbush number; a string starting with digit 0 is translated into a negative Hackenbush number. Without loss of generality, we may assume the string starting with digit 1. From left to right, the first place where the digits changes from 1 to 0 is translated into a decimal point. Let n the number of 1s to the left of the decimal point. Then $n-1$ is the integer part of the Hackenbush number. The digits to the left of the decimal point is the dyadic rational part of the Hackenbush number.

The next proposition gives the solution of Up-Down II.

Proposition 2. *Each colored heap in Up-Down II is a sumber of the form:*

$$S_d = \uparrow(d) + *, \tag{17}$$

where d equals the Hackenbush number of the heap.

Proof. We prove by induction. Assume the claim is true for numbers simpler than d. By induction hypothesis,

$$S_{d^L} = \uparrow(d^L) + * \qquad \text{and} \qquad S_{d^R} = \uparrow(d^R) + *. \tag{18}$$

According to the rule,

$$\begin{aligned}
S_d &= \{S_{d^L}, 0 | 0, S_{d^R}\} \\
&= \{\uparrow(d^L) + *, 0 | 0, \uparrow(d^R) + *\} \\
&= \{\uparrow(d^L), * | *, \uparrow(d^R)\} + * \\
&= \uparrow(d) + *.
\end{aligned} \tag{19}$$

$\square$

Example 2. Consider the heap $WWWBBWW$ (from bottom up). The corresponding binary string of the heap is 1100111. The number of 1's to the left of the first 0 is 2. Thus, the integer part of the number is 1 (=2-1). The rational part is .0111 (in binary notation) which equals 0.4375. This string represents the Hackenbush number 1.4375. According to proposition 2, the heap has the value $\uparrow(1.4375) + *$. □

4   Up-Down Game III

The setup of Up-Down III is the same as Up-Down and Up-Down II, but the legal moves are different.

- When it is L's turn to move, he can choose any one of the heaps and
  1. repeatedly removes the top counter until a white counter is removed, or
  2. removes all the counters from the heap, or
  3. splits the heap into two none-empty heaps.
- When it is R's turn to move, he can choose any one of the heaps and
  1. repeatedly removes the top counter until a black counter is removed, or
  2. removes all the counters from the heap, or
  3. splits the heap into two none-empty heaps.

The player who removes the last counter is the winner.

Let S_d denotes the heap whose Hackenbush number is d. With simple induction, one can show that $S_d > *$ iff $d > 0$, and $S_{d_2} > S_{d_1}$ iff $d_2 > d_1$. At the first glance, it may be seen that both players have many splitting options. But, after a detailed analysis, we know that each player has at most one dominant split option.

1. L will never split S_d when $d > 0$; R will never split S_d when $d < 0$.
2. When L splits $S_d(d < 0)$ into two heaps $-S_b + S_u$, u must be a number greater than b and b must be the minimum among all the possible split options.
3. When R splits $S_d(d > 0)$ into two heaps $S_b - S_u$, u must be a number greater than b and b must be the minimum among all the possible split options.

The next proposition gives the solution of Up-Down III.

Proposition 3. *Each colored heap in Up-Down III is a sumber (or the negation of a sumber) of the form:*

$$S_d = \{S_{d^L}, 0|0, S_{d^R}\} = \uparrow(d) + *, \tag{20}$$

(when R has no dominant split option)

or the form:

$$S_d = \{S_{d^L}, 0|S_b - S_u, S_{d^R}\} = S_b - S_u + \uparrow(m) + *, \tag{21}$$

(when R has a dominant split option)

where $d > 0$ is the Hackenbush number of the heap, $S_b - S_u$ is the dominant split, and m is the simplest number greater than u.

Proof. The proof is by induction. Assume the claim is true for games simpler than S_d. When R has no dominant split option, S_d has the same value as in Up-Down II. When R has a dominant split option, there are two possible cases.

- case $d^L = b$:

$$X(S_d) = \{x : S_{d^L} - S_b + S_u + * < | \uparrow(x)\}$$
$$= \{x : S_u + * < | \uparrow(x)\}$$
$$= \{x > u\} \tag{22}$$

- case $d^L > b$: In this case, R has a dominant split option on d^L. Moreover, by induction hypothesis,

$$S_{d^L} = S_b - S_{u^R} + \uparrow(m') + * , \tag{23}$$

where $S_b - S_{u^R}$ is the dominant split, and m' is the simplest number greater than u^R.

$$X(S_d) = \{x : S_{d^L} - S_b + S_u + * < | \uparrow(x)\}$$
$$= \{x : -S_{u^R} + S_u + \uparrow(m') < | \uparrow(x)\}$$
$$= \{x > u\} \tag{24}$$

In both cases, we have $X(S_d) = \{x > u\}$. By theorem 2,

$$S_d = \{S_{d^L}, 0 | S_b - S_u, S_{d^R}\} = S_b - S_u + \uparrow(m) + * , \tag{25}$$

(when R has no dominant split option)

where m is the simplest number in $X(S_d)$.   □

Example 3. Consider the heap $WWWBBWW$ (from bottom up).
The corresponding Hackenbush number is 1.4375. R can split $WWWBBWW$ into $WW + WBBWW$ whose corresponding Hackenbush numbers are 1 and -1.25 respectively. The simplest number greater than 1.25 is 2. Thus,

$$WWWBBWW = S_{1.4375} = S_1 - S_{1.25} + \uparrow(2) + * . \tag{26}$$

Next, consider the heap $WBBWW = -S_{1.25}$. L can split $WBBWW$ into $WB + BWWW$ of which the corresponding Hackenbush numbers are -1 and 3, respectively. The simplest number greater than 3 is 4. Thus,

$$WBBWW = -S_{1.25} = -S_1 + S_3 - \uparrow(4) + * . \tag{27}$$

Since R has no dominant split option at WW and $BWWW$, we have

$$WW = S_1 = \uparrow(1) + * , \tag{28}$$
$$BWWW = S_3 = \uparrow(3) + * . \tag{29}$$

Finally, putting all together, we have

$$WWWBBWW = S_{1.4375} = \uparrow(2) + \uparrow(3) - \uparrow(4) + * . \tag{30}$$

□

5 Conclusion

We presented three variations of colored heap games. The sumber simplification rule is a powerful tool to analyze these games. In the Up-Down game, the sumbers contain only positive (or negative) ups with integer orders. In the Up-Down II game, the sumbers may contain ups with rational orders. In the Up-Down III game, we find sumbers that contains both positive and negative terms. It is interesting to see how a small change of the rule may produce different game values. The most interesting thing is that all these games can be solved by sumbers.

References

1. E.R. Berlekamp, J.H. Conway, and R.K. Guy. *Winning Ways*, Academic Press, New York, 1982.
2. C.L Bouton. Nim: A Game With a Complete Mathematical Theory, *Annals of Mathematics*, 3:35–39, 1902.
3. J.H. Conway, *On Numbers and Games*, Academic Press, New York, 1976.
4. P.M. Grundy, Mathematics and Games, *Eureka*, 2:6–8, 1939. Reprinted in *Eureka*, 27:9–11, 1964.
5. K.Y. Kao. *On Hot and Tepid Combinatorial Games*, PhD Thesis, University of North Carolina, Charlotte, 1997.
6. K.Y. Kao. Sumbers - Sums of Ups and Downs, *Elec. Journal of Combinatorial Number Theory*, 5, G1, 2005.

An Event-Based Pool Physics Simulator

Will Leckie[1] and Michael Greenspan[2]

[1] Department of Electrical and Computer Engineering,
Queen's University, Kingston, Canada
will.leckie@ece.queensu.ca
[2] Department of Electrical and Computer Engineering, School of Computing,
Queen's University, Kingston, Canada
michael.greenspan@queensu.ca

Abstract. The paper presents a method to simulate the physics of the game of pool. The method is based upon a parametrization of ball motion which allows the time of occurrence of events, such as collisions and transitions between motion states, to be solved analytically. It is shown that the occurrences of all possible events are determined as the roots of polynomials up to fourth order, for which closed-form solutions exist. The method is both *accurate*, returning continuous space solutions for both time and space parameters, and *efficient*, requiring no iterative numerical methods. It is suitable for use within a game tree search, which requires a great many potential shots to be modeled efficiently, and within a robotic pool system, which requires a high accuracy in predicting shot outcomes.

1 Introduction

Pool and its variations and close relatives billiards, carom, and snooker, are collectively classified as cue sports. While the precise origins of pool are unknown, it is an ancient game believed to have evolved from a common ancestor of golf and croquet. Shakespeare made reference to the game, and Marie Antoinette was known to be an enthusiast. There has recently been a resurgence of interest in pool world-wide: pool was recognized as a demonstration sport by the International Olympic Committee at the 1998 Nagano Olympics, and it is estimated that in 2003 over 40.7 million people picked up a cue in the U.S. alone.

The first effort to develop a robotic system to play a cue sport was the SNOOKER MACHINE of Khodabandehloo et al. [17], which was developed in the early 1990s. There are currently a number of such systems under development, including DEEP GREEN [12], the POOL SHARC [2], and others [10,6,7,11,5]. To play pool robotically is a challenging objective; a greater challenge is to achieve a level of play strong enough to compete against a proficient human player. In addition to the purely robotic aspects of the problem such as computer vision and mechatronics, which involve accurate sensing and actuation, there is a significant amount of strategy involved in playing pool. Placement of the cue ball following a shot is considered one of the key elements to successful play, and even moderately accomplished players tend to plan at least 3 shots ahead. In

H.J. van den Herik et al. (Eds.): ACG11, LNCS 4250, pp. 247–262, 2006.
© Springer-Verlag Berlin Heidelberg 2006

this way pool bears a similarity to other games of strategy, such as chess and checkers. One significant difference is that chess and checkers are played on a board with discrete positions so that the number of possible board states is finite, although huge. In contrast, a pool table is a continuous domain, with a truly infinite number of possible table states.

To predict the outcome of a shot accurately (i.e., the final rest locations of all balls) requires a knowledge of physics. Game strategy and physics are therefore intertwined, which is one of the intriguing aspects of the game. Most players have an intuitive understanding of the physics involved, a result of heuristics and many hours of observation. Computational pool, however, requires an explicit physical model of the balls, cue, and table, and their interactions. An early investigation of pool physics was explored by Coriolis [8], and the most thorough treatment to date is that of Marlow [13]. Recent work includes Bayes and Scott [3],Walker [18], Wallace and Schroeder [19], Witters and Duymelinck [20], and Onada [14]; books such as Koehler [9], Petit [15], Alciatore [1]; and a monograph by Shepard [16], who extended aspects of Marlow's work and also explored the statistical basis of certain strategies. Billiards has also been used as a model problem in the field of quantum chaos, although this field is not related to the macroscopic problem of predicting ball motion examined herein.

This paper describes the development of a pool physics simulator that is both numerically accurate and computationally efficient. The simulator has been developed for use within the gaming component of the DEEP GREEN system and was also distributed as the physics model behind the Computational Pool Tournament of the 10th and 11th Computer Olympiads. There currently exist a number of commercial and online pool simulators, each of which have attractive graphics, automatic strategies, and an underlying physics model. While many of these games are based upon seemingly realistic physics models, pool players have varied opinions about the level of realism and usefulness of these simulators, and there are some situations where anomalies occur. All of these simulators are closed systems, with the notable exception of foobilliard [4].

Our objective is to develop a pool physics simulator that is both highly accurate and efficient, which can be used by both the gaming and robotics communities to further the development of computational and robotic pool. For robotic pool it is necessary to have a highly accurate physics simulator due to the continuous nature of the game, so that the outcome of each shot can be predicted accurately. For computational pool, it is important that the physics simulator be efficient, as it may be called repeatedly to evaluate potential shots within a game tree. To satisfy both of these criteria, we have foregone the standard integration method of simulation, which discretizes time, in favor of an event-based method that solves each shot analytically and exactly over a continuous time domain. A further benefit is that the order of events, which may be obscured by the integration method when the time between events is small, will be preserved.

This paper continues in Sect. 2 with a description of the physics model. Section 3 describes the various motion states and events that can occur, and describes a method to predict analytically the time of each event. The method is discussed in Sect. 4 and the paper concludes in Sect. 5 with a description of future work.

2   Pool Physics

In this section we model the physics of the impact of the cue and the ball (2.1), the resulting ball motion (2.2), and the collision of the ball with the rails or other balls (2.3).

2.1   Cue-Ball Impact

Each shot commences with the cue striking the cue ball. Let $(\hat{i}, \hat{j}, \hat{k})$ be the ball-centric coordinate reference frame illustrated in Fig. 1. The central axis of the cue is denoted by a vector $\vec{q}$, which by definition is parallel to the vertical $\hat{j}$-$\hat{k}$ plane. Vector $\vec{q}$ is oriented at an angle θ to the horizontal $\hat{i}$-$\hat{j}$ plane, and we make the simplifying assumption that the cue contacts the ball at a single point $Q \in \vec{q}$, thereby ignoring the geometry of the cue tip. If R is the radius of the ball, a and b are the respective horizontal and vertical displacements of Q from the ball's center, and $c = |\sqrt{R^2 - a^2 - b^2}|$, then $Q = (a, c, b)$. For Q to lie on the ball's surface we must have $a^2 + b^2 \leq R^2$, so the $\hat{j}$ component of Q must always be real and non-negative.

To model the trajectory of the cue ball, its initial velocity $\vec{v}$ immediately after being struck by the cue must be written in terms of the impact parameters a, b, θ and V_0, the magnitude of the velocity of the cue immediately before striking

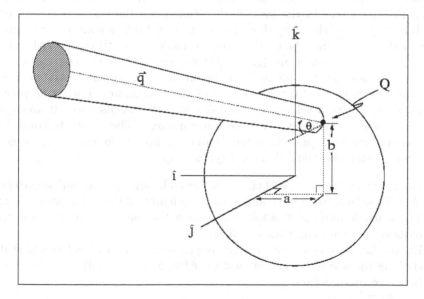

Fig. 1. Cue Impact

the ball. Newton's Second Law $\vec{F} = m\vec{a}$ is integrated to give the velocity of the ball in terms of the force $\vec{F}$ exerted on the ball by the cue impact (cf. [16]). Assuming that the time duration of the collision is very small[1], the force $\vec{F}$ can be treated as a perfectly elastic impulse and integrating Newton's Second Law with respect to time over the small duration $t_f - t_i$ of the collision gives:

$$\int_{t_i}^{t_f} \vec{F}(t)dt = \vec{F} \int_{t_i}^{t_f} \delta(t = t_i)dt = \vec{F} = m \int_{t_i}^{t_f} \vec{a}(t)dt = m\vec{v}(t = t_f), \quad (1)$$

or simply:

$$\vec{F} = m\vec{v}. \quad (2)$$

The post-impact velocity of the ball is therefore expressed in the ball-centric coordinate reference frame as:

$$\vec{v} = \left(0, \frac{-F}{m}\cos\theta, \frac{-F}{m}\sin\theta\right). \quad (3)$$

The magnitude of the force F in terms of the impact parameters is derived by simultaneously solving the equations for conservation of linear momentum and conservation of energy before and after the cue impact to obtain:

$$F = \frac{2mV_0}{1 + \frac{m}{M} + \frac{5}{2R^2}\left(a^2 + b^2\cos^2\theta + c^2\sin^2\theta - 2bc\cos\theta\sin\theta\right)}, \quad (4)$$

where m is the ball mass and M is the cue mass.

The non-zero $\hat{k}$ component of $\vec{v}$ indicates that the ball has a component of initial linear velocity in the vertical direction when the cue is not horizontal, which is the basis of the *jump* shot. The event-based solution method we develop later in this paper could be easily extended to deal with three-dimensional ball motion, but in our model we ignore this $\hat{k}$-component of the cue ball's initial velocity and assume that all balls are confined to movement on the surface of the table, which is true for the majority of shots encountered in actual play.

In addition to linear velocity, the collision with the cue also imparts an angular velocity to the ball whenever a, b, or θ are nonzero. The ability to control the angular velocity (i.e., *spin*) of the ball is an essential skill in mastering the game. Players conceptualize the following 3 types of spin:

- FOLLOW: i.e., *top spin*. FOLLOW is achieved by striking the ball above center ($b > 0$), so that there is a component of spin around the ball's positive $\hat{i}$-axis. The usual desired effect of follow is to have the cue ball continue its forward motion after colliding with an object ball.
- DRAW: i.e., *back spin*. DRAW is the opposite of FOLLOW, and is achieved by striking the ball below center ($b < 0$). After an object ball collision, the cue ball reverses direction.

---

[1] Marlow [13, p.45] empirically determined the collision duration to be $\sim 200\mu sec$ at a speed of 1 m/s.

- ENGLISH: i.e., *side spin*. Left or right ENGLISH is achieved by striking the ball slightly left $(a > 0)$ or right $(a < 0)$ of the center respectively, resulting in a rotation around the ball's $\hat{k}$-axis. There are a number of possible effects of ENGLISH, including altering the angle of incidence of the cue and/or object balls following a collision. ENGLISH can also be communicated to object balls.

Given the moment of inertia of a solid sphere as $I = \frac{2}{5}mR^2$, the post impact instantaneous angular velocity of the ball within the ball frame is:

$$\vec{\omega} = \frac{1}{I}\left(-cF\sin\theta + bF\cos\theta,\ aF\sin\theta,\ -aF\cos\theta\right). \tag{5}$$

It is interesting to note that instances of the above itemized spins can all be achieved with a horizontal cue (i.e., $\theta = 0$), in which case the $\hat{j}$-component $\vec{\omega}_y$ of $\vec{\omega}$ vanishes. When $\vec{\omega}_y = 0$, the ball travels with straight rectilinear motion[2]. There is a fourth type of spin, usually attempted only by more advanced players, wherein the cue is elevated at an extreme angle nearly perpendicular to the table $(\theta \sim 90^o)$. This causes a large $\vec{\omega}_y$ component resulting in a curvilinear ball motion; it is the basis of *curve* and *massé* shots.

2.2 Ball Motion

As the ball moves across the table, the only unbalanced external force acting on it is friction with the table. This friction acts at the point of contact between the ball and the table surface, in a direction opposite to the ball's motion. We approximate this as a single point of contact, although in reality the ball compresses the fibers of the table cloth and sinks slightly into the cloth so that the point of contact is actually a small portion of the ball's spherical surface.

When a ball is struck by the cue, it begins its motion by sliding across the surface of the table. After some time, the interplay between the table friction and the ball's linear and angular velocities causes the ball to start rolling. It takes a very specific combination of the ball's linear and angular velocities to cause it to transition from the sliding to the rolling state. Assume the ball is moving in the $\hat{i}$ direction. Intuitively, the ball will be rolling only when it makes one full revolution about the $\hat{j}$ axis for every distance $2\pi R$ that it travels along the table, equal to the circumference of the ball.

Marlow [13] classified the ball motion as either sliding or rolling by considering the relationship between the linear velocity of the ball and the angular velocity of the ball at the point of contact with the table surface. The *relative velocity* $\vec{u}(t)$ of the point P at the bottom of the ball that is in contact with the table surface is [13]:

$$\vec{u}(t) = \vec{v}(t) + R\hat{k} \times \vec{w}(t). \tag{6}$$

The ball slides across the table surface when $\vec{u}(t) \neq 0$ and rolls across the table surface when $\vec{u}(t) = 0$. Note that the relative velocity $\vec{u}(t)$ has no $\hat{k}$-component

---

[2] Many people believe that the use of ENGLISH necessarily causes the cue ball to travel with curvilinear motion, which is incorrect. The ball travels in a straight trajectory unless struck with an elevated cue stick.

since $\hat{k} \times \hat{k} = 0$. As a result, the angular velocity about the vertical $\hat{k}$ axis does not couple into the translational motion of the ball and must be treated separately as we shall see below.

From Newton's Laws of motion and Eq. 6, the following equations govern the *sliding* ball's state variables for position $\vec{r}(t)$, velocity $\vec{v}(t)$, and angular velocity $\vec{\omega}(t)$ as a function of time t, expressed in a frame of reference attached to the center of the ball such that the $\hat{i}$ axis of this frame is along the ball's direction of motion:

$$\vec{r}_B(t) = \begin{bmatrix} x_B(t) \\ y_B(t) \end{bmatrix} = \begin{bmatrix} v_0 t - \frac{1}{2}\mu_s g t^2 \hat{u}_{0x} \\ -\frac{1}{2}\mu_s g t^2 \hat{u}_{0y} \end{bmatrix} ; \tag{7}$$

$$\vec{v}_B(t) = \vec{v}_0 - \mu_s g t \hat{u}_0 ; \tag{8}$$

$$\vec{\omega}_B(t) = \vec{\omega}_\parallel(t) = \vec{\omega}_0 - \frac{5\mu_s g}{2R} t \left(\hat{k} \times \hat{u}_0 \right) . \tag{9}$$

The "$_B$"-subscripts in the above equations indicate that the equations are expressed in the ball frame. The subtractive terms represent the effect of the table friction slowly bleeding off the ball's speed and spin as time progresses. The "$_0$"-subscripts denote the initial values, and "$_x$" represents the $\hat{i}$-component of a vector, for example. The gravitational constant is represented by g and μ is the dimensionless coefficient of friction, which takes on values between 0.0 and 1.0 and results from the material properties of the ball and table cloth surfaces. Larger values of μ correspond to larger frictional forces which reduce the ball's speed and spin more quickly.

The $\hat{k}$ components of $\vec{r}(t)$ and $\vec{v}(t)$ are assumed to be zero at all times as a result of the earlier assumption that the ball is constrained to movement on the surface of the table. Eq. 9 is a three-dimensional vector equation for the angular velocity, but since the $\hat{k}$-component of $\hat{k} \times \hat{u}_0$ is zero, the $\hat{k}$ component of the angular velocity given by Eq. 9 does not evolve with time. This is a result of the assumption of a single point of contact between the ball and the table cloth. We denote $\vec{\omega}_B(t)$ as $\vec{\omega}_\parallel(t)$ since the ball always lies in the plane of the table, the $\hat{i}$-$\hat{j}$ plane. To calculate the vertical component of the ball's angular velocity as a function of time, denoted by $\omega_\perp$, we use the simple model suggested by Marlow:

$$\omega_\perp(t) = \omega_{\perp 0} - \frac{5\mu_{sp} g}{2R} t . \tag{10}$$

In practice the coefficients of friction governing the sliding and the spinning motion of the ball are denoted by μ_s and μ_{sp}, respectively.

To find the position of the ball in the table frame as a function of time, $\vec{r}(t)$, we first express $\vec{r}_B(t)$ in the table frame using a simple rotation:

$$\vec{r}_T(t) = \begin{bmatrix} \cos\phi & -\sin\phi \\ \sin\phi & \cos\phi \end{bmatrix} \vec{r}_B(t) , \tag{11}$$

where ϕ is the angle between the initial velocity vector of the ball and the $\hat{i}$-axis of the table frame (i.e., the ball's heading relative to the table frame). The position of the ball in the table frame is then $\vec{r}(t) = \vec{r}_0 + \vec{r}_T(t)$. By combining Eqs. 6, 8 and 9, the relative velocity evolves with time according to:

$$\vec{u}(t) = \vec{u}_0 - \frac{7}{2}\mu_s gt\hat{u}_0 \,. \tag{12}$$

The sliding state lasts until $\vec{u}(t) = 0$, so from Eq. 12 the duration τ_S of the sliding state is:

$$\tau_S = \frac{2\,|\vec{u}_0|}{7\mu_s g} \,. \tag{13}$$

The rolling state is defined by the condition $\vec{u}(t) = 0$, whence $|\vec{v}(t)| = R\,|\vec{w}_\parallel(t)|$. When $\vec{u}(t) = 0$ the ball travels one circumference in the $\hat{i}$ direction for every rotation about the $\hat{j}$-axis as shown in Fig. 2. During the sliding state the angular velocity $\vec{\omega}$ is arbitrary and the relative velocity $\vec{u}$ is non-zero. During the rolling state the relative velocity is zero and the angular velocity lies in the $\hat{i}$-$\hat{k}$ plane with $|\vec{\omega}_\parallel| = |\vec{v}|/R$. The force of friction acting on the ball is denoted by $\vec{F}$ and the point of contact between the ball and table is denoted by P in Fig. 2.

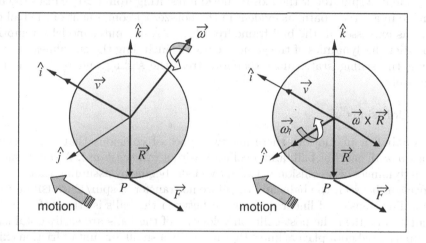

Fig. 2. Ball Dynamics During the Sliding State (left) and Rolling State (right)

Once the relative velocity has vanished and the ball has started rolling, it continues rolling in a straight trajectory because there is no angular velocity around the axis along which the ball is moving. During this *natural roll* state the ball's state variables evolve according to Newton's Law of Motion under the constraint on the angular velocity $\vec{\omega}_\parallel$ imposed by the natural roll condition:

$$\vec{r}(t) = \vec{r}_0 + \vec{v}_0 t - \frac{1}{2}\mu_r gt^2\hat{v}_0 \,; \tag{14}$$

$$\vec{v}(t) = \vec{v}_0 - \mu_r gt\hat{v}_0 \,; \tag{15}$$

$$|\vec{\omega}_{\parallel}(t)| = \frac{|\vec{v}(t)|}{R}. \tag{16}$$

Once again, the subtractive terms in Eqs. 14 and 15 represent the effect of table friction, where μ_r is the coefficient of rolling friction, in general different from $\mu_s{}^3$. By definition, during the rolling state the direction of $\vec{\omega}_{\parallel}$ is perpendicular to the direction of the ball's motion as shown in Fig. 2. During the rolling state the vertical component of the angular velocity, $\vec{\omega}_{\perp}$, continues to evolve according to Eq. 10.

The natural roll state lasts until the table friction reduces the ball's velocity to zero, so from Eq. 15 the duration of the rolling state is:

$$\tau_R = \frac{|\vec{v}_0|}{\mu_r g}. \tag{17}$$

Since $\mu_s \neq \mu_r$, the motion of a moving billiard ball is quantitatively different during the sliding and rolling states. During the natural roll state, the ball always travels in a straight trajectory because the component of spin about the ball's axis of motion is zero by definition. During the sliding state, however, any component of spin about the axis of motion, resulting from $\theta > 0$, causes the ball to move in a curved path, as evident in the non-zero $\hat{j}$-component of the ball position as expressed in the ball frame from Eq. 7. A computer model attempting to predict the dynamics of the game accurately, including the curvilinear motion during the sliding state, must therefore treat the sliding and rolling motions separately.

2.3   Collisions

Once a ball is in motion, there are two types of collisions that can occur: a collision with another ball, and a collision with a rail (*cushion*). A simple model of the dynamics of a collision between two balls begins by assuming that there is no friction between the balls and therefore no transfer of spin/ENGLISH between them. The transfer of linear momentum between the balls is straightforward to calculate and then the post-collision velocities of the balls are easily obtained.

Experienced pool players know that in reality a small amount of friction exists between the two balls when in contact with one another, adding a key dimension to the game. This friction is the source of the important effect known as *throw*, in which the spin of the cue ball is communicated to the object ball during the short time of the collision, causing a deflection of the object ball from the intended post-impact trajectory. The effect of throw is small but must be taken into account for any measure of success on long shots down the table.

Marlow [13] treats the most general case of the collision by including ball-ball friction using linear and angular momentum conservation laws as a starting principle. This results in a system of 16 equations in 16 unknowns which is solved to obtain four equations for the post-impact velocities and angular velocities for

---

[3] Marlow experimentally measured $\mu_{sp}=0.044$, $\mu_r=0.016$ and thereby calculated $\mu_s=0.2$ [13, p. 237].

the two balls. Our model uses Marlow's equations in order to solve the dynamics of the ball-ball collision.

Models of the ball-cushion interaction often involve some model of a damped-spring system. Some of the ball's kinetic energy is absorbed by the cushion, and the angular velocity of the ball is altered by the collision. We use Marlow's [13] damped-spring model to calculate the post-impact linear and angular velocities for the ball-rail collision.

3   Event-Based Simulation

We now consider the problem of simulating the physics of the game. The chosen simulation method must be computationally efficient because it will be used in a game strategy algorithm to expand a search tree, which necessitates invoking the method for a large number of potential shots. We also strive for physical accuracy in the simulation, both to make a simulated game more realistic and to allow its use within a robotic system, wherein slight inaccuracies can result in missed shots.

3.1   Discrete Numerical Integration

The obvious first option for a simulation method is numerical integration. With full knowledge of the state variables for a ball's position $\vec{r}(t)$, velocity $\vec{v}(t)$ and angular velocity $\vec{\omega}(t)$ at a given time t, these state variables are advanced forward in time by a small discrete amount Δt, using some discrete numerical integration technique. The simplest approach equates the acceleration of the ball to the time-derivative of its velocity, and integrates the acceleration equation assuming that the acceleration is constant over Δt. This gives the velocity after Δt in terms of the previous time step's velocity and the (constant) acceleration: $\vec{v}(t + \Delta t) = \vec{v}(t) + \vec{a}\Delta t$. Similarly, integrating the velocity and assuming it is constant over the tiny Δt yields the position $\vec{r}(t + \Delta t) = \vec{r}(t) + \vec{v}\Delta t$. This method is the standard approach for modeling physics in human-interactive computer games of all kinds, where the emphasis is more on apparent realism than on physical accuracy. Computational efficiency is increased as a result of the linear approximations to the actual equations for the state variables $\vec{r}(t)$, $\vec{v}(t)$, and $\vec{\omega}(t)$ as a function of time t. However, this increased efficiency is obtained at the expense of the accuracy of the solution.

In pool, use of the integration method boils down to the problem of the discovery of events *after* they have occurred, and then handling them using the appropriate physics. After each time step, the simulator must consider all ball positions, velocities, and angular velocities to determine whether any events such as ball collisions or rail collisions have occurred during the previous Δt. Due to the continuous measurement domain, it is highly improbable that an event occurs *exactly* at the beginning or end of a time step; it is much more likely that an event occurs sometime within the time step, i.e., at a fractional Δt.

For example, assume that two moving balls collide at time t_c within time step s_i, where the duration of s_i is $[t_{i0}, t_{if}]$; $t_{if} - t_{i0} = \Delta t$. The separation distance D

of the centers of the balls at t_c is exactly equal to $2R$. Prior to t_c, their separation $D > 2R$, and provided Δt is sufficiently small relative to the linear velocities of the balls, then $D < 2R$ within the period $[t_c, t_{if}]$. Upon inspection at time t_{if}, the simulator would realize that a collision must have occurred sometime during s_i, because $D < 2R$. The simulator would then have to back up in time to t_c in order to apply the appropriate equations governing the collision. Backing the model up by a fraction of Δt to calculate the exact values of the state variables at t_c adds complexity and the possibility of lost accuracy. Calculating the exact value of t_c without making any simplifying assumptions that may limit accuracy is itself a challenging problem. Another option is to return iteratively to t_{i0} and simulate forward with a smaller time step, iterating until the time step has been reduced such that $\Delta t \approx t_c - t_{i0}$, in order to advance forward from t_{i0} to t_c with no approximations. This option adds complexity and decreases efficiency.

In addition, a common situation in pool is that of *frozen balls*, where two or more balls are stationary and in contact before being struck by another moving ball. The discrete integration method would have difficulty simply recognizing and handling the events in the correct order, because the time between events is so small. Modeling a break shot this way is difficult, and computer-pool games often use either a heavily-approximated solution, or some randomized selection from an encyclopedia of pre-computed break shots, instead of attempting to generate a new break shot each time.

Therefore, there are several intertwined challenges associated with the integration method applied to modeling pool physics. Choosing a reasonable time step Δt is a difficult trade-off between speed and accuracy, and dealing with the fact that events usually occur within a time step necessitates complexity and approximations. A smaller time step results in a more accurate model for two reasons: (1) the assumption that the speed of the ball is constant over the time step is more valid as Δt is reduced, and (2) with a smaller time step any approximations needed to back the model up by a portion of a time step will be more accurate. However, with a larger time step the model can predict the eventual outcome much faster since fewer computations are necessary. Given these challenges and drawbacks, we looked to another approach to model the outcome of a shot.

3.2 Continuous Event-Based Simulation

Since the equations developed in Subsect. 2.2 provide the value of the ball-state variables at any time t, we chose to take an event-oriented approach that allows us to solve the ball-dynamics problem exactly, in the continuous domain in both time and space.

Motion States and Events. The approach is based upon the observation that a ball can be in one of four possible *motion states* at any time t:

- SLIDING, defined by $\vec{v}(t) \neq 0$ and $\vec{u}(t) \neq 0$, during which the ball slides across the table;

- ROLLING, defined by $\vec{v}(t) \neq 0$, and $\vec{u}(t) = 0$, during which the ball rolls across the table;
- SPINNING, defined by $\vec{v}(t) = \vec{u}(t) = 0$ and $\vec{\omega}(t) = \pm |\vec{\omega}(t)| \hat{k}$, during which the ball spins in place about the vertical axis;
- and STATIONARY, defined by $\vec{v}(t) = \vec{u}(t) = \vec{\omega}(t) = 0$, when the ball is completely stationary.

An *event* is defined as the transition of a ball from one motion state to another. A *motion transition* event occurs simply through the passage of time as a ball's speed and spin evolve under the influence of the table friction. In addition, there are a number of *collision* events that occur when two objects impact, resulting in a change of the ball's linear and angular velocities and therefore a transition to a different motion state. Events are classified into the following five categories:

- CUE-STRIKE, in which the cue ball is struck by the cue stick;
- BALL-COLLISION, in which two balls collide with each other;
- RAIL-COLLISION, in which a given ball collides with a cushion on the table;
- POCKET-COLLISION, in which a given ball is pocketed;
- and MOTION-TRANSITION, in which a given ball transitions from one motion state to another through the passage of time.

Event Time Prediction. Using this framework, the simulation problem boils down to one of *prediction* of when the next event occurs. When modeling a shot, we calculate the time τ_E at which the next event E occurs and advance each ball's state variables $\vec{r}(t)$, $\vec{v}(t)$ and $\vec{\omega}(t)$ forward in time to $t = \tau_E$ using the appropriate equations from Subsect. 2.2. We then apply the appropriate physics for the event, for example calculating the post-collision instantaneous linear and angular velocities of two balls following a BALL-COLLISION event. This is an attractive solution method because the dynamics of the moving balls are calculated completely analytically, with no approximations, and are subject only to floating point (double precision) round off error. A summary of the event-based algorithm is presented in Fig. 3.

The problem of predicting τ_E for the collision of two balls with one or both of them moving is one of the more interesting problems that the event-based method must handle. The event will occur at the time τ_E at which the pair of balls have a distance of separation D between their centers equal to $2R$. The separation of the two balls, denoted by subscripts 1 and 2, respectively, is a time-dependent vector $\vec{d}(t)$ given by:

$$\vec{d}(t) = \vec{r}_2(t) - \vec{r}_1(t) = \begin{bmatrix} a_x t^2 + b_x t + c_x \\ a_y t^2 + b_y t + c_y \end{bmatrix}, \tag{18}$$

where the coefficients a, b, and c are obtained by collecting terms appropriately from the expansion of $\vec{r}_2(t) - \vec{r}_1(t)$ using the parametrization for $\vec{r}(t)$ given by Eq. 7 or 14. In the expansion, the value of μ used is either μ_s if the ball is in the SLIDING state, μ_r if it is in the ROLLING state, or 0 otherwise. The expansion is written in component form in Eq. 18 with a_x denoting the $\hat{i}$-component of the t^2 coefficient of $\vec{d}(t)$, for example.

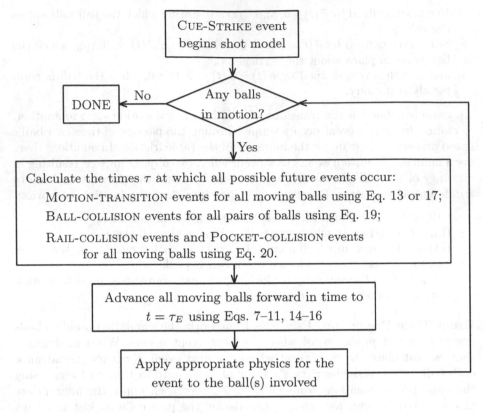

Fig. 3. Event-based Algorithm for Modeling Pool Physics

The time of collision τ_E that we are interested in finding corresponds to the solution of $D^2 = |\vec{d}(t)|^2 = 4R^2$. Expanding Eq. 18 we obtain:

$$(a_x^2 + a_y^2)t^4 + (2a_xb_x + 2a_yb_y)t^3 + (b_x^2 + 2a_xc_x + 2a_yc_y + b_y^2)t^2$$
$$+ (2b_xc_x + 2b_yc_y)t + c_x^2 + c_y^2 - 4R^2 = 0. \tag{19}$$

This quartic polynomial is solved in closed form using standard analytical methods. The four unique and possibly complex roots of Eq. 19 represent possible values of τ_E and must be carefully interpreted to determine which among them is physically possible. If the balls do indeed collide, then the correct root corresponding to τ_E is the smallest positive real root of Eq. 19 that results in both ball centers lying somewhere on the table surface (i.e., within the rail bounds) within the lifetime of both balls' present motion states. If no roots are found that meet these criteria, then it is concluded that this pair of balls do not collide on the table surface within their current motion states and that some other event (such as a MOTION-TRANSITION or RAIL-COLLISION) must precede the collision.

At this point it is worth noting that relaxing the earlier assumption that all balls are constrained to movement on the surface of the table by including $\hat{k}$-components of position $\vec{r}(t)$ and velocity $\vec{v}(t)$ would result in a different functional form of the coefficients a, b, and c in Eqs. 18 and 19 but that, importantly, these coefficients would still be time-independent. As a result, Eq. 19 would still be fourth-order and a solution for the time of collision could still be obtained using analytical methods.

A similar approach is used to determine the time at which a given ball collides with a rail or a pocket. The ball trajectory is parameterized as a function of t and the straight line representing the rail or pocket mouth as a function of s. Since the ball's position vector $\vec{r}(t)$ refers to its center, when predicting rail collisions we look for the intersection of the ball's center and a horizontal line parallel to the rail but one ball radius toward the table center. For example, for a horizontal rail and a ball in the natural roll state, we would solve:

$$\begin{bmatrix} s + R \\ 0 \end{bmatrix} = \begin{bmatrix} x_0 + v_{0x}t - \frac{1}{2}\mu g t^2 \hat{v_{0x}} \\ y_0 + v_{0y}t - \frac{1}{2}\mu g t^2 \hat{v_{0y}} \end{bmatrix}, \tag{20}$$

where the left hand side is the parameterization of the rail and the right hand side is the parameterization of the ball's trajectory $\vec{r}(t)$ in the rolling state, as given by Eq. 14. The right hand side is written in component form and the variables are expressed in the table frame. Again, the appropriate value of μ is used, depending on the ball's motion state. Other rails and the six pockets are parameterized in a similar way.

Eq. 20 is easily solved using simple algebra and the quadratic formula, which returns two possibly complex roots corresponding to the time of collision τ_E. Again, the two roots must be carefully interpreted to determine whether either correspond to a physically meaningful solution.

4   Discussion

Since the event-based algorithm we have presented is completely analytical and uses no approximations in determining the time of the next event and in advancing the state of the table forward in time to the next event, it has the potential to be more accurate than the numerical integration method. There is no need to choose the time step Δt carefully and there is no loss of accuracy arising from approximations to back the model up by a fraction of a time step in order to handle an event. In addition, there is no loss of accuracy due to the approximation that the velocity is constant during the time step. The only limitations on the accuracy of our method are (1) floating-point (double) precision, (2) the values of physical constants such as μ used, and (3) any approximations used to model the physics of a collision event. All of these limitations are also present in the integration method.

The efficiency of the event-based method is another interesting topic. During many billiards shots relatively long time periods can pass in which no events occur at all. For example, on a long cue ball-object ball shot down the table, one or two seconds can pass while the cue ball rolls down the table before it strikes the object ball, during which time no events occur whatsoever. Assume that there are n balls in motion at some time $t = \tau$ (where $n_{max} = 16$ since there are at most 16 balls on the pool table), and assume that time $\Delta\tau$ passes during which none of the moving balls transition to another motion state or collide with another ball or rail. Let the integration-style method use a time step Δt such that there are $N = \Delta\tau/\Delta t$ time steps in $[\tau, \tau + \Delta\tau]$. To simulate the event-free ball dynamics for this time period $\Delta\tau$, the integration-style model must perform:

- $3nN$ operations to step the moving balls' state variables ahead, since there are N time steps, n moving balls, and 3 state equations for each moving ball;
- $12nN$ operations to test for ball-rail/pocket collisions, since there are 6 rails, 6 pockets, N time steps, and n moving balls;
- and $N\left(n\left(n-1\right)/2 + n\left(n_{max} - n\right)\right) = N\left(31n - n^2\right)/2$ operations to test for ball-ball collisions, since there are $n\left(n-1\right)/2$ tests for pairs of moving balls, $n\left(n_{max} - n\right)$ tests for moving-stationary ball collisions, and N time steps.

The total number of operations required is $N\left(61n - n^2\right)/2$, with the number of operations increasing linearly with decreasing time step Δt.

Contrast this result with the requirements for the event-based method:

- n operations to predict the next motion state transition for all of the moving balls;
- $24n$ operations to predict the next ball-rail/pocket collision, since there are 6 rails, 6 pockets, and 2 operations are required for each quadratic solution;
- $19(n\left(n-1\right)/2 + n\left(n_{max} - n\right)) = 19\left(31n - n^2\right)/2$ operations to predict the next ball-ball collision, since there are $n\left(n-1\right)/2$ tests for pairs of moving balls, $n\left(n_{max} - n\right)$ tests for moving-stationary collisions, and 19 operations needed to solve each quartic equation;
- and $3n$ operations to advance all of the moving balls' state variables ahead to the time of the next event.

The total number of operations required is $\left(645n - 19n^2\right)/2$, independent of the amount of time $\Delta\tau$ until the next event.

With three moving balls and $\Delta\tau = 1$ second, for example, the integration method with a rather coarse time step of $\Delta t = 1$ millisecond needs 87,000 operations to solve this one second of the dynamics, while the event-based method needs just 882. A typical shot sees perhaps five to ten events occur, while a perfectly executed shot involving only the cue ball, one object ball and a pocket may see even fewer than five events occur. It is clear that the workload of the event-based approach is not very heavy, even when modeling more complicated dynamics.

There are several other advantages to using the event-based approach. The dynamics and end result of a given shot can be completely and precisely stored by recording only the initial table state (i.e., ball locations) just before the shot as well as a time-stamped list of the events that occur during a shot. A shot can then be played back/animated after the fact using only the recorded initial table state and the list of events, with no need to re-model all of the physics of the shot.

Since the event-based approach needs to solve polynomial equations up to order four, its accuracy depends heavily on the numerical accuracy of the polynomial solver and the interpretation of the results. Closed form analytical formulae exist for the solutions of polynomials up to fourth-order, but care must be taken when implementing these formulae. For example, a fourth-order equation often has at least one pair of complex-conjugate roots. One challenge lies in determining how "real" or how "complex" a given root is, since floating-point numbers are never exactly equal to zero. We search the set of roots returned by our quartic solver to eliminate complex-conjugate pairs, then use a simple thresholding on the imaginary part to determine if any remaining roots are really "real".

5 Conclusion

We described a predictive event-based method to simulate the physics of pool. The method is based upon a parametrization of the separation of balls, which allows event times to be obtained analytically. The solution requires no granular time step and is accurate to floating point precision in both time and space. It is also efficient, requiring no iterative numerical methods.

In future work, we will develop a game strategy program that uses the simulation to model shot outcomes and automatically select shots in order to play a game of 8 Ball. Following that, the physical parameters of the DEEP GREEN system will be measured and calibrated, and the physics simulation and game strategy will be integrated into the DEEP GREEN system to compete against a human opponent.

Acknowledgements

We would like to thank the Natural Sciences and Engineering Research Council of Canada (NSERC) for their support, and David Levy and Jonathan Schaeffer for their suggestions and encouragement of this work.

References

1. D. Alciatore. *The Illustrated Principles of Pool and Billiards*. Sterling Publishers, New York, NY, 2004. ISBN 1402714289.
2. M. Ebne Alian, S. Bagheri Shouraki, M.T. Manzuri Shalmani, P. Karimian and P. Sabzmeydani. Robotshark: A Gantry Pool Player Robot. *ISR 2004: 35th Intl. Sym. Rob*, 2004.
3. J. Bayes and W. Scott. Billiard Ball Collision Experiment. *Am. Jour. Physics*, 3(31):197-200, 1963.
4. F. Berger. http://foobillard.sunsite.dk, 2000.
5. B.R. Cheng, J.T. Li and J.S. Yang. Design of the Neural-Fuzzy Compensator for a Billiard Robot. *IEEE Intl. Conf. Networking, Sensing & Control*, pages 909–913, 2004.

6. S.C. Chua, E.K. Wong, A.W.C. Tan, and V.C. Koo. Decision Algorithm for Pool Using Fuzzy System. *iCAiET 2002: Intl. Conf. AI in Eng. & Tech*, pages 370–375, 2002.

7. S.C. Chua, E.K. Wong, and V.C. Koo. Pool Balls Identification and Calibration for a Pool Robot. *ROVISP 2003: Proc. Intl. Conf. Robotics, Vision, Information and Signal Processing*, pages 312–315, 2003.

8. G.G. Coriolis. *Théorie Mathématique des Effets du Jeu de Billard*. Jacques Gabay, 1835 (republished in 1990).

9. J.H. Koehler. *The Science of Pocket Billiards*. Sportology Publications, Laguna Hills, CA, 1989. ISBN 0962289027.

10. L.B. Larsen, M.D. Jensen, and W.K. Vodzi. Multi Modal User Interaction in an Automatic Pool Trainer. *ICMI 2002: 4th IEEE Intl. Conf. Multimodal Interfaces*, pages 361–366, 2002.

11. Z.M. Lin, J.S. Yang, and C.Y. Yang. Grey Decision-Making for a Billiard Robot. *IEEE Intl. Conf. Systems, Man and Cybernetics*, pages 5350–5355, 2004.

12. F. Long, J. Herland, M.-Ch. Tessier, D. Naulls, A. Roth, G. Roth, and M. Greenspan. Robotic Pool: An Experiment in Automatic Potting. *IROS 2004: IEEE/RSJ Intl. Conf. Intell. Rob. Sys*, pages 361–366, 2004.

13. W.C. Marlow. *The Physics of Pocket Billiards*. Marlow Advanced Systems Technologies, Palm Beach Gardens, FL, 1995. ISBN 0964537001.

14. G. Onada. Comment on "Analysis of Billiard Ball Collisions in Two Dimensions". *Am. Jour. Physics*, 57(5):476-478, 1989.

15. R. Petit. *Billard. Théorie du Jeu*. Chiron Editeur, Saint-Quentin, France, 1997. ISBN 2702705731.

16. R. Shepard. *Amateur Physics for the Amateur Pool Player*. 3d edition. Self-published, 1997.

17. S.W. Shu. *Automating Skills Using a Robot Snooker Player*. PhD Thesis, Bristol University, 1994.

18. J. Walker. The Physics of the Draw, the Follow, and the Masse (in Billiards and Pool). *Scientific American*, 249(1):124-129, 1983.

19. R.E. Wallace and M. Schroeder. Analysis of Billiard Ball Collisions in Two Dimensions. *Am. Jour. Physics*, 56(9):815-819, 1988.

20. J. Witters and D. Duymelinck. Rolling and Sliding Resistive Forces on Balls Moving on a Flat Surface. *Am. Jour. Physics*, 54(1):80-83, 1988.

Optimization of a Billiard
Player – Position Play*

Jean-Pierre Dussault and Jean-François Landry

Département d'Informatique,
Université de Sherbrooke, Sherbrooke (Québec), Canada
{Jean-Pierre.Dussault, Jean-Francois.Landry2}@USherbrooke.CA

Abstract. The paper describes optimization principles to produce a
computer pool player. A good player has technical and planning abilities.
Technically, he[1] sinks balls with precision, and controls the position of
the cue ball after the shot. He uses his technical abilities to devise a game
plan, sinking the balls in a winning order.

We propose to use the optimization techniques in such a way that it
simulates an excellent player. In this paper, we focus on the technical
abilities. We provide optimization models to compute the shots to not
only sink a given ball, but bring the cue ball at a specified target. Some
hints on planning optimization strategies are given.

1 Introduction

The paper presents a key computation to develop an optimal billiard player,
namely the skill to sink a ball while achieving a good position for the next
one. It is a rather basic skill. Actually, in the game sometimes named carambol
(*Billard Français*, three cushions), the only aim is to hit the two balls with the
cue ball, thus positioning the cue ball is most important in this game. For a start,
we limit ourselves to optimizing one shot to (1) sink an object ball by hitting it
with the cue ball, (2) to reach a given target point on the table, and (3) to stay
on a given target point on the table. Although the problem has to be addressed
to make an artificial player, to our knowledge, no reference has described any
solution to this problem.

In order to develop our optimization model, we need to understand deeply
(1) the behavior of balls rolling on a cloth stretched on a table, (2) the effect
of the collision between two balls, and (3) the effect of hitting the cue ball with
the leather tip of the cue stick. The optimization model to be developed will
benefit from any improvement on the physical model. As a physical model, we
use a sufficiently complete model to illustrate the potential of our optimization
modeling approach. We base our physical model on many sources (see below).

We construct a quite general model by first exposing simplified situations that
are made progressively more realistic. We start addressing the simple situation

---

* This research was partially supported by NSERC grant OGP0005491.
[1] For brevity, we use 'he' ('his') where 'he or she' ('his or her') is meant.

H.J. van den Herik et al. (Eds.): ACG11, LNCS 4250, pp. 263–272, 2006.
© Springer-Verlag Berlin Heidelberg 2006

of prescribing the initial speed and spin to the cue ball so that it reaches a given target, given a shoot direction. This allows to introduce a model of ball movement largely inspired by [1]. However, the ideas are common, and are well explained for example in [2,3,4,5].

Next, we consider the situation in which (1) the cue ball is targeted to a collision point, (2) hit an object ball in a prescribed direction, and (3) rebounds toward a target. This introduces simple elastic collisions without ball-ball friction, whose treatment comes from [5].

In a third step, we consider adding the ball-ball friction. Since the ball-ball friction modifies the trajectory of the object ball, now the collision target is no longer known in advance, but must be optimized. Our treatment of the ball-ball friction comes from [2], but similar developments are to be found in [4].

Finally (for this paper!), we address the problem of optimizing an actual shot, identifying (a) the speed, (b) the direction of the cue stick and (c) the contact point on the cue ball. This is simply a change of variables since angular speed of the cue ball is expressed in terms of the shot parameters. For this part, we assume that the ball is struck with a level cue stick (*i.e.* no massé) which imposes a natural limit on the magnitude of the spin with respect to the speed.

Below, we will discuss aspects of the solution of our models, and point out many stimulating perspectives.

2   Cue-Ball Positioning

Two aspects of accuracy are important for the pool player: accurately sink the object ball, and accurately position the cue ball. Computing the target ensuring to sink the object ball is trivial (but nevertheless difficult to perform for an amateur player), so we will concentrate on the computation of the shot in order to position the cue ball on a specified target, or at least as close as possible to the target.

After hitting the object ball, the cue ball is left with some linear ($\mathbf{v}_0$) and angular ($\boldsymbol{\omega}_0$) velocity. Throughout the paper, vectors will be denoted by boldface letters (like $\mathbf{v}_0$)), their norm by italic letters ($v_0 = \|\mathbf{v}_0\|$) or regular symbols ($\omega_0 = \|\boldsymbol{\omega}_0\|$) and finally the normalized vectors will inherit a hat ($\hat{\mathbf{v}}_0 = \frac{\mathbf{v}_0}{v_0}$).

We will first present an optimization formulation to compute suitable velocities in order for the cue ball to reach, and stop on a given target. Since the cue ball was shot by the player, we will then present another optimization problem to compute the angular and linear velocities required to have the cue ball hit the object ball suitably to sink it, and to reach and stop at the target. Finally, we will translate the initial angular and linear velocities required to a specification of the cue tip contact point on the cue ball, and the speed of the cue tip.

2.1   Aiming the Cue Ball at a Target

In this sub section, we simply compute the initial velocities required to bring the cue ball from some point (the origin) to a specified target. The motion of a ball is known to be first of a sliding nature, and after some point, become of a pure rolling nature.

Let the table be in the plane xy, and the vector $\mathbf{e}_3 = (0\ 0\ 1)^t$ point upward the table. The sliding movement is characterized by a non vanishing relative velocity of the contact point of the ball with the cloth, i.e., $0 \neq \mathbf{v}_r(t) = \mathbf{v}(t) + R\mathbf{e}_3 \times \boldsymbol{\omega}(t)$. Then, we may assume that the frictional force is constant and given by the relation $\mathbf{f} = -mg\mu_s\hat{\mathbf{v}}_r(0)$. We denote by I the inertia tensor for the sphere, $I = \frac{2mR^2}{5}$, m is the ball's mass, R its radius, and g is the gravitational constant. Also, μ_s is the sliding friction coefficient, ~ 0.2. We may now deduce the equation for the linear and angular velocities:

$$\mathbf{v}(t) = \mathbf{v}(0) - g\mu_s\hat{\mathbf{v}}_r(0)t \qquad \text{and} \tag{1}$$

$$\boldsymbol{\omega}(t) = \boldsymbol{\omega}(0) - \frac{R\mathbf{e}_3 \times \mathbf{f}}{I}t. \tag{2}$$

Similarly, the linear and angular positions are given by:

$$\mathbf{p}(t) = \mathbf{p}(0) + \mathbf{v}(0)t - \frac{g\mu_s\hat{\mathbf{v}}_r}{2}t^2 \qquad \text{and} \tag{3}$$

$$\mathbf{a}(t) = \mathbf{a}(0) + \boldsymbol{\omega}(0)t - \frac{R\mathbf{e}_3 \times \mathbf{f}}{I}t^2. \tag{4}$$

Using those developments, we may deduce that the relative speed will become null at a time:

$$\tau_r = \frac{2}{7g\mu_s}v_r(0), \tag{5}$$

after which the ball will be in a rolling motion. In rolling motion, the force is $\mathbf{f} = -mg\mu_r\hat{\mathbf{v}}$ and the friction coefficient $\mu_r \sim 0.01$. Since the side spin does not influence trajectories, we may consider the two first components of $\boldsymbol{\omega}$, so that the angular quantities depend on the linear ones:

$$\boldsymbol{\omega}(t) = \frac{1}{R}\mathbf{e}_3 \times \mathbf{v}(t), \tag{6}$$

$$\mathbf{v}(t) = \mathbf{v}(\tau_r) - g\mu_r\hat{\mathbf{v}}_r(t - \tau_r), \tag{7}$$

$$\mathbf{p}(t) = \mathbf{p}(\tau_r) + \mathbf{v}(\tau_r)(t - \tau_r) - \frac{g\mu_r\hat{\mathbf{v}}_r}{2}(t - \tau_r)^2, \quad \text{and} \tag{8}$$

$$\mathbf{a}(t) = \mathbf{a}(\tau_r) + \frac{1}{R}\mathbf{e}_3 \times (\mathbf{v}(\tau_r)(t - \tau_r) - \frac{g\mu_r\hat{\mathbf{v}}_r}{2}(t - \tau_r)^2. \tag{9}$$

Moreover, we deduce that the ball will stop at a time $\tau_f = \frac{v(\tau_r)}{g\mu_r}$.

Now we have all the ingredients to set up an optimization problem to compute $\mathbf{v}(0)$ and $\boldsymbol{\omega}(0)$ in order for the cue ball to reach and stop at the target $\mathbf{s}$. We assume that $\mathbf{v}(0) = \alpha\mathbf{d}_0$ where $\mathbf{d}_0$ is an imposed initial direction of movement. The decision variables for our optimization problem are then α and ω_1, ω_2, and ω_3. Let $x = (\alpha\ \omega_1\ \omega_2\ \omega_3)^t$. The developments above allow to define a function $F(x)$ which computes $\mathbf{p}(\tau_f)$, and we simply solve

$$\min_x \frac{1}{2}\|F(x) - \mathbf{s}\|. \tag{10}$$

As defined, the solution always yields exactly the target. Of course, the initial velocities (particularly angular) are not realistic, but adding lower (l) and upper (u) bound constraints of the form $l \leq x \leq u$ helps ruling out unrealistic solutions; then, the target is not always reachable, for instance if situated backward (with respect to $\mathbf{d}_0$) and far, but then, the solution brings the cue ball as close as possible given the bounds on x.

Figures 1 and 2 show two cases where the target is attained. In Fig. 1 strong spin is used. The curve represents the trajectory while the straight lines are a sampling of the spin velocity $\boldsymbol{\omega}$. Back spin is showed by the lines on the right of the curve while natural roll is on the left. Figures 3 and 4 show two cases in which the target is not reachable because of spin restrictions.

2.2   Aiming the Cue Ball After a Collision

In this section, we set up an optimization problem to take into account that the cue ball must reach a target collision point $\mathbf{c}$ (collision target), then obey the collision laws, and finally try to reach and stop at $\mathbf{s}$ (stop target).

The collision calculations are simplified by the fact that the object ball is at rest when the player executes his shot. We will first develop the model assuming no friction between the balls, and then generalize the model to the realistic case.

Frictionless and Elastic Ball–Ball Collisions. For frictionless ball-ball collisions, the target point $\mathbf{c}$ is easy to compute. Assume the object ball is at $\mathbf{o}$, and the target in the pocket at $\mathbf{b}$. Then, the desired direction for the object ball is $\boldsymbol{\delta} = \mathbf{b} - \mathbf{o}$, and the target $\mathbf{c} = \mathbf{o} - 2R\hat{\boldsymbol{\delta}}$. Since the angular speed may affect the trajectory, we now leave the vector $\mathbf{d}_0$ (previously imposed) free to the optimization process; we choose to use the vector $\mathbf{v}_0$ as the optimization variable. From the developments above, there must be a time value $\bar{t}$ (depending on $\mathbf{v}_0$ and $\boldsymbol{\omega}_0$) for which $\mathbf{p}(\bar{t}) = \mathbf{c}$. From this point, we define a second trajectory denoted with barred quantities: we set $\bar{\mathbf{v}}_0 = \Phi(\mathbf{v}(\bar{t}))$, the resulting direction for the cue ball after the collision, $\bar{\mathbf{p}}_0 = \mathbf{p}(\bar{t})$, $\bar{\boldsymbol{\omega}}_0 = \boldsymbol{\omega}(\bar{t})$, $\bar{\mathbf{a}}_0 = \mathbf{a}(\bar{t})$. We then compute, as in Subsection 2.1, $\bar{\mathbf{p}}(\bar{\tau}_f)$. Our decision variables are now $\mathbf{v}_0$, $\boldsymbol{\omega}_0$, and the unknown time $\bar{t}$ to reach the impact point. Since $\alpha \mathbf{d}_0 = \mathbf{v}_0$, we thus solve

$$\min_{\mathbf{v}_0, \boldsymbol{\omega}_0, \bar{t}} \frac{1}{2}(\|\mathbf{p}(\bar{t}) - \mathbf{c}\|^2 + \|\bar{\mathbf{p}}(\bar{\tau}_f) - \mathbf{s}\|^2). \tag{11}$$

In the above formulation, one should be aware that the first term, $\|\mathbf{p}(\bar{t}) - \mathbf{c}\|^2$ is actually a *constraint*, i.e. must vanish for the computation to make any sense. If the solution is such that $\mathbf{p}(\bar{t})$ is not sufficiently close to $\mathbf{c}$, its weight in the objective function should be increased.

To compute the velocity of the cue ball after the impact, we may define the function Φ as follows:

$$\bar{\mathbf{v}}_{0x} = \frac{\mathbf{v}(\bar{t})_x \delta_y - \mathbf{v}(\bar{t})_y \delta_x}{\delta_x^2 + \delta_y^2} \delta_y \quad \text{and} \quad \bar{\mathbf{v}}_{0y} = -\frac{\mathbf{v}(\bar{t})_x \delta_y - \mathbf{v}(\bar{t})_y \delta_x}{\delta_x^2 + \delta_y^2} \delta_x. \tag{12}$$

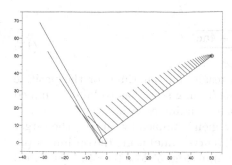

Fig. 1. Example case where the target is attained, using strong spin

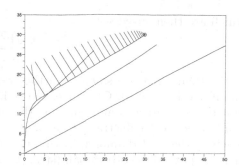

Fig. 2. Example cases where the target is attained, not using strong spin

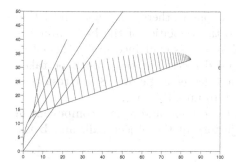

Fig. 3. Example cases where the target is not reachable because of spin restrictions

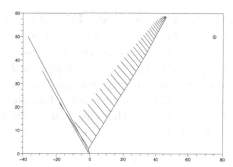

Fig. 4. Example cases where the target is not reachable because of spin restrictions

Linear Shots. In most shots, the cue stick is held horizontally, and thus does not impart curvilinear motion to the cue ball. In such cases, $\boldsymbol{\omega}_0 \perp \mathbf{d}_0$ and we know in advance the direction $\mathbf{d}_0$; the decision variable then reduces to the scalar α. This simplified setting allows to *compute* the value $\bar{t}$ actually, so that the optimization problem is reduced in dimension, as well as in difficulty.

Since we have $\mathbf{p}(t)$ as our target, we can transform the position equations (3 and 8) to get a quadratic equation of the form $\mathbf{a}x^2 + \mathbf{b}x + \mathbf{c}$. In sliding friction:

$$0 = -\mathbf{p}(t) + \mathbf{p}(0) + \mathbf{v}(0)t - \frac{g\mu_s \hat{\mathbf{v}}_r}{2}t^2 \,, \tag{13}$$

$$\text{where} \quad \mathbf{c} = -\mathbf{p}(t) + \mathbf{p}(0)\,, \qquad \mathbf{b} = \mathbf{v}(0)\,, \qquad \mathbf{a} = -\frac{g\mu_s \hat{\mathbf{v}}_r}{2}\,. \tag{14}$$

In rolling friction:

$$0 = -\mathbf{p}(t) + \mathbf{p}(\tau_r) + \mathbf{v}(\tau_r)(t - \tau_r) - \frac{g\mu_r \hat{\mathbf{v}}_r}{2}(t - \tau_r)^2 \,, \tag{15}$$

$$\text{where} \quad \mathbf{c} = -\mathbf{p}(t) + \mathbf{p}(\tau_r)\,, \qquad \mathbf{b} = \mathbf{v}(\tau_r)\,, \qquad \mathbf{a} = -\frac{g\mu_r \hat{\mathbf{v}}_r}{2}\,, \tag{16}$$

which are then solved with:

$$\frac{-\mathbf{b} \pm \sqrt{\mathbf{b}^2 - 4\mathbf{ac}}}{2\mathbf{a}}. \tag{17}$$

We then select the smallest of the two results for our time (or the positive one if a value is < 0). Of course, the sliding/rolling state has to be determined first by calculating τ_r and applying the corresponding equation.

Figures 5 and 6 illustrate the attainable region (sampled by pluses), the target reached (the black diamond) by the cue ball (curved line) using the optimization, and the path of the object ball (straight line): in Fig. 5 close to a 45° cut aimed at the top left pocket and in Fig. 6, an almost direct shot to the side. The rectangle represents the table. Figures 7 and 8 show two unreachable targets: the optimization does its best, and get as close as possible to the target.

Ball-Ball Collision with Friction. As it happens, there exists a small friction coefficient between two balls, imparting to the velocities of the balls after the collision an effect named *collision-induced throw*. As explained in [2], this effect is slight, changing the trajectories of 2 to 4 degrees, but there must be taken care of, in particular for long shots since a 4 degree angle over 8 feet results in a deviation of 6.7 inches, more than enough to miss the shot.

In order to take this effect into account, we must add a new component in the objective function. Let $\tilde{\mathbf{p}}$ be the trajectory of the object ball, and $\mathbf{b}$ the

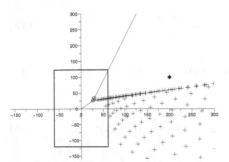

Fig. 5. A close to a 45° cut aimed at the top left pocket

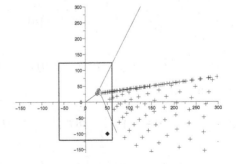

Fig. 6. An almost direct shot to the side

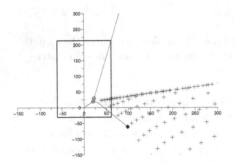

Fig. 7. Unreachable target

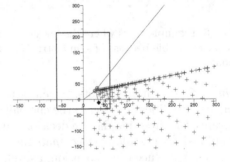

Fig. 8. Unreachable target

target position in the pocket. At some still unspecified time $\tilde{t}$, one must have $\tilde{\mathbf{p}}(\tilde{t}) \approx \mathbf{b}$. Moreover, the cue ball is no longer aimed at a specific target, but must impact the object ball in such a way that it reaches the pocket. $\mathbf{c}$ now represents the center of the object ball and $\mathbf{p}(\bar{t}) - \mathbf{c}$ is no more constant, but subject to optimization. The objective function becomes

$$\min_{\mathbf{v}_0, \boldsymbol{\omega}_0, \mathbf{c}, \bar{t}, \tilde{t}} \frac{1}{2}(\|\, \|\mathbf{p}(\bar{t}) - \mathbf{c}\|^2 - 4 * R^2\|^2 + \|\bar{\mathbf{p}}(\bar{\tau}_f) - \mathbf{s}\|^2 + \|\tilde{\mathbf{p}}(\tilde{t}) - \mathbf{b}\|^2) \quad (18)$$

Here again, the first term, slightly more complicated than before, is a constraint. It expresses the fact that the cue ball will actually touch (hit) the object ball at time $\bar{t}$. Since the so called 90° rule is not exactly obeyed to under realistic friction and elasticity assumptions, the exact contact point is no longer known in advance, but depends on the actual speed and spin of the cue ball at the impact. In practice, the contact point will be close to the contact point as used before, and should be used as an initial guess to the optimization algorithm.

The last term $\|\tilde{\mathbf{p}}(\tilde{t}) - \mathbf{b}\|^2$ also represents a constraint, and could be modeled differently. As a term to minimize, one seeks to reach as close as possible the center of the pocket. One could also specify bounds, required to be satisfied in order to sink the object ball.

The trajectory $\tilde{\mathbf{p}}$ may be computed as above, using the fact that its initial velocity is given by the formulæ

$$\tilde{\mathbf{v}}_{0x} = \mathbf{v}(\bar{t})_x - \bar{\mathbf{v}}_{0x} \quad \text{and} \quad \tilde{\mathbf{v}}_{0y} = \mathbf{v}(\bar{t})_y - \bar{\mathbf{v}}_{0y}. \quad (19)$$

As mentioned above, we must now take into account the collision-induced throw. As it happens, all we have to do is to add a (small) correction to the velocity vectors $\tilde{\mathbf{v}}_0$ and $\bar{\mathbf{v}}_0$. The correction is a component in the direction $\bar{\mathbf{v}}_0$ and obeys the law $v_{by} = \cos(\gamma)\mu_{bb}\tilde{v}_0$, where μ_{bb} is the ball-ball friction, α is the cut angle and

$$\cos(\gamma) = \frac{\left(\sin(\alpha) + \frac{R\omega_z(\bar{t})}{v(\bar{t})}\right)}{\left(\left(\sin(\alpha) + \frac{R\omega_z(\bar{t})}{v(\bar{t})}\right)^2 + \left(\cos(\alpha)\frac{R\omega_{xy}(\bar{t})}{v(\bar{t})}\right)^2\right)^{1/2}}. \quad (20)$$

We then correct the after collision velocities by adding $v_{by}\hat{\tilde{\mathbf{v}}}$ to $\tilde{\mathbf{v}}_0$ and subtracting it from $\bar{\mathbf{v}}_0$.

Figures 9 and 10 show the effect of collision induced throw. In Fig. 9 we see the throw trajectory (dark line) of the object ball without correction of the target. In Fig. 10 we see the modified attainable region when correcting the target. The effect of the throw is slight and has a negligible effect on the repositioning of the cue ball (sampled by crosses), but is important for the sinking of the object ball in the pocket.

Side Spin. The side spin of the cue ball affects the throw. The simple rolling-sliding model presented in Subsection 2.1 does not propagate side spin in its form since its equations only depend on the force components in the $z = 0$ plane. In contrast, the ω_z component may well persist after the cue ball is stopped in

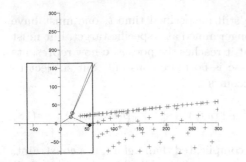

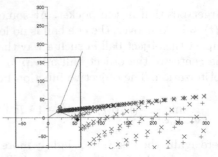

Fig. 9. No correction of the target

Fig. 10. the modified attainable region when correcting the target

extreme situations, as one observes on real tables. We must add a damping to simulate the attenuation of the side spin.

3   Shot Parameters

We consider a parameter transformation from $\mathbf{v}_0$, the velocity of the cue stick, and $\mathbf{h}_0$, the hit point on the cue ball, to $\mathbf{v}_0$, $\boldsymbol{\omega}_0$. We assume that the cue stick is horizontal. In this context, there is only a combination of back—top spin and side spin.

Following [2], let $\mathbf{h}_0 = (0\ h_y\ h_z)$ represent the point hit on the cue ball. After a suitable change of variables so that $\hat{\mathbf{x}} = \mathbf{v}_0$, $\hat{\mathbf{y}} = \mathbf{e}_3 \times \hat{\mathbf{x}}$ and $\hat{\mathbf{z}} = \mathbf{e}_3$, the angular velocity vector will be orthogonal to $h_y\hat{\mathbf{y}} + h_z\hat{\mathbf{z}}$ and its magnitude will be $\frac{5v_0}{2R^2}$. Moreover, in order to avoid miscue, we must impose a constraint $\|\mathbf{h}_0\| \leq \frac{R}{2}$.

4   Solving the Optimization Models

We observe that the developed optimization models are mostly unconstrained problems. We simply include bounds on the parameters, which reflect reality: there is an actual limit on the speed a pool ball can reach! Similarly, we just saw that the cue stick has to hit the ball within half of its radius in order to safely avoid miscue. We model the point hit by the cue using an angle (the spin direction) and a scalar (the amount of spin), the amount being bounded by $\frac{R}{2}$.

In order to solve the optimization efficiently, we will benefit from some knowledge of the game to provide a good initial approximation, and from bounds on the parameters to be satisfied. It is possible to compute in advance a minimum speed required by the cue ball in order to reach the object ball, and so that the object ball actually reaches the pocket.

Also, we will solve the simpler model neglecting the ball-ball friction, and only thereafter the general model, taking as initial guess the simpler solution, we will refine the solution by taking into account the collision-induced throw, and eventually the corrections due to an inelastic ball-ball collision.

In all the cases where the cue stick is horizontal, we know that the cue ball trajectory is linear, so we may express $\mathbf{v}_0 = \alpha \mathbf{d}_0$ with a value of $\mathbf{d}_0$ being either constant (no ball-ball friction), or expressed in a function of other parameters ($\mathbf{c}$ if the cue ball is at the origin). Moreover, since in the case of no ball-ball friction, the side spin has no effect at all, we may reduce the number of decision variables to only 3, namely α, $\bar{t}$, and h_y (the amount of back–top spin). When adding the ball-ball friction, the collision target $\mathbf{c}$, the side spin h_z, and the time $\tilde{t}$ are added to the decision variables.

4.1 Optimization Algorithms

The resulting optimization models may be solved using many optimization methods. The objective functions are smooth, but their derivatives are difficult to express. We produced the graphics using Scilab [6], which provides a function optim to optimize a function subject to bound constraints. The method used is a limited memory quasi-Newton iterative algorithm, and we employed finite differences to compute the gradient values.

As mentioned above, some terms in the objective function are actually penalization of constraints. It could be preferable to use some constrained optimization method, but the simple penalty approach we described seems to work fine in practice. Our assumption is as follows.

We will incorporate those computations within a player. We then use the open source OPT++ [7] optimization software. Again, we chose a quasi-Newton iterative method to treat the minimization subject to bound constraints.

5 Extensions

We presented the basics of an optimization methodology to achieve good position play. Below we discuss three extensions that make the methodology really powerful.

5.1 Breaking Clusters

Often in actual play, the position play consists in breaking clusters of balls. Our optimization models adapt easily by specifying a target position (the cluster) together with a speed range; above, the speed range was null, i.e., we required the cue ball to stop on the target.

5.2 Using the Rail for Positioning

As it was apparent in the figures, the reachable region extends largely outside the table. In such cases, a simplified model assuming perfect collisions between the ball and the rail would allow to aim the cue ball at the mirror target. However, the energy loss in the collision with the cushion is as large as 70%, so that accurate position play must take into account the friction and inelasticity of the collision.

Building on the above model, the cue ball's trajectory after the collision with the object ball will be separated into $n + 1$ parts; the first part reaches a first cushion, the second part a second cushion, and so on. The presented model treats the special case $n = 0$.

5.3 Using Other Balls for Positioning

Instead of considering the other balls on the table as added difficulties, we may well use them to achieve positioning by having the cue ball bouncing on them. Again, it consists in adding pieces to the path of the cue ball.

6 Conclusion

In this paper we presented the fundamentals of an optimization methodology to give a virtual pool player good sinking and positioning abilities. This constitutes the basic skill on which a good player builds his game.

Still discussing technical skills, we will have to add to our models the presence of other objects (other balls, cushions) which may influence the optimization and the possibility to break a cluster of balls.

We anticipate to building a function estimating the difficulty of a shot. This difficulty coefficient should take into account the difficulty to sink the object ball, but also to reach a point close to the stop target. Moreover, even if the stop target is not well approached, maybe the next ball remains "easy", and thus the shot should not be discarded. A planner should consider all the remaining balls on the table. Probably a dynamic programming model can prove useful for this planning.

Of course, we would like to add spectacular shots, such as massés. Our optimization model can already deal with various angular velocities, it remains to have a model to convert the actual shot ((1) cue stick angle and force, (2) point to hit on the cue ball) into the appropriate v_0, ω_0 initial conditions of the cue ball.

References

1. D. Sénéchal. Mouvement d'une Boule de Billard entre les Collisions. *Unpublished manuscript*, 1999.
2. R. Shepard. Amateur Physics for the Amateur Pool Player. *Self published*, 1997.
3. W.C. Marlow. *The Physics of Pocker Billiards*. MAST, Palm Beach Gardens, Florida, 1995.
4. D.G. Alciatore. *The Illustrated Principles of Pool and Billiards*. Sterling Publishing, 2004.
5. P. Grogono. Mathematics for Snooker Simulation. *Unpublished manuscript*, 1996.
6. *Scilab Group*. Ψlab 3.1, Institut National de Recherche en Informatique et Automatique - Domaine de Voluceau - Rocquencourt - B.P. 105 - 78153 - Le Chesnay Cedex - France, email: Scilab@inria.fr, 2005.
7. OPT++. *An object-Oriented Nonlinear Optimization Library*. http://csmr.ca. sandia.gov/ projects/opt++.

Author Index

Lecture Notes in Computer Science

For information about Vols. 1–4227

please contact your bookseller or Springer

Vol. 4268: G. Parr, D. Malone, M. Ó Foghlú (Eds.), Autonomic Principles of IP Operations and Management. XIII, 237 pages. 2006.

Vol. 4267: A. Helmy, B. Jennings, L. Murphy, T. Pfeifer (Eds.), Autonomic Management of Mobile Multimedia Services. XIII, 257 pages. 2006.

Vol. 4266: H. Yoshiura, K. Sakurai, K. Rannenberg, Y. Murayama, S. Kawamura (Eds.), Advances in Information and Computer Security. XIII, 438 pages. 2006.

Vol. 4265: L. Todorovski, N. Lavrač, K.P. Jantke (Eds.), Discovery Science. XIV, 384 pages. 2006. (Sublibrary LNAI).

Vol. 4264: J.L. Balcázar, P.M. Long, F. Stephan (Eds.), Algorithmic Learning Theory. XIII, 393 pages. 2006. (Sublibrary LNAI).

Vol. 4263: A. Levi, E. Savas, H. Yenigün, S. Balcisoy, Y. Saygin (Eds.), Computer and Information Sciences – ISCIS 2006. XXIII, 1084 pages. 2006.

Vol. 4262: K. Havelund, M. N\'u\~nez, G. Ro\csu, B. Wolff (Eds.), Formal Approaches to Software Testing and Runtime Verification. VIII, 255 pages. 2006.

Vol. 4261: Y. Zhuang, S. Yang, Y. Rui, Q. He (Eds.), Advances in Multimedia Information Processing - PCM 2006. XXII, 1040 pages. 2006.

Vol. 4260: Z. Liu, J. He (Eds.), Formal Methods and Software Engineering. XII, 778 pages. 2006.

Vol. 4259: S. Greco, Y. Hata, S. Hirano, M. Inuiguchi, S. Miyamoto, H.S. Nguyen, R. Słowiński (Eds.), Rough Sets and Current Trends in Computing. XXII, 951 pages. 2006. (Sublibrary LNAI).

Vol. 4257: I. Richardson, P. Runeson, R. Messnarz (Eds.), Software Process Improvement. XI, 219 pages. 2006.

Vol. 4256: L. Feng, G. Wang, C. Zeng, R. Huang (Eds.), Web Information Systems – WISE 2006 Workshops. XIV, 320 pages. 2006.

Vol. 4255: K. Aberer, Z. Peng, E.A. Rundensteiner, Y. Zhang, X. Li (Eds.), Web Information Systems – WISE 2006. XIV, 563 pages. 2006.

Vol. 4254: T. Grust, H. Höpfner, A. Illarramendi, S. Jablonski, M. Mesiti, S. Müller, P.-L. Patranjan, K.-U. Sattler, M. Spiliopoulou, J. Wijsen (Eds.), Current Trends in Database Technology – EDBT 2006. XXXI, 932 pages. 2006.

Vol. 4253: B. Gabrys, R.J. Howlett, L.C. Jain (Eds.), Knowledge-Based Intelligent Information and Engineering Systems, Part III. XXXII, 1301 pages. 2006. (Sublibrary LNAI).

Vol. 4252: B. Gabrys, R.J. Howlett, L.C. Jain (Eds.), Knowledge-Based Intelligent Information and Engineering Systems, Part II. XXXIII, 1335 pages. 2006. (Sublibrary LNAI).

Vol. 4251: B. Gabrys, R.J. Howlett, L.C. Jain (Eds.), Knowledge-Based Intelligent Information and Engineering Systems, Part I. LXVI, 1297 pages. 2006. (Sublibrary LNAI).

Vol. 4250: H.J. van den Herik, S.-C. Hsu, T.-s. Hsu, H.H.L.M. Donkers (Eds.), Advances in Computer Games. XIV, 273 pages. 2006.

Vol. 4249: L. Goubin, M. Matsui (Eds.), Cryptographic Hardware and Embedded Systems - CHES 2006. XII, 462 pages. 2006.

Vol. 4248: S. Staab, V. Svátek (Eds.), Managing Knowledge in a World of Networks. XIV, 400 pages. 2006. (Sublibrary LNAI).

Vol. 4247: T.-D. Wang, X. Li, S.-H. Chen, X. Wang, H. Abbass, H. Iba, G. Chen, X. Yao (Eds.), Simulated Evolution and Learning. XXI, 940 pages. 2006.

Vol. 4246: M. Hermann, A. Voronkov (Eds.), Logic for Programming, Artificial Intelligence, and Reasoning. XIII, 588 pages. 2006. (Sublibrary LNAI).

Vol. 4245: A. Kuba, L.G. Nyúl, K. Palágyi (Eds.), Discrete Geometry for Computer Imagery. XIII, 688 pages. 2006.

Vol. 4244: S. Spaccapietra (Ed.), Journal on Data Semantics VII. XI, 267 pages. 2006.

Vol. 4243: T. Yakhno, E.J. Neuhold (Eds.), Advances in Information Systems. XIII, 420 pages. 2006.

Vol. 4242: A. Rashid, M. Aksit (Eds.), Transactions on Aspect-Oriented Software Development II. IX, 289 pages. 2006.

Vol. 4241: R.R. Beichel, M. Sonka (Eds.), Computer Vision Approaches to Medical Image Analysis. XI, 262 pages. 2006.

Vol. 4239: H.Y. Youn, M. Kim, H. Morikawa (Eds.), Ubiquitous Computing Systems. XVI, 548 pages. 2006.

Vol. 4238: Y.-T. Kim, M. Takano (Eds.), Management of Convergence Networks and Services. XVIII, 605 pages. 2006.

Vol. 4237: H. Leitold, E. Markatos (Eds.), Communications and Multimedia Security. XII, 253 pages. 2006.

Vol. 4236: L. Breveglieri, I. Koren, D. Naccache, J.-P. Seifert (Eds.), Fault Diagnosis and Tolerance in Cryptography. XIII, 253 pages. 2006.

Vol. 4234: I. King, J. Wang, L. Chan, D. Wang (Eds.), Neural Information Processing, Part III. XXII, 1227 pages. 2006.

Vol. 4233: I. King, J. Wang, L. Chan, D. Wang (Eds.), Neural Information Processing, Part II. XXII, 1203 pages. 2006.

Vol. 4232: I. King, J. Wang, L. Chan, D. Wang (Eds.), Neural Information Processing, Part I. XLVI, 1153 pages. 2006.

Vol. 4231: J. F. Roddick, R. Benjamins, S. Si-Saïd Cherfi, R. Chiang, C. Claramunt, R. Elmasri, F. Grandi, H. Han, M. Hepp, M. Hepp, M. Lytras, V.B. Mišić, G. Poels, I.-Y. Song, J.D. Trujillo, C. Vangenot (Eds.), Advances in Conceptual Modeling - Theory and Practice. XXII, 456 pages. 2006.

Vol. 4230: C. Priami, A. Ingólfsdóttir, B. Mishra, H.R. Nielson (Eds.), Transactions on Computational Systems Biology VII. VII, 185 pages. 2006. (Sublibrary LNBI).

Vol. 4229: E. Najm, J.F. Pradat-Peyre, V.V. Donzeau-Gouge (Eds.), Formal Techniques for Networked and Distributed Systems - FORTE 2006. X, 486 pages. 2006.

Vol. 4228: D.E. Lightfoot, C.A. Szyperski (Eds.), Modular Programming Languages. X, 415 pages. 2006.